Cilia and Mucus

Cilia and Mucus

From Development to Respiratory Defense

edited by
Matthias Salathe
University of Miami School of Medicine
Miami, Florida

Section Editors

Kenneth B. Adler
North Carolina State University
Raleigh, North Carolina

Richard C. Boucher
University of North Carolina at Chapel Hill
Chapel Hill, North Carolina

Peter Satir
Albert Einstein College of Medicine
Bronx, New York

MARCEL DEKKER, INC. NEW YORK • BASEL

ISBN: 0-8247-0441-X

This book is printed on acid-free paper.

Headquarters
Marcel Dekker, Inc.
270 Madison Avenue, New York, NY 10016
tel: 212-696-9000; fax: 212-685-4540

Eastern Hemisphere Distribution
Marcel Dekker AG
Hutgasse 4, Postfach 812, CH-4001 Basel, Switzerland
tel: 41-61-261-8482; fax: 41-61-261-8896

World Wide Web
http://www.dekker.com

The publisher offers discounts on this book when ordered in bulk quantities. For more information, write to Special Sales/Professional Marketing at the headquarters address above.

Current printing (last digit):
10 9 8 7 6 5 4 3 2 1

PRINTED IN THE UNITED STATES OF AMERICA

Preface

The first international meeting on cilia, mucus, and mucociliary interactions was held in Jerusalem in February 1997. It was an unprecedented opportunity for basic and clinical scientists to come together and exchange information on the mucociliary system in a comprehensive way. Many participants suggested that similar events should be held in the future. In November 1999, through the generosity of the Terme di Sirmione in Italy, we were able to hold a second international meeting devoted to cilia, mucus, and mucociliary interactions. Again, basic and clinical scientists from around the world attended the meeting and shared their findings on the biology of the mucociliary apparatus. The participants also identified areas that require further research to clarify the role of mucociliary dysfunction in diseases.

Major accomplishments in the field over the last several years include the recognition that cilia and dynein arms, in addition to their transport function in the respiratory tract, serve a pivotal role in development of the mammalian embryo. Also, new experiments have shed light on changes in mucus expression and goblet cell metaplasia under conditions mimicking airway disease. One area requiring our continued efforts is the identification of a method to measure mucus load in the airways in vivo and to assess the ability of the mucociliary system to deal with such mucus loads in a more direct way. Many additional areas of interest are covered in this book. It is therefore a unique collection of papers that will prove a valuable resource for scientists and clinicians.

On behalf of all the scientists who attended the 1999 meeting, I would like to extend our sincerest gratitude to the Prince of Collalto, who made the meeting possible, and to the people of the Terme di Sirmione, who showed us such generosity and provided smooth, flawless logistics for the meeting. Without them, this book would not have been possible. I would also like to thank Boehringer Ingelheim, who supported this publication with an unrestricted grant. And last but not least: let's do it again.

Matthias Salathe

Contents

Contributors

William M. Abraham Division of Pulmonary and Critical Care Medicine, University of Miami at Mount Sinai Medical Center, Miami Beach, Florida

Kenneth B. Adler College of Veterinary Medicine, North Carolina State University, Raleigh, North Carolina

John E. Agnew Departments of Thoracic Medicine and Medical Physics, Royal Free and University College Medical School, London, United Kingdom

Nancy J. Akley College of Veterinary Medicine, North Carolina State University, Raleigh, North Carolina

Ross Albright Department of Biomedical Engineering, University of Nevada, Reno, Nevada

Serge Amselem INSERM U468, Hôpital Henri-Mondor, Créteil, France

Ronald Anderson Department of Immunology, MRC Unit for Inflammation and Immunity, Pretoria, South Africa

Sandra D. Anderson Department of Respiratory Medicine, Royal Prince Alfred Hospital, Sydney, Australia

P. W. Andrew Department of Microbiology and Immunology, University of Leicester, Leicester, United Kingdom

David J. Asai Department of Biological Sciences, Purdue University, West Lafayette, Indiana

William D. Bennett Division of Pulmonary Medicine, Department of Medicine, University of North Carolina at Chapel Hill, Chapel Hill, North Carolina

J. R. Blake School of Mathematics and Statistics, University of Birmingham, Birmingham, United Kingdom

James C. Bonner National Institute of Environmental Health Sciences, National Institutes of Health, Research Triangle Park, North Carolina

Richard J. Bookman Department of Molecular and Cellular Pharmacology, University of Miami School of Medicine, Miami, Florida

Brian W. Booth College of Veterinary Medicine, North Carolina State University, Raleigh, North Carolina

Richard C. Boucher Division of Pulmonary Medicine, Department of Medicine, University of North Carolina at Chapel Hill, Chapel Hill, North Carolina

Alex Braiman Department of Chemistry, Ben-Gurion University of the Negev, Beer-Sheva, Israel

Anne-Marie Bridoux INSERM U468, Hôpital Henri-Mondor, Créteil, France

Peter T. P. Bye Department of Respiratory Medicine, Royal Prince Alfred Hospital, Sydney, Australia

Valère Cacheux INSERM U468, Hôpital Henri-Mondor, Créteil, France

D. Caput Sanofi Recherche, Labège Innopole, France

Ingemar Carlstedt Department of Cell and Molecular Biology, Lund University, Lund, Sweden

Mary M. J. Chang University of California, Davis, Davis, California

Catherine Chapelin INSERM U468, Hôpital Henri-Mondor, Créteil, France

Yin Chen University of California, Davis, Davis, California

Annick Clément Service de Pneumologie Pédiatrique, Hôpital Armand-Trousseau, Paris, France

Peter Cole Host Defence Unit, National Heart & Lung Institute, London, United Kingdom

Joy Conway Respiratory, Cell, and Molecular Biology Research Division, Southampton General Hospital, Southampton, United Kingdom

Julia R. Davies Department of Cell and Molecular Biology, Lund University, Lund, Sweden

C. William Davis Cystic Fibrosis/Pulmonary Research and Treatment Center, University of North Carolina at Chapel Hill, Chapel Hill, North Carolina

Evangelia Daviskas Department of Respiratory Medicine, Royal Prince Alfred Hospital, Sydney, Australia

Bruce Denney School of Medicine, University of Nevada, and VA Sierra Nevada Health Care System, Reno, Nevada

Scott R. Diehl Molecular Genetic Epidemiology Unit, National Institute of Dental and Craniofacial Research, National Institutes of Health, Bethesda, Maryland

Bénédicte Duriez INSERM U468, Hôpital Henri-Mondor, Créteil, France

Estelle Escudier Service d'Histologie-Embryologie, Groupe Hospitalier Pitié-Salpêtrière, Paris, France

John H. Evans* Department of Physiology, University of Massachusetts Medical School, Worcester, Massachusetts

---

* *Current affiliation*: Department of Pediatrics, National Jewish Medical and Research Center, Denver, Colorado

Cahit Evrensel Department of Biomedical Engineering, University of Nevada, Reno, Nevada

Charles Feldman Department of Medicine, Johannesburg Hospital, and School of Medicine, University of the Witwatersrand, Johannesburg, South Africa

Bernard M. Fischer Division of Pediatric Pulmonary Diseases, Duke University Medical Center, Durham, North Carolina

W. Michael Foster* School of Hygiene and Public Health, The Johns Hopkins University, Baltimore, Maryland

E. A. Gaffney School of Mathematics and Statistics, University of Birmingham, Birmingham, United Kingdom

Michel Goossens INSERM U468, Hôpital Henri-Mondor, Créteil, France

Amir Hasani Departments of Thoracic Medicine and Medical Physics, Royal Free and University College Medical School, London, United Kingdom

Amgad Hassan Department of Biomedical Engineering, University of Nevada, Reno, Nevada

Robert A. Hirst Department of Molecular Biosciences, Adelaide University, Adelaide, Australia

Michael E. J. Holwill Physics Department, King's College London, London, United Kingdom

O. Houcine Laboratoire de Cytophysiologie et Toxicologie Cellulaire, Université Paris 7, Paris, France

Sang Won Hyun Department of Pharmaceutical Sciences, School of Pharmacy, University of Maryland, Baltimore, Maryland

Beom Tae Kim Department of Pharmaceutical Sciences, School of Pharmacy, University of Maryland, Baltimore, Maryland

* *Current affiliation*: Duke University Medical Center, Durham, North Carolina

Kwang Chul Kim Department of Pharmaceutical Sciences, School of Pharmacy, University of Maryland, Baltimore, Maryland

Malcolm King Pulmonary Research Group, University of Alberta, Edmonton, Canada

Michael Knowles Division of Pulmonary Medicine, Department of Medicine, University of North Carolina at Chapel Hill, Chapel Hill, North Carolina

Alon Korngreen* Department of Chemistry, Ben-Gurion University of the Negev, Beer-Sheva, Israel

Peter Krumpe School of Medicine, University of Nevada, and VA Sierra Nevada Health Care System, Reno, Nevada

Alison B. Lansley† Department of Physiology, University of Massachusetts Medical School, Worcester, Massachusetts

J. Laoukili Laboratoire de Cytophysiologie et Toxicologie Cellulaire, Université Paris 7, Paris, France

Eduardo Lazarowski Cystic Fibrosis/Pulmonary Research and Treatment Center, University of North Carolina at Chapel Hill, Chapel Hill, North Carolina

Martin M. Lee‡ Pulmonary Research Group, University of Alberta, Edmonton, Canada

Min Ki Lee Department of Pharmaceutical Sciences, School of Pharmacy, University of Maryland, Baltimore, Maryland

Ching Li Sinica Academy, Nankang, Taiwan

Yuehua Li College of Veterinary Medicine, North Carolina State University, Raleigh, North Carolina

---

* *Current affiliation*: Abteilung Zellphysiologie, Max-Planck Institut für medizinische Forschung, Heidelberg, Germany

† *Current affiliation*: Department of Pharmacy, University of Brighton, Brighton, United Kingdom

‡ *Current affiliation*: Channing Lab, Brigham & Women's Hospital, Harvard Medical School, Boston, Massachusetts

Erik Lillehoj Department of Pharmaceutical Sciences, School of Pharmacy, University of Maryland, Baltimore, Maryland

Weiyuan Ma Department of Chemistry, Ben-Gurion University of the Negev, Beer-Sheva, Israel

Mariangela Macchione College of Veterinary Medicine, North Carolina State University, Raleigh, North Carolina, and Faculdade de Medicina, Universidade de São Paulo, São Paulo, Brazil

F. Marano Laboratoire de Cytophysiologie et Toxicologie Cellulaire, Université Paris 7, Paris, France

Linda D. Martin College of Veterinary Medicine, North Carolina State University, Raleigh, North Carolina

Daoud Meerzaman Department of Pharmaceutical Sciences, School of Pharmacy, University of Maryland, Baltimore, Maryland

Anil Mehta Tayside Institute of Child Health, Ninewells Hospital and Medical School, Dundee, United Kingdom

K. Million Laboratoire de Cytophysiologie et Toxicologie Cellulaire, Université Paris 7, Paris, France

T. J. Mitchell Division of Infection and Immunity, Institute of Biomedical and Life Sciences, University of Glasgow, Glasgow, United Kingdom

H. M. Mitchison Department of Paediatrics, Royal Free and University College Medical School, London, United Kingdom

R. Muimo Tayside Institute of Child Health, Ninewells Hospital and Medical School, Dundee, United Kingdom

X. Mwimbi Tayside Institute of Child Health, Ninewells Hospital and Medical School, Dundee, United Kingdom

Jay A. Nadel Departments of Medicine and Physiology, University of California, San Francisco, San Francisco, California

Peadar G. Noone Division of Pulmonary Medicine, Department of Medicine, University of North Carolina at Chapel Hill, Chapel Hill, North Carolina

C. O'Callaghan Department of Child Health, University of Leicester, and Leicester Royal Infirmary, Leicester, United Kingdom

Thomas O'Riordan Division of Pulmonary and Critical Care Medicine, State University of New York at Stony Brook, Stony Brook, New York

Jose T. Orozco Section of Molecular and Cell Biology, University of California, Davis, Davis, California

Lawrence E. Ostrowski Division of Pulmonary Medicine, Department of Medicine, University of North Carolina at Chapel Hill, Chapel Hill, North Carolina

Anthony M. Paradiso Cystic Fibrosis/Pulmonary Research and Treatment Center, University of North Carolina at Chapel Hill, Chapel Hill, North Carolina

Jacek Pawlik Pediatric Division, Institute of Tuberculosis and Lung Diseases, Rabka, Poland

Konan Peck Sinica Academy, Nankang, Taiwan

Gaëlle Pennarun INSERM U468, Hôpital Henri-Mondor, Créteil, France

E. Perret Sanofi Recherche, Labège Innopole, France

Zvi Priel Department of Chemistry, Ben-Gurion University of the Negev, Beer-Sheva, Israel

Scott H. Randell Cystic Fibrosis/Pulmonary Research and Treatment Center, University of North Carolina at Chapel Hill, Chapel Hill, North Carolina

Carla M. Pedrosa Ribeiro Cystic Fibrosis/Pulmonary Research and Treatment Center, University of North Carolina at Chapel Hill, Chapel Hill, North Carolina

Michael Robinson Department of Respiratory Medicine, Royal Prince Alfred Hospital, Sydney, Australia

Gilles Roger Service d'Oto-Rhino-Laryngologie, Hôpital Armand-Trousseau, Paris, France

Duncan F. Rogers Department of Thoracic Medicine, National Heart & Lung Institute (Imperial College), London, United Kingdom

Lesilee S. Rose Section of Molecular and Cell Biology, University of California, Davis, Davis, California

Ewa Rutkiewicz Institute of Human Genetics, Poznań, Poland

Matthias Salathe Division of Pulmonary and Critical Care Medicine, University of Miami School of Medicine, Miami, Florida

Michael J. Sanderson Department of Physiology, University of Massachusetts Medical School, Worcester, Massachusetts

Jonathan M. Scholey Section of Molecular and Cell Biology, University of California, Davis, Davis, California

Dawn Signor Section of Molecular and Cell Biology, University of California, Davis, Davis, California

Shai D. Silberberg Department of Life Sciences and Zlotowski Center for Neuroscience, Ben-Gurion University of the Negev, Beer-Sheva, Israel

Alinka K. Smith Department of Thoracic Medicine, National Heart & Lung Institute (Imperial College), London, United Kingdom

S. L. Spiden Department of Paediatrics, Royal Free and University College Medical School, London, United Kingdom

Eiichi Sudo* Pulmonary Research Group, University of Alberta, Edmonton, Canada

Cui-e Sun National Institute of Dental and Craniofacial Research, National Institutes of Health, Bethesda, Maryland

Annette J. Theron Department of Immunology, MRC Unit for Inflammation and Immunity, Pretoria, South Africa

---

* *Current affiliation*: Hospital of Printing Bureau of Ministry of Finance, Tokyo, Japan

Frédéric Tournier Laboratoire de Cytophysiologie et Toxicologie Cellulaire, Université Paris 7, Paris, France

Natalya Uzlaner Department of Chemistry, Ben-Gurion University of the Negev, Beer-Sheva, Israel

Judith A. Voynow Division of Pediatric Pulmonary Diseases, Duke University Medical Center, Durham, North Carolina

Elizabeth M. Wagner Department of Medicine, School of Medicine, The Johns Hopkins University, Baltimore, Maryland

Yue-Fen Wang National Institutes of Health, Bethesda, Maryland

Karen P. Wedaman Section of Molecular and Cell Biology, University of California, Davis, Davis, California

Robert Wilson Host Defence Unit, National Heart & Lung Institute, London, United Kingdom

Michał Witt Institute of Human Genetics, Poznań, Poland

Reen Wu Department of Internal Medicine, University of California, Davis, Davis, California

Diego F. Wyszynski* National Institute of Dental and Craniofacial Research, National Institutes of Health, Bethesda, Maryland

Pan-Chyr Yang Department of Internal Medicine, National Taiwan University Hospital, Taipei, Taiwan

Chong-Jen Yu Department of Internal Medicine, National Taiwan University Hospital, Taipei, Taiwan

Jerzy Żebrak Pediatric Division, Institute of Tuberculosis and Lung Diseases, Rabka, Poland

Yu Hua Zhao University of California, Davis, Davis, California

---

* *Current affiliation*: Genetics Program, Department of Medicine, Boston University School of Medicine, Boston, Massachusetts

Part I

Cilia

This is an exciting time to be working on cilia, their motility, and the consequences of ciliary activity and inactivity for human disease. New techniques in molecule genetics, bioinformatics, macromolecular analysis, and imaging are expanding our knowledge of the structure, biochemistry, and physiology of cilia at a rapid rate. Among the new insights are:

1. The appreciation of the diversity of molecular motors involved in ciliary motility and ciliary biogenesis, including the cloning and localization of specific axonemal and cytoplasmic dyneins and ciliary kinesins.
2. The ability to grow human cilia from normal and PCD (primary ciliary dyskinesia) patients in culture in amounts adequate to analyze protein composition or, alternatively, to utilize knowledge of gene orthologous between unicellular ciliated eukaryotes and man coupled with new information on the human genome to identify loci involved in human ciliary motility.
3. The ability to record ciliary motility at high speed, so as to simultaneously map changes that are involved in the signal transduction pathways that control ciliary beat frequency, a critical parameter for mucociliary clearance.
4. The role of cilia in determining left-right asymmetry in mammalian embryos, thus relating situs inversus (SI) to PCD, as in Kartagener's syndrome.

We have here a series of chapters that delineates and builds on these new findings from basic biophysical modeling of the dynein arm activity in the axoneme to the regulation of ciliary beat frequency (CBF) by Ca^{2+} to identifying genes for PCD and Kartagener's syndrome. The reader will encounter two specific aspects of results several times in this volume.

THE EFFECT OF Ca^{2+} ON CBF

In these chapters, Sanderson et al. (Chapter 4) and Salathe and Bookman (Chapter 5) clearly document a rapid rise in CBF that follows within about 100 ms of a rise in intracellular Ca^{2+}. It is generally assumed, from *Chlamydomonas* mutant studies, that CBF is primarily controlled by the mechanochemical cycle of the outer arm dynein (OAD). In papers presented at Sirmione, Italy, but absent from this collection, Hamasaki et al. made the case that CBF is directly proportional to microtubule sliding velocity produced by OAD, which in turn is proportional to the phosphorylation of an OAD regulatory light chain. Light-chain phosphorylation is controlled by activation of a PKA both in protozoa and in mammalian cilia. In *Tetrahymena*, addition of cAMP and Ca^{2+} act synergistically to increase phosphorylation of the critical light chain in axonemes over PKA alone, while Ca^{2+} alone has no effect. Here, Braiman et al. (Chapter 6) show that activation of PKA and/or PKG increases CBF of rabbit tracheal cilia by about 30% over baseline in the absence of Ca^{2+} elevation, while CBF more than doubles when PKA/PKG activation and Ca^{2+} elevation both occur. Elevation of Ca^{2+} alone does not increase CBF. Braiman et al. (Chapter 6) suggest that Ca^{2+} exerts its effect via a Ca^{2+}–calmodulin complex. There are several Ca^{2+}-binding proteins in the axoneme, including calmodulin and centrin, the latter of which is a subunit constituent of inner arm dynein (IAD). Could Ca^{2+} act to make dynein more accessible to light-chain phosphorylation in the presence of activated PKA by altering its conformation? Light-chain phosphorylation–dephosphorylation would be consistent with the fast-slow changes in duty cycle that are postulated in the Salathe-Bookman chapter and that have been demonstrated as fast and slow microtubule sliding as part of the spoke control mechanism in the axoneme in a series of studies from Sale's laboratory (e.g., Ref. 1). Also, phosphorylation level is independent of immediate changes in Ca^{2+} or cAMP and, if operating in the presence of a constitutive phosphatase, would be expected to produce a hysteresis loop of the type described by Sanderson et al. (Chapter 4).

OAD MUTANTS AND PCD

Witt et al. (Chapter 9), Spiden and Mitchison (Chapter 10), and Pennurum et al. (Chapter 11) provide compelling evidence that PCD is genetically heterogeneous, as might be expected if the genetic defects that produce ciliary paralysis or abnormal beating in lower organisms have orthologs in the human genome. The coupling between PCD and situs inversus (SI) would then depend on exactly which genes are mutated. The present understanding of the development of left-right asymmetry in the mammalian embryo is that normal development requires motile embryonic nodal cilia (2). Nodal cilia are evidently unique in being 9 + 0, but motile, because of the presence of IADs. No other known epithelial cilia have this phenotype, but it is characteristic of certain sperm studied in some detail by Baccetti et al. (3) which have a helical, rather than planar beat. It would not be surprising to find mutations that affected OADs only and would not affect the nodal cilia, so that SI would be absent, but somatic 9 + 2 cilia would be dysfunctional. Witt et al. (Chapter 9) describe such cases, which result in families where affected individuals exhibit ciliary dyskinesia only without situs inversus (CDO families).

The default condition where nodal cilia are immotile is not considered to be SI, but rather the potential for symmetry randomization, so that with all other developmental processes functioning correctly, about 50% of affected individuals will have SI and 50% will have normal asymmetries. However, randomization implies that conditions for normal or inverted asymmetries will not always be strictly concordant, so that intermediates will occur, which probably will not survive, particularly because of abnormal heart development. It is possible that the frequency of PCD affecting left-right asymmetry is considerably underestimated because only successful embryos exhibit totally normal or totally reversed organ development.

Since abnormal development of the heart would result from incomplete asymmetry determination, this work suggests that certain ciliary mutations may be the underlying cause of fetal mortality in an undetermined number of situations.

Peter Satir

REFERENCES

1. G Habermacher, WS Sale. Regulation of flagellar dynein by phosphorylation of a 138-Kd inner arm dynein intermediate chain. J Cell Biol 136:167–176, 1997.
2. S Nonaka, Y Tanaka, Y Okada, S Takeda, A Harada, Y Kanai, M Kido, N Hirokawa. Randomization of left-right asymmetry due to loss of nodal cilia generating leftward flow of extraembryonic fluid in mice lacking KIF3B motor protein. Cell 95:829–837, 1998.
3. B Baccetti, AG Burrini, R Dallai, V Pallini. The dynein electrophoretic bands in axonemes naturally lacking the inner or the outer arm. J Cell Biol 80:334–340, 1979.

1

Cytoplasmic Dynein-2 and Ciliogenesis

David J. Asai
Purdue University
West Lafayette, Indiana

Dynein was first isolated from *Tetrahymena* cilia nearly four decades ago (1,2). This discovery of a microtubule motor together with the realization that dynein produces directed movements along microtubules revolutionized the thinking about how cells achieve microtubule-based movements (3–5).

Dynein is the high molecular weight microtubule molecular motor that powers ciliary beating and nonaxonemal movements (reviewed in Refs. 6–8). A dynein particle transduces the free energy of ATP hydrolysis into mechanical force that is applied to the surface of the microtubule, enabling the dynein to translocate toward the proximal or minus-end of the microtubule. The dynein cross-bridge cycle includes the tight attachment of the dynein to the microtubule, the dissociation of the dynein from the microtubule upon dynein binding to ATP, the conformational change in the dynein structure that depends on ATP hydrolysis and subsequent release of the hydrolysis products, and the reattachment of the dynein to the microtubule. Under the appropriate conditions, a molecular cargo tethered to the dynein will be carried from one part of the cell to another along a microtubule track. The cargo may be a membrane-bounded organelle or vesicle, the centrosome, a kinetochore, or another microtubule.

THE DYNEIN HEAVY CHAIN

In situ, dynein is a complex oligomer comprised of up to one dozen polypeptides of different sizes. These subunits include several light chains ($M_r < 25$ kDa), up to four intermediate chains (M_r 60–140 kDa), and one, two, or three heavy chains ($M_r > 500$ kDa). The light and intermediate chains, together with nondynein proteins, including the dynactin complex, contribute to the regulation of dynein activity and mediate the tethering of the dynein to its molecular cargo (e.g., Ref. 9). The actual motor activity of dynein resides in each heavy chain. An isolated heavy chain, separated from the other dynein constituents, is able to produce microtubule translocation in vitro at a velocity similar to that produced by intact dynein (10). Depending on the source of the dynein, the dynein particle comprises a single heavy chain, a homodimer of two identical heavy chains, a heterodimer, or a heterotrimer.

Favorable electron micrographs reveal that each heavy chain forms a tail, a globular head, and a short ''antenna'' that extends from the head (11). The tail interacts with other proteins to tether the dynein to its cargo, and in dyneins comprising multiple heavy chains, the heavy chains are gathered together by their tail domains to form a ''bouquet'' structure. Controlled proteolysis can separate the head from the tail and has revealed that the globular head domain contains the motor activity and the tail tethers the dynein to its cargo (12,13). The ''antenna'' is the B-link that is thought to be the ATP-sensitive microtubule-binding site required for movement.

Over the last decade, the complete sequences of nearly 20 dynein heavy chains have been reported. These include dyneins of all recognized functional classes—axonemal (ciliary) outer and inner arms, and nonaxonemal types 1 and 2—and from a wide variety of organisms. The examination of these sequences reveals several features found in all dyneins:

1. The dynein heavy chain is very large, approximately 4600 residues in length.
2. The N-terminal ~1300 residues form the tail domain. The sequence of this domain is the most divergent portion of the heavy chain, which is consistent with its role in specifying the cargo to which the dynein is tethered.
3. The central catalytic domain contains four evenly spaced phosphate-binding P-loops. The sequence of the first P-loop (P-1) is invariant—GPAGTGKT—and the first P-loop is the site of $MgATP^{2-}$ hydrolysis, which is required for movement. The function(s) of the other P-loops is not understood, but one dynein heavy chain can bind up to four molecules of ATP and the dynein activity may be modulated by the binding of nucleotides to these other sites (14–16).

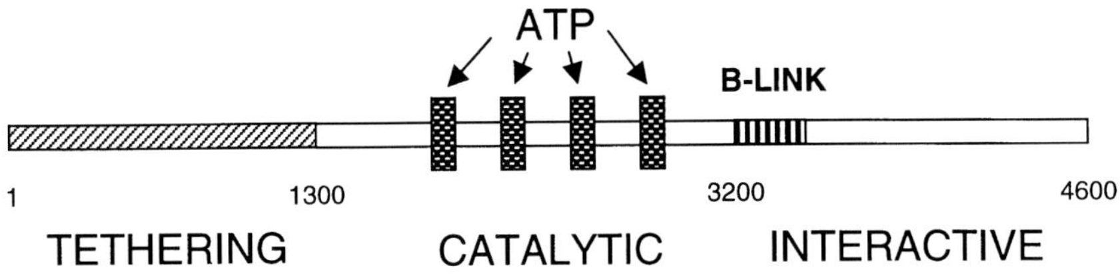

Figure 1 Three domains in the dynein heavy-chain sequence. The N-terminal 1300 residues form the short tail that interacts with other dynein subunits to tether the dynein to its molecular cargo. The catalytic domain is in the middle of the sequence and includes the four evenly spaced phosphate-binding loops (P-loops). The C-terminal portion of the heavy chain forms the interactive domain that includes the coiled-coil region that is thought to form the B-link.

4. The C-terminal interactive domain contains a short region whose sequence predicts two alpha-helices interrupted by a short flexible loop. Based on this predicted secondary structure, it was proposed that this region forms an intramolecular coiled-coil and is the B-link (17), and genetic and biochemical experiments support this hypothesis (18–21).

The important features of the dynein heavy chain are summarized in Figure 1.

FUNCTIONAL SPECIALIZATION OF DYNEIN

Well before the cloning and sequencing of any dynein heavy chain, it was evident that many organisms contain multiple dynein heavy-chain isoforms (reviewed in Ref. 22). For example, mutations in the green alga *Chlamydomonas* showed that axonemal outer and inner dynein arms are composed of distinct heavy chains, and biochemical studies were able to separate and analyze some of these isoforms (23–25). Each of the heavy chains is located in a precise place along the axoneme, and each isoform makes a unique contribution to the overall movement of the cilium or flagellum (26–28).

Organisms with cilia or flagella express a family of approximately 14 dynein heavy chains. Of these, as many as 12 are axonemal, and the remaining two or three are nonaxonemal or "cytoplasmic." Each dynein heavy chain isoform is encoded by a separate gene, and the number and sequences of the heavy chain isoforms are highly conserved across widely divergent organisms (29). Essentially the same family of dyneins is found in a wide spectrum of organisms, including the ciliated protozoan *Tetrahymena*, the sea urchin, the green alga *Chla-*

mydomonas, and the rat. The cumulative alignment of the catalytic domains of heavy chains from several model organisms illustrates this sequence conservation and assorts the dyneins into functional types (Fig. 2). This grouping of sequences is consistent with the experimental evidence that different isoforms produce distinct forces. Further, the sequence conservation observed across widely diverged organisms indicates that the gene duplications that gave rise to the multiple dynein genes occurred rapidly and prior to the divergence of the species. Once

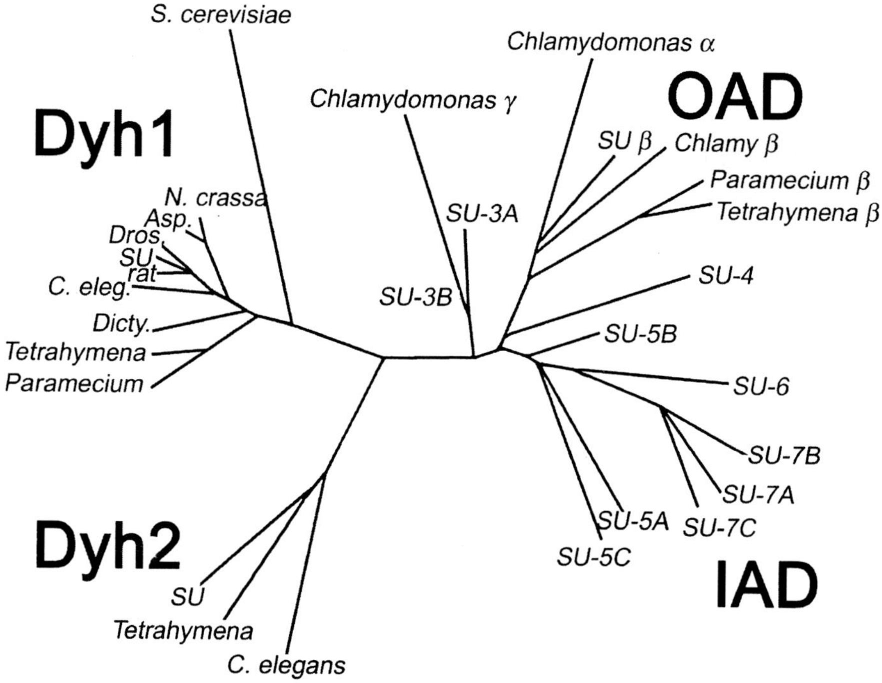

FIGURE 2 Four classes of dynein sequences. The ~ 1200-residue catalytic domains of several dynein heavy chains were aligned by CLUSTAL (GCG, University of Wisconsin) and the results of the alignment displayed in this unrooted tree. The sequences cluster into four groups: (1) cytoplasmic dynein Dyh1, (2) cytoplasmic dynein Dyh2, (3) axonemal outer arm dyneins (OADs), and (4) axonemal inner arm dyneins (IADs). Abbreviations: Scere, *S. cerevisiae*; Ncrass, *Neurospora*; Asp, *Aspergillus*; Dros, *Drosophila*; SU, sea urchin; Celeg, *Caenorhabditis*; Dicty, *Dictyostelium*; Tet, *Tetrahymena*; Para, *Paramecium*; Chlamy, *Chlamydomonas*. (Figure by Nathan Pankratz.)

duplicated, further sequence divergence was constrained, presumably because of the high level of functional specialization of the different dynein isoforms.

CYTOPLASMIC DYNEIN VERSUS AXONEMAL DYNEIN

Soon after the discovery of axonemal dynein, nonaxonemal or ''cytoplasmic'' dynein-like activities were described. The mitotic apparatus is an obvious and important structure whose movement may involve a dynein (30–32). Studies with antibodies to axonemal dynein found cross-reacting antigen in the mitotic apparatus, often distributed throughout the spindle (e.g., Ref. 33). These observations led to the idea that dynein might be responsible for the sliding apart of the two half-spindles during anaphase. These early ideas are probably too simple because a minus-end-directed motor acting on microtubules between the spindle poles (e.g., in the zone of overlap between the two half-spindles) would cause the spindle poles to be drawn together rather than separated. However, later studies using other specific antibodies and the tools of molecular genetics confirmed the finding that dynein is present and functioning in the mitotic apparatus. There is good evidence from different experimental systems that dynein is required for the formation of the bipolar spindle during prophase, mediates the attachment and prometaphase movements of the chromosomes on the kinetochore microtubules, and contributes to the separation of the spindle poles during anaphase (34–41).

With the growing appreciation of the roles of dynein in nonaxonemal motility, it was important to isolate and characterize the putative cytoplasmic dynein: Are the dyneins from the axoneme and from the cytosol the same enzyme or not? One favored experimental system was the unfertilized sea urchin egg because of the massive quantities of cytosolic proteins that can be obtained. Excellent progress by several laboratories was made in the purification and analysis of sea urchin egg dynein (42–48). The later stages of the sea urchin embryo are ciliated and the unfertilized egg is a storehouse of proteins used by the embryo after fertilization. Thus, the sea urchin egg provides an exceptional opportunity to explore the possibility that there are two different classes of dynein—ciliary and nonciliary—that might be distinguished from one another at the protein level. In the unfertilized sea urchin egg, two kinds of dyneins are immunologically distinct and bind with different affinities to taxol-induced microtubules: 20S soluble dynein, a ciliary precursor, and HMr-3, a dynein that functions in the cytosol and is excluded from the cilia (49,50).

Neuronal tissues are also a rich source of cytoplasmic dynein. Since the first successful temperature-dependent in vitro assembly of brain tubulin, high-speed extracts of vertebrate brain are the favored material from which to purify microtubules. Under standard assembly conditions, the α- and β-tubulins co-assemble with microtubule-associated proteins (MAPs), including tau and the high molecular weight MAPs. Among the high molecular weight MAPs is MAP1C, which is conventional cytoplasmic dynein (51). The biochemical analysis of MAP1C revealed that cytoplasmic dynein has a high CTPase activity, whereas axonemal dynein does not (52).

DYH1 AND DYH2—TWO CYTOSOLIC DYNEINS

The biochemical studies summarized in the previous paragraphs demonstrated the existence of at least one cytoplasmic dynein that is distinct from axonemal dyneins. The cloning and sequencing of dynein heavy chain genes provided a more precise understanding of the relationship between cytoplasmic and axonemal dyneins. The first dynein heavy-chain sequence to be completed was that of the sea urchin axonemal outer arm β heavy chain (53,54). This was soon followed by the determination of several other sequences, including those of the genes encoding cytoplasmic dyneins. There are conserved sequence differences in the catalytic domain that may distinguish cytoplasmic dynein from axonemal dynein (55).

An RNA-directed polymerase chain reaction (PCR) strategy provided the vehicle for the more extensive analysis of the entire dynein heavy chain family expressed in a given cell, tissue, or organism (56). The method uses degenerate oligonucleotide primers in the polymerase chain reaction to amplify short regions of the dynein heavy chain gene encoding the highly conserved P-1 region. This strategy was first applied to sea urchin embryos (57). Among the 14 sea urchin dynein genes are two that conform to the putative cytoplasmic dynein sequence motif. These two genes were originally called DYH1A and DYH1B; here they are referred to as DYH1 and DYH2, respectively. The DYH2 gene has also been identified in several other organisms, including *Tetrahymena*, *Chlamydomonas*, *Caenorhabditis*, rat, and human. As shown in the unrooted tree in Figure 2, the catalytic domain sequences of Dyh1 and Dyh2 are distinct.

DYH1 encodes the Dyh1 cytoplasmic dynein heavy-chain isoform that is found in all eukaryotes, including those organisms that apparently only have a single dynein (yeast, filamentous fungi, and slime mold). Dyh1 is the same as MAP1C, DHC1a, and cDHC1. DYH2 encodes the Dyh2 cytoplasmic dynein heavy-chain isoform that is found in all organisms examined that have more than one dynein. The gene encoding Dyh2 is expressed in many tissues including those whose cells are unciliated and an antibody to Dyh2 stained elements of the Golgi apparatus in fibroblasts (58,59). Dyh2 is the same as DHC1b, cDHC2, and DLP4. In this paper, the two cytoplasmic dyneins are referred to as Dyh1 and Dyh2.

The discovery that in many organisms two cytoplasmic dyneins coexist in the same cell led to studies aimed at dissecting the functions performed by each dynein. Where the contribution of a specific isoform has been investigated, the evidence is that Dyh1 and Dyh2 perform separate functions. In organisms with only one dynein, Dyh1 is used to position nuclei during and after mitosis and to organize the microtubule cytoskeleton. In cells with both cytoplasmic dyneins, Dyh1 is required for the normal chromosome segregation during mitosis and participates in the intracellular trafficking of membrane-bounded organelles and vesicles that are part of the secretory and endocytic pathways (39,60). In contrast, Dyh2 appears not to be directly involved in mitosis or endocytosis and exocytosis.

DYH2 IN CILIOGENESIS

So what is the function of Dyh2? The first clue came from the sea urchin study that originally discovered Dyh2 (57). In organisms with cilia or flagella, the expression of the genes that encode axonemal structures—more than 200 separate genes—is induced in response to deciliation or deflagellation. Ciliated sea urchin embryos were deciliated twice and then allowed to regenerate their cilia. At different times during reciliation, the expression patterns of individual dynein heavy-chain genes were measured by Northern blotting. As expected, the expression of tubulin and the axonemal outer arm dynein β heavy chain were upregulated during reciliation of the sea urchin embryo. Further, as expected, the expression of the conventional cytoplasmic dynein gene DYH1 was not changed. The surprise was that DYH2 expression also increased during reciliation. Thus, the sea urchin study revealed an interesting paradox: the sequence of Dyh2 suggests that it is a cytoplasmic dynein, but its expression pattern is more similar to that of bona fide axonemal dyneins. This finding led to the hypothesis that Dyh2 is involved in the formation but not the motility of the cilium. Other experiments have revealed a similar upregulation in DYH2 expression during ciliogenesis in *Tetrahymena* and rat tracheal epithelial cells (39,61).

The hypothesis that Dyh2 is not part of the ciliary motility apparatus but is involved in the assembly of the axoneme was examined in primary cultures of rat ciliated tracheal epithelial (RTE) cells (61). In these cultures, ciliogenesis was initiated by the establishment of an air/liquid interface. Northern blots revealed that the expression of DYH2 increased during ciliogenesis. If the cultures remained submerged, the DYH2 was expressed at a constant low level. If growth factors were withdrawn from the medium, a treatment that accelerates ciliogenesis, the expression of DYH2 was correspondingly stimulated. Isoform-specific antibodies revealed that Dyh2 was mostly cytoplasmic and appeared to accumulate at the apical end of the ciliated RTE cells. This location of Dyh2 is consistent with the idea that Dyh2 carries ciliary precursors from the sites of their synthesis to the site of ciliary assembly; the minus-ends of the microtubules are at the apical end of the ciliated epithelial cells. In nonciliated cells, anti-Dyh2 staining was weak and diffused. In contrast, an antibody to axonemal dynein brightly stained the cilia. These results are summarized in Figure 3.

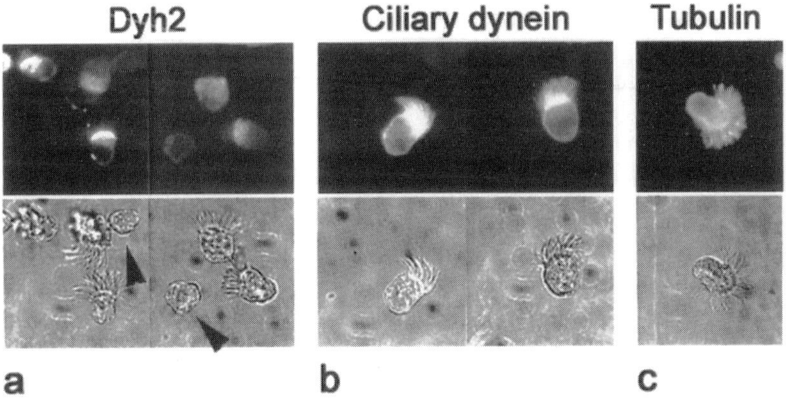

FIGURE 3 Dyh2 accumulates at the apical ends of rat ciliated epithelial cells. Rat tracheal epithelial cells were processed for indirect immunofluorescence microscopy. The upper panels are the fluorescence images and the lower panels are the corresponding brightfield images. (a) Affinity-purified anti-Dyh2 antibody brightly stained a crescent-shaped region at the apical ends of ciliated cells, but only faintly stained the cilia. Unciliated cells (arrowheads) were only weakly stained with this antibody. (b) Antibody against axonemal dynein brightly stained the cilia. (c) Antibody against tubulin brightly stained the cilia and much of the cell body. This figure is similar to one previously published (61).

DYH2 IN DIFFERENT CELLS

A powerful strategy that should lead to the understanding of the precise role of Dyh2 is to disrupt the expression of only the DYH2 gene and then to evaluate the resulting phenotype. This strategy has been pursued in two model organisms and has produced two different results. Insertional mutations of the gene encoding Dyh2 in *Chlamydomonas* led to stumpy flagella and the accumulation of protein raft complexes in the axoneme (62,63). The *Chlamydomonas* phenotype is consistent with the idea that Dyh2 is the motor responsible for retrograde intraflagellar transport (IFT) (64). This function for Dyh2 is further demonstrated in the analysis of IFT in the sensory cilium of the nematode (65).

A very different result was obtained when the DYH2 gene was disrupted by targeted homologous gene replacement in *Tetrahymena* (39). Transformants lacking the DYH2 gene were missized and misshapen. Unlike in *Chlamydomonas*, Dyh2 is not required for ciliogenesis in *Tetrahymena*. In reciliation experiments, the DYH2 transformants regenerated their cilia. However, instead of the normal ordered rows of cilia, the cilia on the DYH2 transformants emerged in no discernible pattern. Thus, in *Tetrahymena*, Dyh2 is required for the proper organization of the cortical microtubule cytoskeleton but not for cilia formation.

DYH2 AND DYNAMIC MICROTUBULE ORGANIZATION

Taken together, these results suggest an intriguing hypothesis: Dyh2 is involved in the organization of dynamic arrays of microtubules. The function of Dyh2 is manifested in different ways depending upon the organism examined. In *Chlamydomonas*, the flagellar microtubules are extremely dynamic because the flagella rapidly grow or shrink in response to environmental cues. In contrast, *Tetrahymena* cilia are terminal and therefore static structures, but the *Tetrahymena* cortical cytoskeleton is dynamic. The cortical cytoskeleton is disassembled and then reassembled at every cell division (in the laboratory, *Tetrahymena* divides every 2.5 hours) and also must be reorganized after environmental insult (e.g., the high Ca^{2+} used to deciliate the cells). An important problem for further study is to understand how Dyh2 organizes microtubules.

The apparent functional diversity of Dyh2 is reminiscent of the variety of tasks performed by kinesin-II (66). In both cases, the specific job performed by the microtubule motor appears to depend on the cellular context in which it functions. Perhaps the combination of Dyh2 with different dynein subunits—a different heavy chain or accessory protein—leads to functional specialization. In this regard, it is interesting to note the existence of multiple Dyh2-like isoforms (67).

SUMMARY

Organisms with cilia or flagella express approximately 14 different dynein heavy-chain genes, and each gene encodes a separate dynein heavy-chain isoform. The individual isoforms produce different forces leading to an exquisite level of functional specialization among the dyneins. There are at least two nonciliary or cytoplasmic dynein heavy chains, Dyh1 and Dyh2. Dyh1 functions in mitosis and organelle trafficking. In contrast, Dyh2 appears to be responsible for organizing dynamic arrays of microtubules. An important role for Dyh2 occurs in cells programmed to produce cilia—including tracheal epithelial cells—in which Dyh2 is involved in ciliogenesis.

ACKNOWLEDGMENTS

Our studies of cytoplasmic dyneins are supported by grants from the National Science Foundation and the American Cancer Society.

REFERENCES

1. IR Gibbons. Studies on the protein components of cilia from *Tetrahymena pyriformis*. Proc Natl Acad Sci USA 50:1002–1010, 1963.
2. IR Gibbons, AJ Rowe. Dynein: A protein with adenosine triphosphatase activity from cilia. Science 149:424–426, 1965.

3. P Satir. Studies on cilia. II. Examination of the distal region of the ciliary shaft and the role of the filaments in motility. J Cell Biol 26:805–834, 1965.
4. KE Summers, IR Gibbons. Adenosine triphosphate-induced sliding of tubules in trypsin-treated flagella of sea-urchin sperm. Proc Natl Acad Sci USA 68:3092–3096, 1971.
5. WS Sale, P Satir. Direction of active sliding of microtubules in *Tetrahymena* cilia. Proc Natl Acad Sci USA 74:2045–2049, 1977.
6. ELF Holzbaur, RB Vallee. Dyneins: Molecular structure and cellular function. Annu Rev Cell Biol 10:339–372, 1994.
7. ME Porter. Axonemal dyneins: assembly, organization, and regulation. Curr Opin Cell Biol 8:10–17, 1996.
8. SM King. The dynein microtubule motor. Biochim Biophys Acta 1496:60–75, 2000.
9. SJ King, TA Schroer. Dynactin increases the processivity of the cytoplasmic dynein motor. Nature Cell Biol 2:20–24, 2000.
10. WS Sale, LA Fox. Isolated β-heavy chain subunit of dynein translocates microtubules in vitro. J Cell Biol 107:1793–1797, 1988.
11. U Goodenough, J Heuser. Structural comparison of purified proteins with in situ dynein arms. J Mol Biol 180:1083–1118, 1984.
12. K Ogawa. Studies on flagellar ATPase from sea urchin spermatozoa. II. Effect of trypsin digestion on the enzyme. Biochim Biophys Acta 293:514–525, 1973.
13. G Mocz, IR Gibbons. ATP-insensitive interaction of the amino-terminal region of the β heavy chain of dynein with microtubules. Biochemistry 32:3456–3460, 1993.
14. G Mocz, IR Gibbons. Phase partition analysis of nucleotide binding to axonemal dynein. Biochemistry 35:9204–9211, 1996.
15. S Kinoshita, T Miki-Noumura, CK Omoto. Regulatory role of nucleotides in axonemal function. Cell Motil Cytoskel 32:46–54, 1995.
16. E Hirakawa, H Higuchi, YY Toyoshima. Processive movement of single 22S dynein molecules occurs only at low ATP concentrations. Proc Natl Acad Sci USA 97: 2533–2537, 2000.
17. DJ Asai, CJ Brokaw. Dynein heavy chain isoforms and axonemal motility. Trends Cell Biol 3:398–402.
18. ME Porter, JA Knott, LC Gardner, DR Mitchell, SK Dutcher. Mutations in the SUP-PF-1 locus of *Chlamydomonas reinhardtii* identify a regulatory domain in the β-dynein heavy chain. J Cell Biol 126:1495–1507, 1994.
19. MP Koonce. Identification of a microtubule-binding domain in a cytoplasmic dynein heavy chain. J Biol Chem 272:19714–19718, 1997.
20. MA Gee, JE Heuser, RB Vallee. An extended microtubule-binding structure within the dynein motor domain. Nature 390:636–639, 1997.
21. MP Koonce, I Tikhonenko. Functional elements within the dynein microtubule-binding domain. Mol Biol Cell 11:523–529, 2000.
22. G Piperno. Functional diversity of dyneins. Cell Motil Cytoskel 17:147–149, 1990.
23. G Piperno, Z Ramanis, EF Smith, WS Sale. Three distinct inner dynein arms in *Chlamydomonas* flagella: Molecular composition and location in the axoneme. J Cell Biol 110:379–389, 1990.
24. O Kagami, R Kamiya. Translocation and rotation of microtubules caused by multiple species of *Chlamydomonas* inner-arm dynein. J Cell Sci 103:653–664, 1992.

25. R Kamiya. Exploring the function of inner and outer dynein arms with *Chlamydomonas* mutants. Cell Motil Cytoskel 32:98–102, 1995.
26. CJ Brokaw, R Kamiya. Bending patterns of *Chlamydomonas* flagella: IV. Mutations with defects in inner and outer dynein arms indicate differences in dynein arm function. Cell Motil Cytoskel 8:68–75, 1987.
27. CJ Brokaw. Control of flagellar bending: A new agenda based on dynein diversity. Cell Motil Cytoskel 28:199–204.
28. DJ Asai. Multi-dynein hypothesis. Cell Motil Cytoskel 32:129–132, 1995.
29. DJ Asai. Functional and molecular diversity of dynein heavy chains. Sems Cell Devel Biol 7:311–320, 1996.
30. D Mazia, RR Chaffee, RM Iverson. Adenosine triphosphatase in the mitotic apparatus. Proc Natl Acad Sci USA 47:788–790, 1961.
31. R Weisenberg, EW Taylor. Studies on ATPase activity of sea urchin eggs and the isolated mitotic apparatus. Exp Cell Res 53:372–384, 1968.
32. T Miki. The ATPase activity of the mitotic apparatus of the sea urchin egg. Exp Cell Res 29:92–101, 1963.
33. H Mohri, T Mohri, I Mabuchi, I Yazaki, H Sakai, K Ogawa. Localization of dynein in sea urchin eggs during cleavage. Develop Growth Differ 18:391–397, 1976.
34. CM Pfarr, M Coue, PM Grissom, TS Hays, ME Porter, JR McIntosh. Cytoplasmic dynein is localized to kinetochores during mitosis. Nature 345:263–265, 1990.
35. ER Steuer, L Wordeman, TA Schroer, MP Sheetz. Localization of cytoplasmic dynein to mitotic spindles and kinetochores. Nature 345:266–268, 1990.
36. EA Vaisberg, MP Koonce, JR McIntosh. Cytoplasmic dynein plays a role in mammalian mitotic spindle formation. J Cell Biol 123:849–858, 1993.
37. CL Rieder, SP Alexander. Kinetochores are transported poleward along a single astral microtubule during chromosome attachment to the spindle in new lung cells. J Cell Biol 111:81–95, 1990.
38. CJ Echeverri, BM Paschal, KT Vaughan, RB Vallee. Molecular characterization of the 50-kDa subunit of dynactin reveals function for the complex in chromosome alignment and spindle organization during mitosis. J Cell Biol 132:617–633, 1996.
39. S-w Lee, JC Wisniewski, WL Dentler, DJ Asai. Gene knockouts reveal separate functions for two cytoplasmic dyneins in *Tetrahymena thermophila*. Mol Biol Cell 10:771–784, 1999.
40. ER Hildebrandt, MA Hoyt. Mitotic motors in *Saccharomyces cerevisiae*. Biochim Biophys Acta 1496:99–116, 2000.
41. DJ Sharp, HM Brown, M Kwon, GC Rogers, G Holand, JM Scholey. Functional coordination of three mitotic motors in *Drosophila* embryos. Mol Biol Cell 11:241–253, 2000.
42. S Hisanaga, H Sakai. Cytoplasmic dynein of the sea urchin egg. I. Partial purification and characterization. Develop Growth Differ 22:373–384, 1980.
43. MM Pratt. The identification of a dynein ATPase in unfertilized sea urchin eggs. Dev Biol 74:364–378, 1980.
44. JM Scholey, B Neighbors, JR McIntosh, ED Salmon. Isolation of microtubules and a dynein-like MgATPase from unfertilized sea urchin eggs. J Biol Chem 259:6516–6525, 1984.

45. PJ Hollenbeck, F Suprynowicz, WZ Cande. Cytoplasmic dynein-like ATPase cross-links microtubules in an ATP-sensitive manner. J Cell Biol 99:1251–1258, 1984.
46. DJ Asai, L Wilson. A latent activity dynein-like cytoplasmic magnesium adenosine triphosphatase. J Biol Chem 260:699–702, 1985.
47. ME Porter, PM Grissom, JM Scholey, ED Salmon, JR McIntosh. Dynein isoforms in sea urchin eggs. J Biol Chem 263:6759–6771, 1988.
48. KR Foltz, DJ Asai. Ionic strength-dependent isoforms of sea urchin egg dynein. J Biol Chem 263:2878–2883, 1988.
49. DJ Asai. An antiserum to the sea urchin 20S egg dynein reacts with embryonic ciliary dynein but it does not react with the mitotic apparatus. Dev Biol 118:416–424, 1986.
50. PM Grissom, ME Porter, JR McIntosh. Two distinct isoforms of sea urchin egg dynein. Cell Motil Cytoskel 21:281–292, 1992.
51. RB Vallee, JS Wall, BM Paschal, HS Shpetner. Microtubule-associated protein 1C from brain is a two-headed cytosolic dynein. Nature 332:561–563, 1988.
52. V Pallini, C Mencarelli, L Bracci, M Contorni, P Ruggiero, A Tiezzi, R Manetti. Cytoplasmic nucleoside-triphosphatases similar to axonemal dynein occur widely in different cell types. J Submicrosc Cytol 15:229–235, 1983.
53. IR Gibbons, BH Gibbons, G Mocz, DJ Asai. Multiple nucleotide-binding sites in the sequence of the dynein β heavy chain. Nature 352:640–643, 1991.
54. K Ogawa. Four ATP-binding sites in the midregion of the β heavy chain of dynein. Nature 352:643–645, 1991.
55. DJ Asai, SM Beckwith, KA Kandl, HH Keating, H Tjandra, JD Forney. The dynein genes of *Paramecium tetraurelia*: Sequences adjacent to the catalytic P-loop identify cytoplasmic and axonemal heavy chain isoforms. J Cell Sci 107:839–847, 1994.
56. DJ Asai, PS Criswell. Identification of new dynein heavy chain genes by RNA-directed PCR. Meth Cell Biol 47:579–585, 1995.
57. BH Gibbons, DJ Asai, W-JY Tang, TS Hays, IR Gibbons. Phylogeny and expression of axonemal and cytoplasmic dynein genes in sea urchins. Mol Biol Cell 5:57–70, 1994.
58. Y Tanaka, Z Zhang, N Hirokawa. Identification and molecular evolution of new dynein-like protein sequences in rat brain. J Cell Sci 108:1883–1893, 1995.
59. EA Vaisberg, PM Grissom, JR McIntosh. Mammalian cells express three distinct dynein heavy chains that are localized to different cytoplasmic organelles. J Cell Biol 133:831–842, 1996.
60. A Harada, Y Takei, Y Kanai, Y Tanaka, S Nonaka, N Hirokawa. Golgi vesiculation and lysosome dispersion in cells lacking cytoplasmic dynein. J Cell Biol 141:51–59, 1998.
61. PS Criswell, LE Ostrowski, DJ Asai. A novel cytoplasmic dynein heavy chain: Expression of DHC1b in mammalian ciliated epithelial cells. J Cell Sci 109: 1891–1898, 1996.
62. ME Porter, R Bower, JA Knott, P Byrd, W Dentler. Cytoplasmic dynein heavy chain 1b is required for flagellar assembly in *Chalmydomonas*. Mol Biol Cell 10:693–712, 1999.
63. GJ Pazour, BL Dickert, GB Witman. The DHC1b (DHC2) isoform of cytoplasmic dynein is required for flagellar assembly. J Cell Biol 144:473–481, 1999.

64. JL Rosenbaum, DG Cole, DR Diener. Intraflagellar transport: The eyes have it. J Cell Biol 144:385–388, 1999.
65. D Signor, KP Wedaman, JT Orozco, ND Dwyer, CI Bargmann, LS Rose, JM Scholey. Role of a class DHC1b dynein in retrograde transport of IFT motors and IFT raft particles along cilia, but not dendrites, in chemosensory neurons of living *Caenorhabditis elegans*. J Cell Biol 147:519–530, 1999.
66. JR Marszalek, LSB Goldstein. Understanding the functions of kinesin-II. Biochim Biophys Acta 1496:142–150, 2000.
67. PS Criswell, DJ Asai. Evidence for four cytoplasmic dynein heavy chain isoforms in rat testis. Mol Biol Cell 9:237–247, 1998.

2

Dynein Motor Activity During Ciliary Beating

Michael E. J. Holwill
King's College London
London, United Kingdom

INTRODUCTION

Elsewhere in this volume some of the biological and biochemical challenges offered by ciliary motion have been discussed. Here I propose to address certain physical aspects of the movement, with a view to examining the constraints placed on the system in respect of the microtubule displacements produced, the forces developed, and the energy expended during ciliary bending. A major objective is to use the observed motion of a cilium to deduce how the assemblies of dynein motor molecules behave within the axoneme, relating in particular the motor activity to microtubule sliding, ciliary bend shape, and beat frequency. After a brief discussion of the relevant biophysical background, the numbers of active arms required to generate different physical characteristics of the motion will be estimated. It will emerge that the numbers required to satisfy the geometry of the movement are different from those required to provide the force and energy, and the reconciliation of this difference provides information about the behavior of the arms within the system.

BIOPHYSICAL BACKGROUND

A cilium undergoes cyclic activity, usually with a frequency in the range of 5–50 Hz, in which a power, or effective, stroke generates force to induce relative motion between the cell and its surrounding fluid; examples include motion of the mucous layer in the human lung and free-swimming protozoa. The effective stroke is followed by a recovery stroke, generally not the reverse of, and usually dissipating less power than, the effective stroke, to return the cilium to the beginning of its cycle. The power for the motion is provided by the dynein motors, or arms, which utilize the chemical fuel ATP as a source of energy. The arms also have a cyclic action, with a force-generating, or duty, phase as one of the components; in other phases of the cycle, the arm undergoes mechanochemical changes to prepare it for the next duty phase. A 5-μm-long cilium contains about 4000 inner and outer arms. There is evidence (e.g., Ref. 1) that the inner and outer arms perform different functions in the axoneme, with the inner arms considered to be important in controlling bend shape, while the outer arms control beat frequency. The inner arms, with three distinct types repeating every 96 nm along a microtubule, are more complex in structure and arrangement than the outer arms, which are all of the same type and spaced at 24 nm intervals (2). For the purposes of the current discussion, it will be assumed that the outer arms dominate the bending process and that the contribution of the inner arms to this process is not significant. This is clearly an oversimplification, but it affects neither the main thrust of the argument nor the conclusions drawn.

The forces exerted by the motors produce the relative microtubule sliding between the peripheral doublets of the axoneme (3), which leads to ciliary bending. In a number of cells, e.g., *Paramecium*, an alteration in the speed or direction of locomotion in response to an external stimulus is achieved through a changed pattern of ciliary activity; studies of the relationships between the changes indicate that ciliary motion is controlled by the cell. An appropriate physical model for the cilium is therefore a controlled, active, mechanochemical oscillator. Such models are not, of course, novel. For example, in 1958, Machin (4) represented the flagellum as a vibrating elastic beam and demonstrated that energy must be introduced along the length of an organelle to produce the observed wave patterns with sustained, and sometimes increasing, amplitude, and thereby anticipated the discovery of dynein. Other workers, e.g., Brokaw (5) and Holwill and Miles (6), introduced parameters into the modeling to reflect the mechanochemistry of the system.

The available evidence (e.g., Ref. 7) suggests that dynein generates force in one direction only, with the arms on one microtubule (number N) in the axoneme pushing the neighboring microtubule ($N + 1$) tipwards. One consequence of this is that if a ciliary beat is planar, motor activity alternates between the two halves of the axoneme separated by the beat plane. Transition between the effective and recovery strokes is achieved by a switching mechanism (e.g., Ref. 8), which will have a frequency equal to that of the cilium. The dynein arm action is therefore subject to considerable control, the mechanism of which is essential to understand in a comprehensive model of ciliary behavior.

MICROTUBULE SLIDING DISPLACEMENT

At the end of a typical effective stroke, the angle between the tangents to the ends of the major bend on the cilium is about 100°. From this, it is straightforward to calculate that the maximum distance one microtubule has slid relative to its neighbor is about 100 nm, assuming that the cilium was straight at the beginning of the stroke. The amount of relative sliding is, of course, different for each pair of neighboring microtubules, being greatest for the pair lying most nearly parallel to the beat plane of the cilium. Estimates of the distance, or step size, through which an individual outer dynein arm can push a microtubule in one arm cycle vary from 4 nm (9) to 40 nm (10), with values of 8 nm or 16 nm currently being favored. Given a value of 8 nm, a sliding displacement of 100 nm could be achieved by the action of about 12 dynein arms. This is the minimum number of arms that need to be active on each doublet. Because of the unidirectional character of the dynein motors, a minimum number of 48 arms (12 on each of 4 microtubules) will be active within the whole cilium to produce the complete effective stroke. During the effective stroke, a maximum of 1000 outer arms could be active, so that, based on the geometry of the system, only about 5% need be active to produce the observed sliding.

FORCE GENERATION

To produce ciliary beating, the shearing forces generated by the dynein motors must overcome both external and internal resistances to motion. External resistances arise from the interaction between the cilium and its liquid environment and, since the system operates in the low Reynolds number regime of hydrodynamics, are dominated by viscosity. Several authors (e.g., Refs. 11–14) have investigated the propulsive thrust developed by cilia, and the interaction between the organelles and the surrounding medium is well understood. If the precise beat form of a cilium is known, the forces generated and energy dissipated at any stage of its beat can be calculated, but since the patterns of ciliary beating are generally nonuniform, it is more useful to estimate mean values of force and energy production. Using the expressions derived by Sleigh and Holwill (15) and the ciliary force coefficients derived by Lighthill (16), it can be shown that, at every instant during its cycle, a force of the order of 10 pN is exerted by a beating cilium on the liquid environment.

The internal resistances to ciliary movement arise from the elasticity of the system and, possibly, its internal viscosity. There are two components to the local elasticity, one, assumed constant, derived from stable structures such as the microtubules, the other due to transient linkages, such as those made by the dynein arms, and consequently fluctuating throughout the motion. The Young's modulus for microtubules has recently been measured as about 5×10^6 N m^{-2} (17), but insufficient information is available about the transient links to make a quantitative estimate of the variable elasticity. The microtubules and other structures are surrounded by an aqueous medium, contained within the ciliary membrane, which will oppose their motion, but the detailed nature of the resistance is not known, so that a quantitative determination of these resistive forces cannot be made. As noted by Brokaw (18), it is unlikely that the internal resistances will exceed the external forces, and, for the purposes of the current discussion, the two will be assumed to have the same value of 10 pN.

The force generated by extracted outer dynein arms has been measured and found to be in the range 1–5 pN for a single arm. Since the system operates at low Reynolds number, so that inertial forces have a negligible effect, the active and resistive forces are balanced at all stages of the motion. This allows the equation of motion to be established (e.g., Ref. 4) and, further, allows an estimate to be made of the minimum number of dynein arms that are required to maintain the motion. Since the total resistive force is about 20 pN, between 4 and 20 arms must be active in the cilium *at all times*. If the outer arms dominate the force-generation process, only about 1% of them need to be active at any one time, considerably less than the total number available.

ENERGY DISSIPATION

To estimate the rate at which the arms must be active, it is necessary to consider the energy dissipation of a cilium, which depends on the beat frequency, bend shape, elasticity of ciliary structures, and the viscosity of the surrounding liquid. A typical value for a 5-μm-long cilium is 10^{-16} J per beat (e.g., Ref. 15). The energy available from outer-arm dynein in vitro is about 10^{-19} J per arm cycle (11). This figure is similar to the value of 0.2×10^{-19} J for the work done by a dynein molecule calculated from the in vitro values of displacement (8 nm) and force (5 pN). If an energy supply of 10^{-19} J per arm cycle is available in the cilium, it implies that about 1000 arms, i.e., 50% of the outer arms, are active during one beat of a cilium.

DISCUSSION AND CONCLUSIONS

In this chapter, different physical characteristics associated with ciliary movement have been used to estimate the numbers of dynein arms required to produce that feature of the motion. Consideration of the sliding geometry indicates that a complete effective stroke can be formed by the action of a maximum of 48 outer dynein arms. Calculation of force production indicates that between 4 and 20 arms must be generating force at all times during a beat, while energy requirements indicate that 1000 arms must be active during a complete beat cycle. It is important to show that these different numbers reflect a consistent pattern of behavior for the action of dynein motors in the axoneme.

The number of arms required for force and energy production can be used to estimate the proportion of the arm cycle occupied by the duty phase. Suppose the 4–20 arms needed to sustain the force at any instant act synchronously, while successive groups are introduced sequentially, so that there is no overlap between the duty phase of one set of arms and the next set of become active. Because the force must be applied continuously, the end of the duty phase for the first set of arms will coincide with the beginning of the duty phase of the next set. The fraction of the cycle occupied by the duty phase is then found by dividing the number of arms needed to sustain the force by the number needed to satisfy the energy requirements of one beat; this fraction will therefore be between 4/1000 and 20/1000, or 0.4% and 2%. A typical ciliary beat occupies about 33 ms, which is roughly the period of the dynein arm cycle in vitro. The force-generation (or duty) phase of dynein is therefore predicted to be a small fraction of the complete cycle, supporting a similar earlier conclusion by Hamasaki et al. (19) based on a different argument.

On the basis of microtubule displacement, it is predicted that a relatively small number (maximum 48) of arms is required to generate the complete effective stroke, assuming that each motor has a step size of 8 nm. One way in which the apparent discrepancy between this figure and that based on force and energy can be resolved is by using the prediction that the dynein arm is active for a small fraction of its own cycle, and also of the ciliary beat. To sustain the continuous force needed to produce a smooth effective stroke, at least one arm should be active on each microtubule at all times during the stroke. This has the consequence that, regardless of the nature of the coordination between the motors, several hundred arms must enter their duty phase during this stroke. A significant implication is then that each dynein arm does not produce the displacement of 8 nm assumed earlier. This is not unreasonable, since the estimates of step size are made on in vitro preparations under conditions of negligible loading. Within the axoneme, the load conditions are different and could result in a displacement smaller than 8 nm for each arm. If 100 arms produce the 100 nm relative displacement between adjacent microtubules, each arm generates, on average, a displacement of 1 nm. A step size of 8 nm is attractive in terms of the mechanism, since it is a multiple of the 4 nm diameter of the tubulin monomer, and the arms are themselves spaced at 24 nm. However, the large size of the dynein motors (~10 nm diameter) and the flexibility of the arms suggest that the microtubule dimension might not place a rigid constraint on the system.

Another scenario is that groups of dynein arms act simultaneously, with each group moving the microtubule by 8 nm; in this way the number of arms could be made sufficient to satisfy the energy requirements of the system. If the force-generating phase of the dynein cycle remained at about 1% of the cycle time, the beat would occur in about 12 discrete phases. Support for this type of mechanism is provided by observations of quantized beating in cilia (20). A longer force-generating phase of 10% of the cycle time would allow smooth beating to occur with this model; a ratio of this magnitude is not precluded by the observations of Hamasaki et al. (19).

In any modeling of the action of motors within a cilium (e.g., Ref. 2), it is essential that the known requirements of ciliary force generation, energy dissipation, and geometry are satisfied. Suggestions are made in this paper as to ways in which the different numbers of arms needed for these physical processes can be reconciled, but further theoretical modeling coupled with experimental observation is required to provide a complete description of motor activity within the cilium.

REFERENCES

1. CJ Brokaw, R Kamiya. Bending patterns of *Chlamydomonas* flagella: IV. Mutants with defects in inner and outer dynein arms indicate differences in dynein arms function. Cell Motil Cytoskel 8:68–75, 1987.
2. HC Taylor, P Satir, MEJ Holwill. Assessment of inner dynein arm structure and possible function in ciliary and flagellar axonemes. Cell Motil Cytoskel 43:167–177, 1999.
3. P Satir. Mechanisms and controls of microtubule sliding in cilia. Symp Soc Exp Biol 35:172–201, 1982.
4. KE Machin. Wave propagation along flagella. J Exp Biol 35:796–806, 1958.
5. CJ Brokaw. Computer simulation of flagellar movement IV. Properties of an oscillatory two-state cross-bridge model. Biophys J 16:1029–1041, 1976.
6. MEJ Holwill, CA Miles. A mechanochemical model of flagellar activity. Biophys J 11:851–859, 1971.
7. WS Sale, P Satir. Direction of active sliding microtubules in Tetrahymena cilia. Proc Nat Acad Sci 74:2045–2049, 1977.
8. P Satir. Switching mechanisms in the control of ciliary motility. Mod Cell Biol 4: 1–46, 1985.
9. S Kamimura, R Kamiya. High-frequency vibration in flagellar axonemes with amplitudes reflecting the size of tubulin. J Cell Biol 116:1443–1454, 1992.
10. CJ Brokaw. Cross-bridge behavior in a sliding filament model for flagella, in: S Inoue and R Stephens, eds. Molecules and Cell Movement. New York: Raven Press, 1975, pp 165–179.
11. MEJ Holwill, P Satir. Generation of propulsive forces by cilia and flagella, in: J

Bereiter-Hahn, OR Anderson, and W-E Reif, eds. Cytomechanics. Berlin Heidelberg: Springer-Verlag, 1987, pp 120–130.
12. JR Blake. Infinite models for ciliary propulsion. J Fluid Mech 49:209–222, 1971.
13. RE Johnson, CJ Brokaw. Flagellar hydrodynamics: A comparison between resistive-force theory and slender body theory. Biophys J 25:113–127, 1979.
14. J Lighthill. Reinterpreting the basic theorem of flagellar hydrodynamics. Journal of Engineering Mathematics 30:25–34, 1996.
15. MA Sleigh, MEJ Holwill. Energetics of ciliary movement in Sabellaria and Mytilus. J Exp Biol 50:733–743, 1969.
16. J Lighthill. Flagellar Hydrodynamics. SIAM Review 18:161–230, 1976.
17. JE Schoutens. Prediction of Elastic Properties of Sperm Flagella. J Theor Biol 171: 163–177, 1994.
18. CJ Brokaw. Spermatozoan Motility: a biophysical survey. Biol J Linn Soc 7:423–439, 1975.
19. T Hamasaki, MEJ Holwill, K Barkalow, P Satir. Mechanochemical aspects of axonemal dynein activity studied by in vitro microtubule translocation. Biophys J 69: 2569–2579, 1995.
20. SS Baba. Regular steps in bending cilia during the effective stroke. Nature 282:717–720, 1979.

3

Ciliary Growth by Intraciliary Trafficking

Dawn Signor, Karen P. Wedaman,
Jose T. Orozco, Lesilee S. Rose,
and Jonathan M. Scholey
University of California, Davis
Davis, California

INTRODUCTION

Cilia are microtubule-based eukaryotic organelles that function in motility and sensory transduction. The basic design of both motile and sensory cilia consist of a membrane-bounded cylinder encompassing nine outer doublet microtubules, surrounding either a central pair apparatus (motile cilia, 9 + 2) or no inner microtubules (sensory cilia, 9 + 0; except in nematodes where sensory cilia typically contain a variable number of inner singlet microtubules) (1,2). The coordinated action of accessory structures such as dynein arms, radial spokes, and nexin links enable motile cilia to beat in a concerted manner to provide force for locomotion or for moving a fluid environment over the cell. In contrast, sensory cilia do not beat due to their relatively simple structure lacking these accessory structures, but instead incorporate sensory receptors and other components required for sensory signal transduction (3).

The assembly of ciliary axonemes involves the active transport of preassembled protein complexes (axonemal precursors) from the cell body where they are synthesized to the base of the axoneme, followed by their translocation to the distal tip of the cilia where they are incorporated (1,2,4). This process, called intraflagellar transport (IFT), is required for both ciliary/flagellar synthesis and maintenance and is characterized as the bidirectional transport of electron-dense macromolecular protein complexes called raft particles along the outer doublet axonemal microtubules and in association with the membrane (5,6). Recent work has demonstrated roles of kinesin and dynein microtubule-based motors in driving bidirectional IFT in both motile cilia/flagella and immotile sensory cilia (6). For example, heteromeric kinesin-II is responsible for driving the anterograde transport of IFT rafts from the base to the tip of the axoneme, an activity that is essential for the assembly and maintenance of cilia and flagella on sea urchin embryos and *Chlamydomonas*, respectively (7–10). Conversely, the retrograde transport of IFT rafts in *Chlamydomonas* is facilitated by a cytoplasmic dynein of the DHC1b class (11,12).

Sea urchin kinesin-II is the founding member of a subfamily of kinesin-related motor complexes called the heteromeric kinesins (13,14). This protein is heterotrimeric in structure, consisting of two nonidentical motor subunits, KRP_{95} and KRP_{85}, and a non–motor accessory polypeptide KAP and is an anterograde transport motor moving toward the plus-ends of microtubules in vitro at a rate of ~0.5 μm/sec (14–16).

The DHC1b class of dyneins is placed phylogenetically between the axonemal and cytoplasmic classes of dynein heavy chains (DHC) and includes divergent isotypes, which are present in both ciliated and nonciliated cells (17–20). DHC1b dyneins are retrograde transport motors moving toward the minus-ends of microtubules and may play roles in Golgi organization as well as retrograde IFT (19).

Elegant biochemical studies performed in *Chlamydomonas* have demonstrated that the basic subunit of IFT raft particles is a 16S multiprotein complex consisting of a minimum of 15 polypeptides (8,21,22). Two of these polypeptides share sequence homology with the OSM-1 and OSM-6 proteins that are required for sensory cilia function in the nematode *Caenorhabditis elegans* (8,22). Interestingly, *C. elegans* have no motile cilia, but possess only sensory cilia that comprise the terminal endings of a subset of neurons (60 out of a total of 302 neurons) in the adult worm (3,23). Of these 60 ciliated neurons, 26 are thought to be chemosensory, as their terminal sensory cilia are in contact with the external environment through pores in the body wall cuticle (3,23). These chemosensory neurons include the amphid and inner labial neurons of the head and the phasmid neurons of the tail. Like other sensory cilia, these chemosensory cilia contain a rudimentary axoneme that lack all accessory proteins required for motility and thus function to concentrate receptors and other signaling molecules responsible for detecting environmental stimuli that control at least five distinct behaviors including egg-laying, mating, chemotaxis, osmotic avoidance, and dauer larvae formation (3,24). These chemosensory neurons are similar in structure and function to sensory neurons in higher organisms that control sight (phototransducing rod cells of the retina), sound (kinocilia in the ear), and smell (olfactory neurons).

Due to many appealing features of these simple nematodes, including well-characterized genetics, a completely sequenced genome, a well-defined nervous system, their transparent nature, and numerous mutants available that demonstrate defects in sensory structure/function, we have begun using *C. elegans* as a model system to study the microtubule-based transport pathways required to build a sensory cilium/sensory neuron (25,26). To this end we have developed a novel in vivo transport assay, which allows us to directly label and visualize the movement of microtubule-motors and their cargoes within a living *C. elegans* neuron (25). As we will describe here, we have used this assay to characterize the in vivo transport properties of *C. elegans* homologs of the IFT transport motor heterotrimeric kinesin-II (27) and putative IFT raft particle components OSM-1 and OSM-6. These three proteins are expressed in chemosensory neurons and concentrate in sensory cilia. Using time-lapse fluorescence microscopy, we have demonstrated bidirectional transport of kinesin-II and IFT raft particle components OSM-1 and OSM-6 along cilia (IFT) and dendrites (intradendritic transport) (26). These proteins moved at the same rate in both the anterograde and retrograde direction, suggesting that kinesin-II drives the transport of IFT rafts from their site of synthesis in the cell body, out along dendritic microtubules, to their site of action in sensory cilia. Additionally, their similar retrograde transport rates suggest that they are retrieved by a common retrieval pathway. By using this assay to exploit existing mutants with chemosensory cilia defects, we have demonstrated that CHE-3 DHC1b cytoplasmic dynein is responsible for the retrograde transport of kinesin-II and IFT rafts within cilia but is not required for their retrograde retrieval along the dendrite back to the cell body (26). Using this assay we hope to completely dissect the active-transport pathways required for building a ciliated sensory neuron.

EXPERIMENTAL DESIGN

To visualize the movement of microtubule motors and putative cargo molecules in living *C. elegans* cells, heritable lines of transgenic worms are isolated that express these proteins labeled with the green fluorescence protein (GFP) (26). The in vivo transport of these fusion proteins is then monitored using time-lapse fluorescence microscopy (Fig. 1A). Briefly, genomic sequence encoding microtubule motors or cargo molecules and their upstream regulatory sequences isolated using standard molecular biology techniques. These sequences are placed upstream and in frame with the GFP gene in the *C. elegans* transformation vector

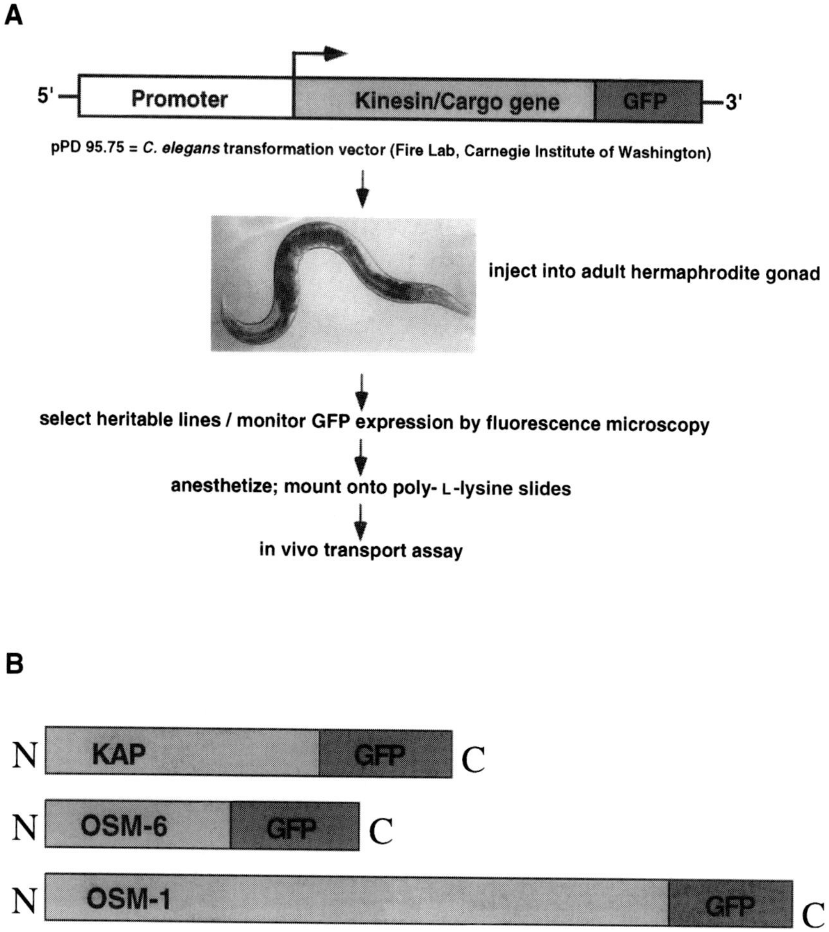

Figure 1 Experimental design of our in vivo transport assay. (A) Flow diagram describing the methods used to create transgenic worms and visualize fusion protein movement. *C. elegans* transformation vectors containing the gene of interest upstream from the GFP gene are microinjected into adult hermaphroditic worms. Heritable lines of worms expressing the GFP-fusion protein of interest are isolated, and the in vivo transport of the GFP-fusion protein is visualized using time-lapse fluorescence microscopy. (B) Schematic depicting the three GFP-fusion proteins monitored for movement in this study: kinesin-II :: GFP (labeled through its associated accessory polypeptide KAP) and IFT raft components OSM-6 :: GFP and OSM1 :: GFP.

pPd 95.77 (kindly provided by the Fire Lab, Carnegie Institute of Washington, Baltimore, MD). This vector drives the expression of a motor/cargo::GFP fusion protein. Once constructed, the transformation vector is microinjected into the adult hermaphrodite gonad with a coinjection marker, and heritable lines expressing the fusion protein of interest are isolated and maintained by self-fertilization. To monitor transport, worms are anesthetized in 1 or 10 mM levamisole, mounted on poly-L-lysine–coated slides, and time-lapse images of moving fluorescent particles are collected at 100× magnification every 0.5–1.0 seconds. To investigate the involvement of other genes in this transport pathway, these constructs were introduced into various mutant background by either microinjection or genetic crosses.

RESULTS AND DISCUSSION

We have thus far characterized the bidirectional in vivo transport velocities of kinesin-II (labeled through its non–motor accessory polypeptide KAP), OSM-1, and OSM-6 (Fig. 1B). Four classes of transport have been studied (Fig. 2): anterograde intradendritic transport, characterized as the transport from the cell body, along the dendrite to the base of the cilium; anterograde IFT, characterized as transport from the transition zone at the base of the cilium, along the axoneme toward the tip; retrograde IFT, from the tip of the cilium back to the transition zone; and retrograde intradendritic transport from the transition zone at the base of the cilium back toward the cell body. By monitoring the movement of individual fluorescence "dots" over multiple frames, average transport velocities were calculated for all classes of transport investigated (Table 1). It is clear that both the IFT motor kinesin-II and putative IFT raft particle proteins OSM-1 and OSM-6 move at the same velocity along both dendrites and cilia. The simplest interpretation of these data is that kinesin-II drives the transport of IFT raft particles from the cell body where they are synthesized and preassembled, out along the dendrite to the base of the transition zone (where they accumulate), then from the transition zone along the axoneme to the tip of the cilia. However, because kinesin-II is a plus-end–directed, anterograde transport motor, these proteins presumably must be retrieved by a minus-end–directed transport motor. The fact that both kinesin-II and IFT raft proteins OSM-1 and OSM-6 share the same retrograde transport velocities suggests that these proteins are retrieved by a common retrograde retrieval pathway.

We have investigated if the DHC1b cytoplasmic dynein, CHE-3 (28), is responsible for the retrograde retrieval of both the anterograde motor kinesin-II and its putative IFT raft particle cargo, by expressing these GFP fusion proteins in a *che-3* mutant background and assaying for both in vivo transport and localiza-

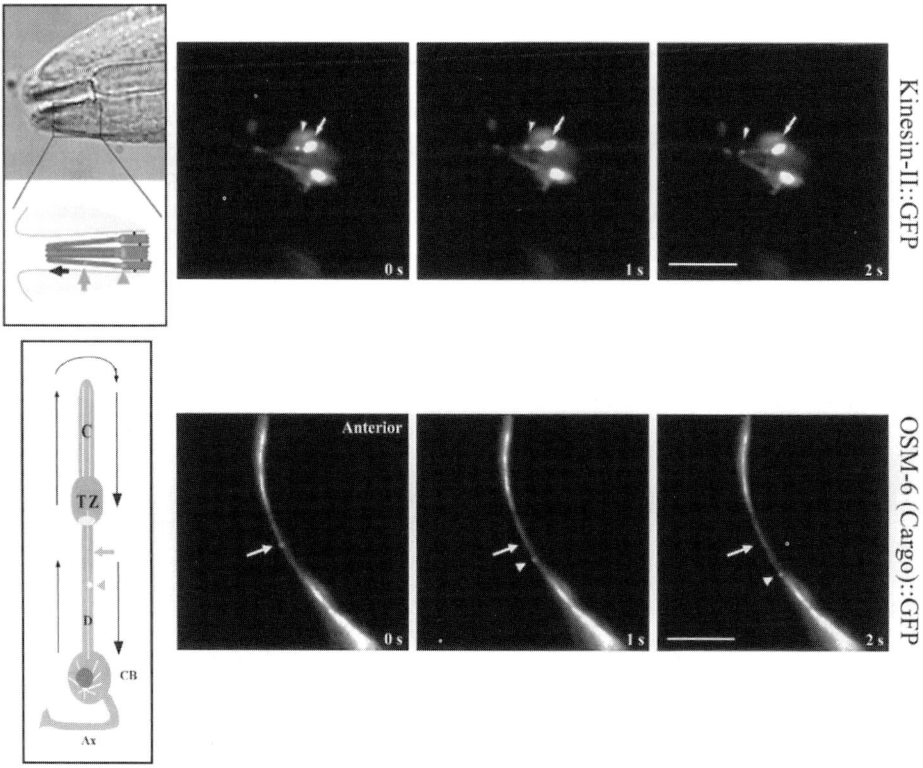

Figure 2 Bidirectional transport of kinesin-II :: GFP and IFT raft components OSM-1 :: GFP and OSM-6 :: GFP along sensory cilia and dendrites. These three GFP-fusion proteins can be seen moving in amphid neurons of the head along the sensory cilia (top panel) and associated dendrites (bottom panel). Shown are time-lapse panels of kinesin-II :: GFP and OSM-6 :: GFP, demonstrating the movement of these fluorescent particles in vivo. Top panel shows the anterograde IFT transport of kinesin-II :: GFP particles moving from the base of the cilium (transition zone) toward the tip of the cilium; lower panel shows the retrograde intradendritic transport of OSM-6 :: GFP particles moving from the base of the cilium back toward the neuronal cell body. Cartoon images are presented for orientation. Arrows point to regions of fixed fluorescence (i.e., the accumulation of fluorescence at the transition zone in the upper panel); arrowheads point to moving fluorescent particles. C = Cilia, TZ = transition zone, D = dendrite, CB = cell body, and Ax = axon. Bar, 5 μm.

TABLE 1 Bidirectional Transport Velocities of IFT Components Along Sensory Cilia and Dendrites

	IFT (transport in cilia), average velocity (μm/s)		Intradendritic transport, average velocity (μm/s)	
Transgenic strain	Anterograde	Retrograde	Anterograde	Retrograde
OSM-1::GFP	0.67 ± 0.10	1.10 ± 0.14	0.71 ± 0.07	1.04 ± 0.08
OSM-6::GFP	0.68 ± 0.11	1.08 ± 0.16	0.75 ± 0.10	1.02 ± 0.10
Kinesin-II::GFP	0.67 ± 0.09	1.07 ± 0.10	0.71 ± 0.08	1.01 ± 0.14

tion. In the *che-3* mutant background, we fail to see any retrograde IFT movement of kinesin-II::GFP, OSM-1::GFP, or OSM-6::GFP along cilia, although the anterograde transport of these fusion proteins occurs normally. Accordingly, when these transgenic worms were examined by confocal microscopy, we observed an accumulation of kinesin-II::GFP, OSM-1::GFP, and OSM-6::GFP at the distal tips of the slightly truncated cilia in *che-3* mutants, compared to the normal enrichment of these fusion proteins in the transition zones of wild-type worms (Fig. 3A). Coupled with a lack of in vivo transport of these proteins, this accumulation at ciliary tips is presumably due to the continuous anterograde movement of kinesin-II and IFT rafts from the transition zone to the tip of the cilia but defective retrieval in the absence of CHE-3 cytoplasmic dynein function. However, CHE-3 dynein is not required for the retrograde transport of kinesin-II, OSM-1, and OSM-6 from the transition zone back to the cell body because the retrograde transport of these GFP-fusion proteins occurs normally in the dendrite of *che-3* mutants (Fig. 3B).

CONCLUSIONS

Our data argue that kinesin-II drives the transport of IFT rafts in the anterograde direction from the neuronal cell body, out along the dendrite, to its final destination at the tip of the sensory cilium. Then, CHE-3 cytoplasmic dynein retrieves both the anterograde motor, kinesin-II, and its associated IFT raft cargo from the ciliary tip back to the transition zone to replenish the available pools of anterograde motor and raft particles. The fact that kinesin-II, OSM-1, and OSM-6 all accumulate at the transition zone suggests that this region serves as a ''loading dock'' for IFT rafts and the specific cargo that they may be carrying. The retrieval of IFT rafts and kinesin-II from the transition zone back toward to the cell body likely depends on an as-yet-unidentified retrograde motor, such as another cytoplasmic dynein isoform or a minus-end–directed kinesin-related motor.

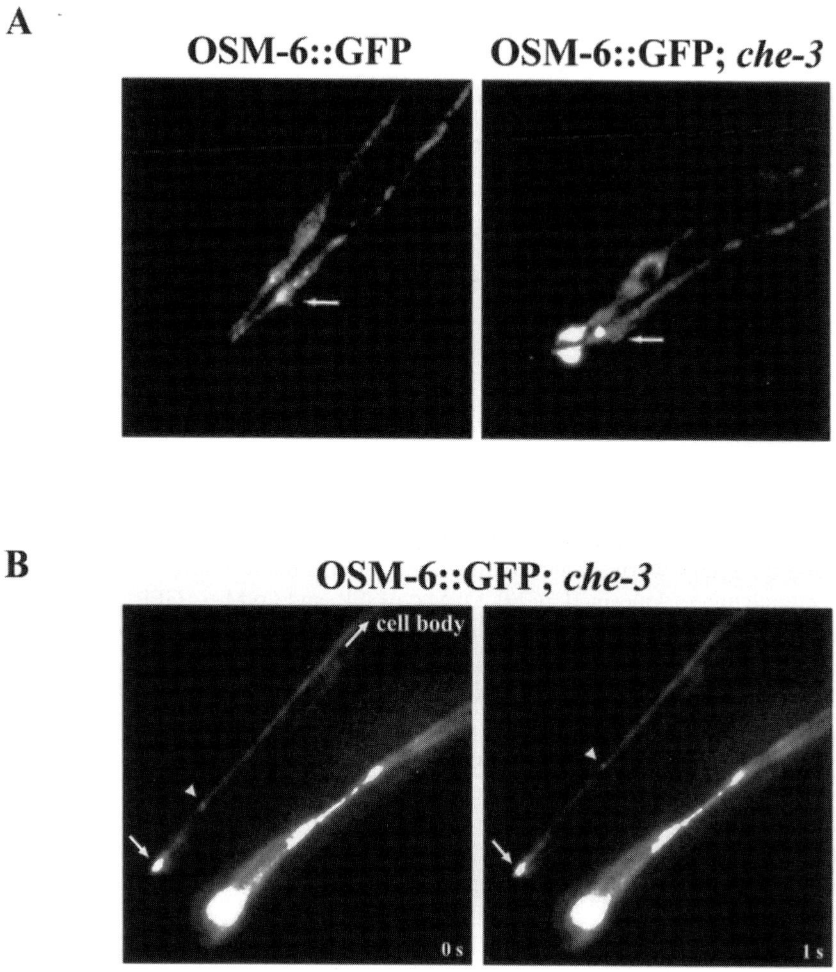

Figure 3 CHE-3 cytoplasmic dynein is required for the retrograde transport of kinesin-II and IFT raft particles along cilia, but not dendrites. (A) Confocal image of phasmid chemosensory neurons of OSM-6::GFP expressing transgenic worms. In wild-type worms (left image), OSM-6::GFP accumulates normally at the base of the cilium in the transition zone (arrow) and localizes diffusely along the length of the sensory cilium. In *che-3* dynein mutant worms (right image), OSM-6::GFP accumulates dramatically at the tips of the cilia in large bulbous structures. (B) Although we see no retrograde movement of kinesin-II::GFP, OSM-6::GFP, or OSM-1::GFP along cilia in *che-3* mutants, these fusion proteins demonstrate normal retrograde transport in the associated dendrite, suggesting a specific subcellular role of CHE-3 cytoplasmic dynein in mediating retrograde IFT. Shown is a time-lapse panel OSM-6::GFP moving retrogradely from the transition zone back toward the cell body in the *che-3* mutant background. Arrows point to the accumulation of OSM-6::GFP at ciliary tips, and arrowheads point to moving OSM-6::GFP particles.

It is plausible that IFT rafts are general transport vehicles used to convey key neuronal components from their site of synthesis in the cell body, out along the dendrite, to their site of action at the distal tip of the sensory cilium. Such components may include axonemal components for sensory cilium synthesis and maintenance, sensory receptors, and other signaling molecules that are necessarily concentrated at the ciliary tip and required to detect and transduce signals from the environment. The fact that mutations in the *osm-1*, *osm-6*, and *che-3* genes result in worms with defective chemosensory behavior and ciliary structural defects (3,22) supports this hypothesis. Because there is typically no vesicle transport within cilia (5), membrane-associated proteins (i.e., receptors) could be transported to the tip of the cilium by IFT rafts, which in *Chlamydomonas* occurs in association with the flagellar membrane. These IFT rafts may associate with such cargoes either in the cell body or at the transition zone.

We have thus far used this in vivo transport assay to investigate the roles of two microtubule-based motors in driving bidirectional transport required for sensory ciliogenesis in *C. elegans* chemosensory neurons. We are now using this assay to exploit a number of other existing mutants in *C. elegans* that demonstrate defects in chemosensory behavior and/or sensory cilia structure to further identify the specific molecules required in this pathway. Additionally, we are using a reverse genetic approach to isolate kinesin-II mutant worms to show unequivocally that kinesin-II transports IFT raft particle components such as OSM-1 and OSM-6.

REFERENCES

1. KA Johnson. Keeping the beat: form meets function in the *Chlamydomonas* flagellum. Bioessays 17:847–854, 1995.
2. RE Stevens. Ciliogenesis in sea urchin embryos—a subroutine in the program of development. Bioessays 17:331–340, 1995.
3. LA Perkins, EM Hedgecock, JN Thomson, JG Culotti. Mutant sensory cilia in the nematode *Caenorhabditis elegans*. Dev Biol 117:456–487, 1986.
4. KA Johnson, JL Rosenbaum. Polarity of flagellar assembly in *Chlamydomonas*. J Cell Biol 19:1605–1611, 1992.
5. KG Kozminski, KA Johnson, P Forscher, JL Rosenbaum. A motility in the eukaryotic flagellum unrelated to flagellar beating. Proc Natl Acad Sci USA 90:5519–5523, 1993.
6. JL Rosenbaum, DG Cole, DR Diener. Intraflagellar transport: the eyes have it. J Cell Biol 144(3):385–388, 1999.
7. RL Morris, JM Scholey. Heterotrimeric kinesin-II is required for the assembly of motile 9 + 2 ciliary axonemes in sea urchin embryos. J Cell Biol 138:1009–1022, 1997.

8. DG Cole, DR Diener, AL Himelblau, PL Beech, JC Fuster, JL Rosenbaum. Chlamydomonas kinesin-II-dependent intraflagellar transport (IFT): IFT particles contain proteins required for ciliary assembly in *Caenorhabditis elegans* sensory neurons. J Cell Biol 141(4):993–1008, 1998.
9. Z Walther, M Vashishtha, JL Hall. The *Chlamydomonas* Fla10 gene encodes a novel kinesin homologous protein. J Cell Biol 126:175–188, 1994.
10. KG Kozminski, PL Beech, JL Rosenbaum. The *Chlamydomonas* kinesin-like protein FLA10 is involved in motility associated with the flagellar membrane. J Cell Biol 131:1517–1527, 1995.
11. GJ Pazour, BL Dickert, GB Witman. The DHC1b (DHC2) isoform of cytoplasmic dynein is required for flagellar assembly. J Cell Biol 144(3):473–481, 1999.
12. ME Porter, R Bower, JA Knott, P Byrd, W Dentler. Cytoplasmic dynein heavy chain 1b is required for flagellar assembly in *Chlamydomonas*. Mol Biol Cell 10:693–712, 1999.
13. DG Cole, WZ Cande, RJ Baskin, DA Skoufias, CJ Hogan, JM Scholey. Isolation of a sea urchin egg kinesin-related protein using peptide antibodies. J Cell Sci 101: 291–301, 1992.
14. DG Cole, SW Chinn, KP Wedaman, K Hall, T Vuong, JM Scholey. Novel heterotrimeric kinesin-related protein purified from sea urchin eggs. Nature 366:268–270, 1993.
15. KP Wedaman, DW Meyer, DJ Rashid, DG Cole, JM Scholey. Sequence and submolecular localization of the 115-kD accessory subunit of the heterotrimeric kinesin-II ($KRP_{85/95}$) complex. J Cell Biol 132:371–380, 1996.
16. DJ Rashid, KP Wedaman, JM Scholey. Heterodimerization of the two motor subunits of the heterotrimeric kinesin, $KRP_{85/95}$. J Mol Biol 252:157–162, 1996.
17. BH Gibbons, DJ Asai, WJ Tang, TS Hays, IR Gibbons. Phylogeny and expression of axonemal and cytoplasmic dynein genes in sea urchins. Mol Biol Cell 5(1):57–70, 1994.
18. PS Criswell, LE Ostrowski, DJ Asai. A novel cytoplasmic dynein heavy chain: expression of DHC1b in mammalian ciliated epithelial cells. J Cell Sci 109(7):1891–1898, 1996.
19. E Holzbaur, RB Vallee. Dyneins: molecular structure and cellular functions. Ann Rev Cell Biol 10:339–372, 1994.
20. EA Vaisberg, PM Grissom, JR McIntosh. Mammalian cells express three distinct dynein heavy chains that are localized to different cytoplasmic organelles. J Cell Biol 133(4):831–842, 1996.
21. G Piperno, K Mead. Transport of a novel complex in the cytoplasmic matrix of *Chlamydomonas* flagella. Proc Natl Acad Sci USA 94:4457–4462, 1997.
22. J Collet, CA Spike, EA Lundquist, JE Shaw, RK Herman. Analysis of osm-6, a gene that affects sensory cilium structure and sensory neuron function in *Caenorhabditis elegans*. Genetics 148:187–200, 1998.
23. S Ward, N Thompson, NJG White, S Brenner. Electron microscopic reconstruction of the anterior sensory anatomy of the nematode *Caenorhabditis elegans*. J Comp Neurol 160:313–317, 1975.
24. TA Starich, RK Herman, CK Kari, WH Yeh, WS Schackwitz, M Schulyler, J Collet,

J Thomas, D Riddle. Mutations affecting chemosensory neurons of *Caenorhabditis elegans*. Genetics 139:171–188, 1995.

25. JT Orozco, KP Wedaman, D Signor, H Brown, L Rose, JM Scholey. Movement of motor and cargo along cilia. Nature 398:674, 1999.
26. D Signor, KP Wedaman, JT Orozco, H Brown, L Rose, JM Scholey. Movement of motor and cargo along cilia. Nature 398:674, 1999.
27. D Signor, KP Wedaman, LS Rose, JM Scholey. Two heteromeric kinesin complexes in chemosensory neurons and sensory cilia of *Caenorhabditis elegans*. Mol Biol Cell 10:345–360, 1999.
28. SR Wicks, CJ deVries, H van Luenen, R Plasterk. Personal communication. Amsterdam: Netherlands Cancer Institute, Div. Of Mol. Biol.

4

The Regulation of Airway Ciliary Beat Frequency by Intracellular Calcium

Michael J. Sanderson, Alison B. Lansley,* and John H. Evans†
University of Massachusetts Medical School
Worcester, Massachusetts

INTRODUCTION

During breathing, the airways transport into the lungs large quantities of air that is contaminated with a variety of pollutants, particles, and bacteria. To prevent the accumulation of these contaminants, the airways have developed a defense mechanism of mucociliary clearance that relies on ciliary activity (Sanderson, 1997; Wanner et al., 1996). The airways, from the trachea to the terminal bronchioles, are lined with a ciliated epithelium. The cilia of each cell project towards the airway lumen and are bathed in a "watery" ionic solution called the periciliary layer (Fig. 1a). Each cilium performs a repetitive beat cycle consisting of a rest, recovery, and effective stroke. During the effective stroke the cilium makes contact with the overlying mucus and transports it, together with entrapped particles, along the airways for expulsion at the esophagus.

**Current affiliation*: University of Brighton, Brighton, United Kingdom.
†*Current affiliation*: National Jewish Medical and Research Center, Denver, Colorado.

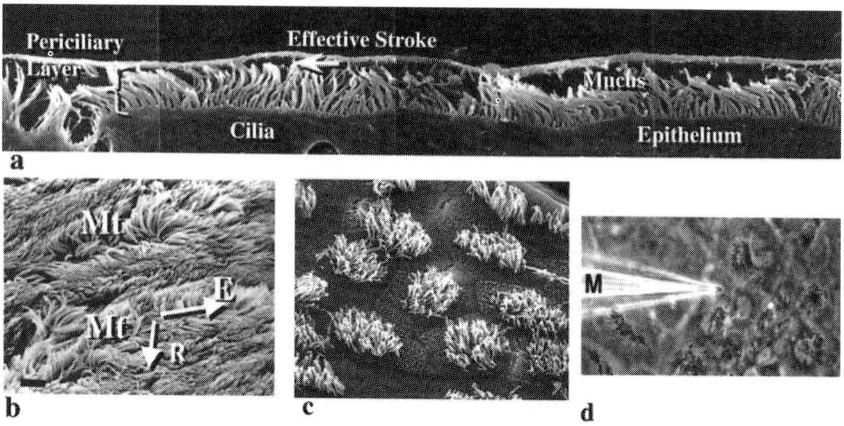

FIGURE 1 The ciliated epithelium of the rabbit airways. (a) A scanning electron micrograph (SEM) of a profile section of the tracheal ciliated epithelium showing the organization of the mucociliary interface. Numerous cilia, arising from the epithelial cells, are observed at various stages of their beat cycle. The cilia beat within the periciliary layer, and their effective stroke is directed towards the left. The overlying mucus is supported by the periciliary layer at the tips of the cilia and is transported when the ciliary tips engage the mucus. (b) An SEM of tracheal ciliated epithelium observed from above. The cilia are densely packed and individual cells are not distinguishable. The ciliary activity is organized into metachronal patterns (Mt) to minimize interference. The effective stroke (E) directed to the upper right, while the recovery stroke and the associated metachronal (Mt) wave are directed toward the bottom. (c) An SEM of an outgrowth of the ciliated epithelium in tissue culture. The cilia of individual cells are centralized in each cell, and the boundaries of each cell can be identified. (d) A phase-contrast light micrograph of a cultured outgrowth similar to that shown in c. Again, individual cells and their cilia can be identified. A glass microprobe (M) is aligned just above the cells and is used to mechanically stimulate a single cell.

To avoid interference and to enhance cooperative activity, the densely packed cilia beat with a metachronal organization coordinated by hydrodynamic forces acting through the periciliary fluid that surrounds the cilia (Fig. 1b). Because mucus is a viscous, elastic secretion, it cannot be moved by individual cilia, but with metachronal coordination the cooperative efforts of multiple cilia can be recruited for mucus transport. By relying on the recovery stoke of the cilia to initiate and propagate metachrony, many cilia can be recruited into the process of mucus transport before the slowing and disorganization effects of cilia-mucus contact occur (Sanderson and Sleigh, 1981).

Although the cilia provide the driving force for mucus clearance and are vital for maintaining healthy lungs, it is surprising that we do not fully understand the mechanisms controlling airway ciliary activity. The major parameter of ciliary activity that contributes to the efficiency of mucus transport is beat frequency. Mucus transport is driven by interactions between the ciliary tip and the mucus (Fig. 1a). Therefore a faster beat frequency results in a faster ciliary tip velocity, which translates into faster mucus transport. Furthermore, cilia with faster beat frequencies have a greater influence on the surrounding periciliary layer and thereby increase the hydrodynamic coupling between adjacent cilia. Increased hydrodynamic coupling enhances the metachronal organization of ciliary activity, and as a result more cilia can be recruited and coordinated into the process of mucus transport.

It has been well established that the rate of ciliary beat frequency is increased by increases in intracellular calcium concentration ($[Ca^{2+}]_i$), but the details of this mechanism are not well understood (Korngreen and Priel, 1994; Lansley et al., 1992; Salathe et al., 1993; Verdugo et al., 1983). We believe this situation results from the fact that it has been difficult to simultaneously measure ciliary beat frequency and changes in $[Ca^{2+}]_i$ and that the changes in $[Ca^{2+}]_i$ and beat frequency, within and between cells, occur rapidly and with complex temporal and spatial patterns. As a result, we have developed and describe here high-resolution techniques for the simultaneous measurement of ciliary beat frequency and $[Ca^{2+}]_i$ (Lansley and Sanderson, 1999). With these techniques we have gained some insight into the Ca^{2+}-based regulatory mechanisms of airway ciliary activity by discovering that the ciliary beat frequency can be regulated by a frequency-modulation process controlled by oscillations in $[Ca^{2+}]_i$ at the ciliary base (Evans and Sanderson, 1999a, b).

TISSUE CULTURE OF CILIATED EPITHELIAL CELLS

For these studies, the ciliary activity of cultured airway cells was examined. Explants of the ciliated mucosa of the rabbit trachea were placed on collagen-coated coverslips to allow the outgrowth of cells. Cells are usually ready for use between 5 and 10 days. The details of these techniques are described in detail by Dirksen et al. (1995). The epithelial cultures provide a robust experimental preparation in which the individual cells and their cilia can be readily identified—a prerequisite essential for the correlation of ciliary activity with changes in $[Ca^{2+}]_i$ (Fig. 1c). A simple protocol, previously found to initiate increases in beat frequency and Ca^{2+} waves in these cultured cells, was mechanical stimulation of a single cell (Sanderson and Dirksen, 1986; Sanderson et al., 1990). As a result, mechanical stimulation has been used to initiate cell responses to correlate beat frequency and changes in $[Ca^{2+}]_i$. A glass microprobe (tip ~1 μm) driven by a piezoelectric device is used to mechanically stimulate a single cell by briefly (150 msec) dimpling its apical surface (Fig. 1d).

MEASUREMENT OF CILIARY BEAT FREQUENCY WITH HIGH-SPEED VIDEO MICROSCOPY

Beat frequency can be determined from the variation in light intensity of the microscope image that results from ciliary movement (Sanderson and Dirksen, 1985). For this technique, the ciliary activity is observed with phase-contrast optics, which enhances the variation in light intensity as the cilia move. The variation in light intensity has been frequently detected from a few cilia with a photomultiplier to generate an analog waveform that represents the repetitive beat cycles; frequency is obtained by counting the peaks per second. To accelerate this analysis and to compensate for system ''noise'' or irregular signals, the waveforms have been processed using spectrum analyzers, which rely on functions such as fast Fourier transforms. While this approach provides frequency data, the time resolution of the data is reduced to the duration of the sampling window, which is commonly 1–2 seconds (Korngreen and Priel, 1994; Mao and Wong, 1998; Salathe et al., 1993). Unfortunately, these approaches are inadequate to quantify the rapid changes in beat frequency induced by changes in $[Ca^{2+}]_i$.

Because ciliary beat frequency can approach rates of 40 Hz, accurate quantification of the waveform signal requires (according to the Nyquist theorem) a minimum sampling rate of 80 Hz to ensure that signal aliasing does not occur (Inoue and Spring, 1997). If the various phases of the beat cycle are also to be measured or a higher fidelity signal is to be recorded, a faster sampling rate is necessary. As a result, we have implemented a high-speed video system, based on conventional video equipment, to record ciliary activity at 240 frames per second (fps) (Roos and Parker, 1990; Sanderson, 1998; Sanderson and Dirksen, 1995). An additional advantage of using video images is that multiple groups of cilia can be measured simultaneously.

Normally, video images are formed by sequential horizontal (H_{sync}) lines that scan the image (first scan = even field). At the bottom of the image, the scanning process is reinitiated by a vertical synchronization (V_{sync}) pulse (Inoue and Spring, 1997). A second series of horizontal scans is completed (odd field), and these scan lines are interlaced between the lines of the first scan to form a complete image. A V_{sync} pulse initiates the scan of the next image. Each field is completed in 1/60 second and each new video image occurs every 1/30 second. To achieve imaging speeds faster than the normal 30 fps, the scanning protocol of our video camera, a charge coupled device (CCD) was modified. Instead of using the V_{sync} pulse to reset the second scan to be interlaced with the first scan, the scan is reset to occur along the same lines as the original scan. An increase in the frame rate was achieved by the generation of additional V_{sync} pulses (Fig.

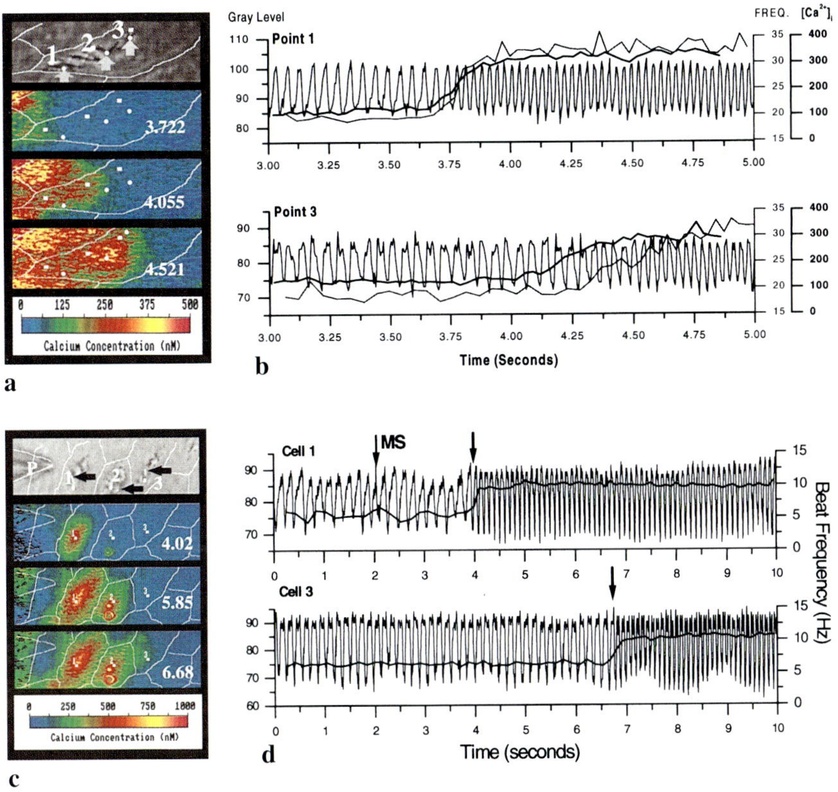

FIGURE 4.6 Intracellular and intercellular Ca^{2+} waves evoke sequential increases in ciliary beat frequency. (a) The changes in $[Ca^{2+}]_i$ within a single cell associated with the propagation of an intracellular wave that is part of an intercellular wave initiated by mechanical stimulation in an adjacent cell at 37°C. (b) Changes in the gray level of the phase-contrast image (left scale and waveform), beat frequency (right scale, gray line and text), and $[Ca^{2+}]_i$ (right scale, black line and text) with respect to time during the passage of the intracellular Ca^{2+} wave across two points within the cell. The $[Ca^{2+}]_i$ is sequentially elevated as the Ca^{2+} wave moves from point 1 to point 3. Following a slight delay, after the passage of the Ca^{2+} wave, the beat frequency rapidly increases within 3–6 beat cycles to a maximal rate. The beat frequency at point 3 remained at a basal rate until the Ca^{2+} wave passed the basal area of the cilia, even though the cilia at point 1 were beating at a maximal rate. (c) An intercellular Ca^{2+} wave induced by mechanical stimulation (MS, arrow indicates time of stimulation) of an adjacent cell, propagating through three cells at 25°C. (d) Changes in the gray level of the phase-contrast image (left scale and waveform) and beat frequency (right scale, black line) with respect to time during the passage of the intercellular Ca^{2+} wave through 3 cells. At 25°C, the basal beat frequency is slower (note difference in time scales between b and d). As the Ca^{2+} wave passes the base of the cilia of each cell, the beat frequency increases (arrows) to a maximal rate within 2–4 beat cycles. The maximal beat frequency rate is also slower at 25°C. (All) A phase-contrast image of the cells is shown at the top of each image panel. The time at which each image was recorded is indicated in each image. The white lines represent cell boundaries. The magnitude of the changes in $[Ca^{2+}]_i$ is indicated by the color scales. The positions at which $[Ca^{2+}]_i$ and ciliary beat frequency were analyzed to generate the graphs (right) are indicated by the large and small white squares, respectively. Beat frequencies are calculated for each beat cycle and plotted at the end of each beat cycle.

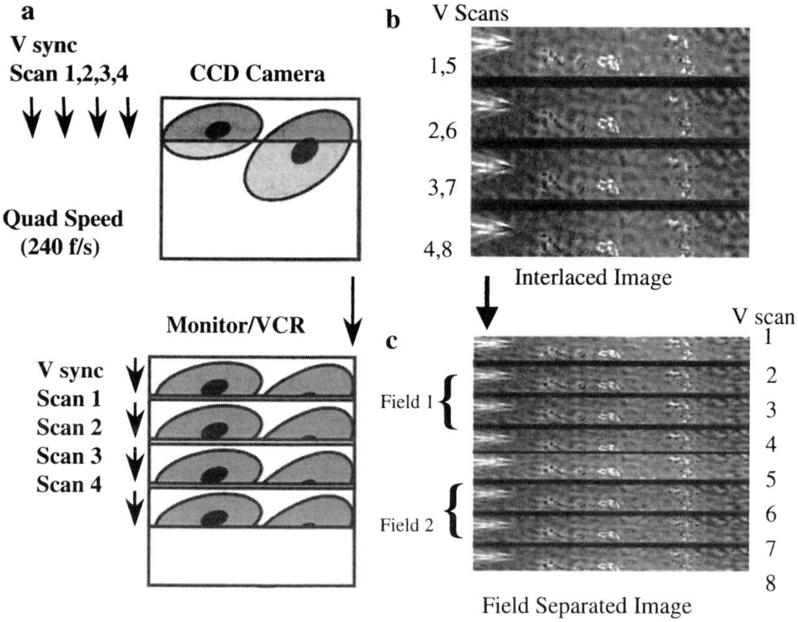

FIGURE 2 The principle of high-speed video for measuring ciliary beat frequency. (a) The generation of the multiple images. The scanning electronics of the CCD camera are modified so that the same part of the camera faceplate is scanned four times during each field (1/60 second). This results in 4 × 1/4 identical images. Two fields are interlaced to generate a recordable video image (b). To view each image in the correct sequence, the first field must be extracted and placed at the top of the image (c), while the second field is placed at the bottom of the image.

2a). After initiating the scan of the image, another V_{sync} pulse is generated when the scan has only covered one fourth of the image. The area scanned represents the top quarter (~100 lines) of the CCD camera faceplate. As a result, identical images are produced at 240 fps. The width and horizontal resolution of the image remain the same, but the image height is reduced to one quarter and, because the scans follow the same lines of one video field and ignores the lines of the interlaced field, there is a reduction (by half) in the vertical spatial resolution. This loss of image resolution is not a problem because the area used to detect the ciliary movement is wider than two video lines. A four-panel image is the typical output of the high-speed video system (Fig. 2b). Because the scan function of the videocassette recorder (VCR) and video monitor were not modified, each panel actually consists of two interlaced images, with a total of eight images, from the same area. To correctly view the temporal sequence of the recorded images, it is necessary to extract and rearrange the interlaced images by digital image processing (Fig. 2c). The even lines (representing images 1–4) are extracted and placed, in order, at the top of the image. The odd lines (representing images 5–8) are placed in order at the bottom of the image. This results in eight sequential images per video frame. However, image reconstruction is not essential for measuring beat frequency. The ciliary beat frequency of multiple groups of cilia is simply extracted from these images by sequentially determining the gray intensity of the image at a site of ciliary movement. By carefully positioning a small detector area over the cilia, a uniform waveform representing beat frequency is obtained (see Fig. 6b and d), from which the beat frequency of each beat cycle can be calculated.

MEASUREMENT OF $[Ca^{2+}]_i$ WITH FAST VIDEO IMAGING OF FURA-2

To measure the $[Ca^{2+}]_i$, cells are loaded with the Ca^{2+}-specific dye fura-2. This dye is available in a cell-permeant form [acetoxymethyl (AM) ester] that readily enters the cell and becomes trapped following the cleavage of the AM group from the dye. Cells are viewed with a fluorescence microscope. Because changes in $[Ca^{2+}]_i$ occur rapidly, it is necessary to sample images quickly if these changes are to be correlated with changes in ciliary beat frequency. However, the technique of ratio imaging with fura-2 to quantify $[Ca^{2+}]_i$ can be slow. Reasons for this include slow filter changes and image acquisition and camera lag. To overcome these problems, we have used a single wavelength recording technique, which allows $[Ca^{2+}]_i$ changes to be recorded at 30 fps (Fig. 3). These measurements are quantified by reference to calibration images, recorded when the time constraints were less important. A detailed description of this technique is provided by Leybaert et al. (1998).

Briefly, the initial starting conditions are recorded (10 frames at wavelengths 340 and 380 nm). From these images and calibration values, the initial $[Ca^{2+}]_i$ can be calculated from the experimental ratios of 340/380 images (Fig. 3). The initial fluorescence values (F_0) are determined. The experimental images are subsequently collected at 30 fps at a single wavelength of 380 nm for approximately 30 seconds. During this time the changes in calcium are proportional to the change in fluorescence of the image, i.e., $[Ca^{2+}]_i \sim F_0/F_t$. After about 30 seconds, a reference image is obtained at 340 nm, and this is used with the 380 nm image to determine the exact $[Ca^{2+}]_i$ at 30 seconds. Images are further recorded for another 30 seconds at 30 fps, and the relative changes are used once again to determine the $[Ca^{2+}]_i$. At the end of the experiment, a final set of reference images were recorded (Fig. 3). The major advantage with ratio imaging is that potential artifacts such as changes in dye concentration, e.g., bleaching, are compensated for. When returning to single wavelength recordings, this potential

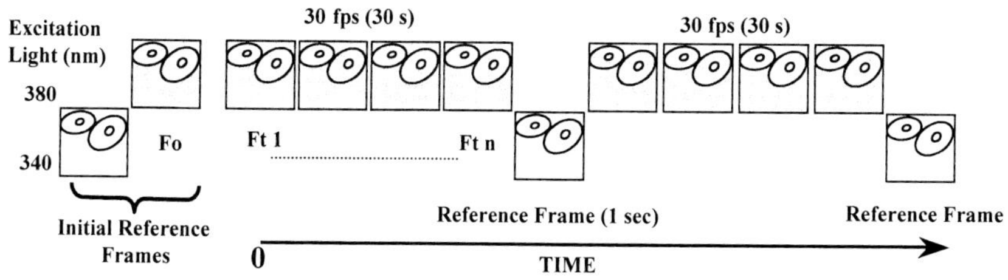

FIGURE 3 The principle of single wavelength recording for the measurement of $[Ca^{2+}]_i$. Initially, a pair of images are recorded using 340 (bottom line) and 380 (top line) nm excitation. This image pair is used to determine the initial $[Ca^{2+}]_i$. The 380 nm image also serves as the reference fluorescence image at time zero (F_o). Fast image recording (30 fps) is performed at 380 nm for 30–60 seconds. A second 340 nm reference image is obtained and fast recording at 380 nm resumed. At the end of the experiment, a final 340 nm image is recorded. The change in fluorescence at 380 nm is determined with reference to F_o and this change in fluorescence is converted into $[Ca^{2+}]_i$ with respect to the initial $[Ca^{2+}]_i$ and F_o.

artifact is more of a concern. However, the reference images can be used to calculate the real $[Ca^{2+}]_i$ at known time points, and these can be used to calibrate the single wavelength calculations.

SIMULTANEOUS MEASUREMENT OF CILIARY BEAT FREQUENCY AND $[Ca^{2+}]_i$

To correlate changes in ciliary beat frequency and $[Ca^{2+}]_i$, it is important to measure each parameter simultaneously. This is achieved by using different light wavelengths to monitor each process (Fig. 4). This method is described in detail by Lansley and Sanderson (1999). To measure $[Ca^{2+}]_i$, excitation light of 340 or 380 nm (for fura-2) is generated by filtration of a Hg arc lamp. This light is directed to the specimen by a 400 nm dichroic mirror. The resulting fluorescent light at 510 nm passes through the 400 nm dichroic mirror but is reflected by a 600 nm dichroic mirror through a 510 nm filter to the silicone intensified camera (SIT). The images are recorded on separate optical memory disc recorders (OMDRs). To measure ciliary activity, white light from a halogen source is filtered at 645 nm to generate a beam that illuminates the specimen and passes through the dichroic mirrors and is reflected to the CCD camera (Fig. 4). The resulting high-speed images are recorded on a second OMDR.

The alignment of the analysis area is critical for the formation of a clear signal representing either ciliary activity or $[Ca^{2+}]_i$ (Fig. 5). The change in ciliary position, and therefore the range in image intensity, is greatest toward the ciliary

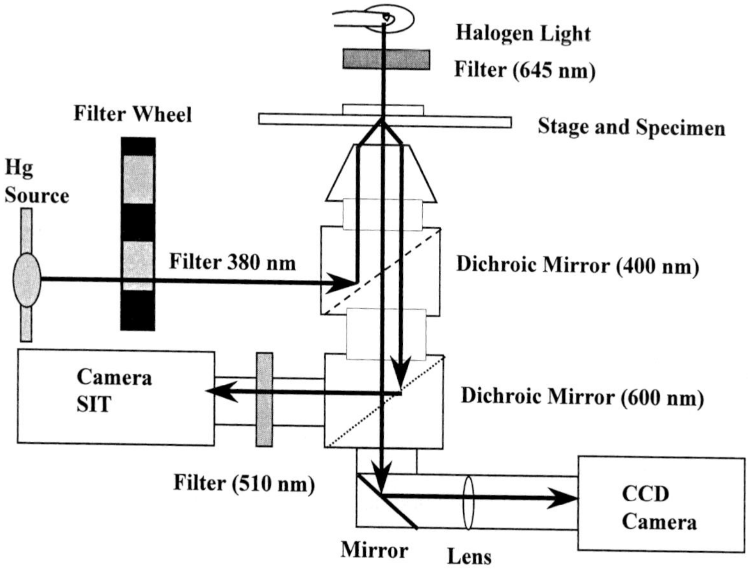

FIGURE 4 A schematic of the microscope optics used for the simultaneous recording of fura-2 fluorescence to measure $[Ca^{2+}]_i$ and the phase-contrast image of the cilia to measure beat frequency. Excitation light (340 or 380 m) is selected by filtration from an Hg bulb and directed to the specimen by a 400 nm dichroic mirror. The resulting fluorescent image, at 510 nm, passes through the 400 nm dichroic mirror and is deflected through a barrier filter to the SIT camera by a 600 nm dichroic mirror. The phase-contrast image is formed using 645 nm light. This light passes through all the dichroic mirrors to reach the CCD camera. A reduction lens equalizes the difference in the magnification that would result from the different faceplate sizes of each camera.

tip. Consequently, to measure beat frequency a small area of interest is selected towards the ciliary tip. The detector area is small so that it detects the nearly synchronous movement of only a few cilia. A larger area extends over a great range of the metachronal activity of the cilia and, as a result, records a more complex waveform. Changes in $[Ca^{2+}]_i$ occur within the cell and are most likely to affect the cilia, at least initially, through the ciliary base. As a result, the area of interest for calcium measurements is aligned with the base of the cilia (Fig. 5). In this case, a slightly larger area is used to compensate for the increase in the signal-to-noise ratio. It is important to point out that the ciliary and calcium events occur at different focal planes. The cilia are above the cell surface while the calcium changes occur within the cell. This difference can be accommodated

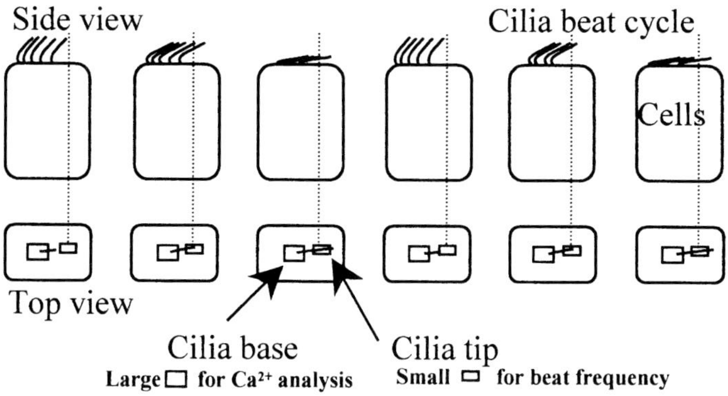

FIGURE 5 A cartoon illustrating the relative alignment positions of the areas of interest used to analyze ciliary activity and changes in $[Ca^{2+}]_i$. The side view (top) emphasizes that cilia move asymmetrically. Because ciliary activity is detected by variations in light intensity, the best location for the small area of interest is near the ciliary tip where the maximum contrast is generated. The area of interest used for $[Ca^{2+}]_i$ measurements is larger and placed at the base of the cilia being monitored.

by focusing the CCD (cilia) and SIT (calcium) cameras at slightly different image planes.

CHANGES IN CILIARY BEAT FREQUENCY INDUCED BY CALCIUM

An intercellular calcium wave, induced by mechanical stimulation of a single cell, was observed as it propagated through several adjacent cells. The sequential elevation of Ca^{2+} across an individual cell is called an intracellular wave, and as an intracellular Ca^{2+} wave propagated across an individual cell, a sequential increase in beat frequency was observed (Fig. 6a and b). As the Ca^{2+} change propagated through adjacent cells as an intercellular wave, a sequential increase in beat frequency of the adjacent cells was also observed (Fig. 6c and d).

In all cells the increase in beat frequency only occurred after the $[Ca^{2+}]_i$ had increased at the base of the cilia. It is important to note that the ciliary beat frequency increase was not exactly proportional to the increase in $[Ca^{2+}]_i$. At low or resting $[Ca^{2+}]_i$ the beat frequency is slow, with some variability (Fig. 7c and d). In fact, the $[Ca^{2+}]_i$ must increase to ~150 nM before increases in beat frequency were observed. At a $[Ca^{2+}]_i$ of approximately 300–500 nM, the ciliary activity reached a maximal beat frequency, even though the $[Ca^{2+}]_i$ continued to increase

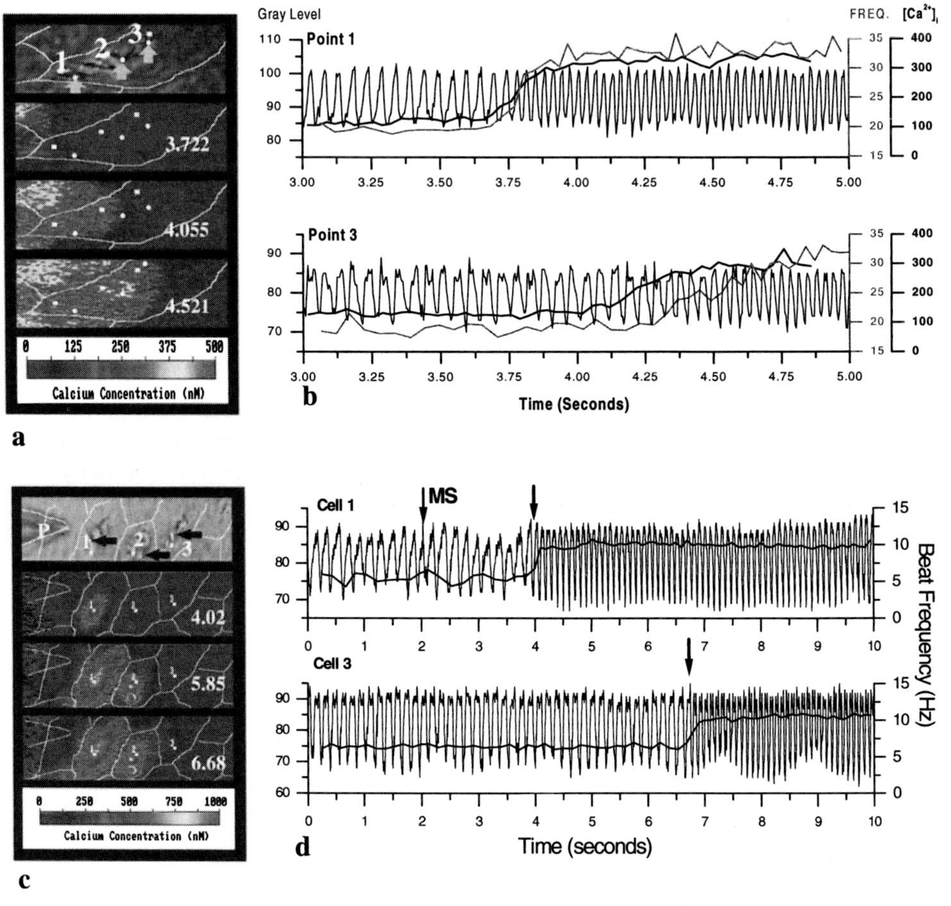

FIGURE 6 Intracellular and intercellular Ca^{2+} waves evoke sequential increases in ciliary beat frequency. (a) The changes in $[Ca^{2+}]_i$ within a single cell associated with the propagation of an intracellular wave that is part of an intercellular wave initiated by mechanical stimulation in an adjacent cell at 37°C. (b) Changes in the gray level of the phase-contrast image (left scale and waveform), beat frequency (right scale, gray line and text), and $[Ca^{2+}]_i$ (right scale, black line and text) with respect to time during the passage of the intracellular Ca^{2+} wave across two points within the cell. The $[Ca^{2+}]_i$ is sequentially elevated as the Ca^{2+} wave moves from point 1 to point 3. Following a slight delay, after the passage of the Ca^{2+} wave, the beat frequency rapidly increases within 3–6 beat cycles to a maximal rate. The beat frequency at point 3 remained at a basal rate until the Ca^{2+} wave passed the basal area of the cilia, even though the cilia at point 1 were beating at a maximal rate. (c) An intercellular Ca^{2+} wave induced by mechanical stimulation (MS, arrow indicates time of stimulation) of an adjacent cell, propagating through three cells

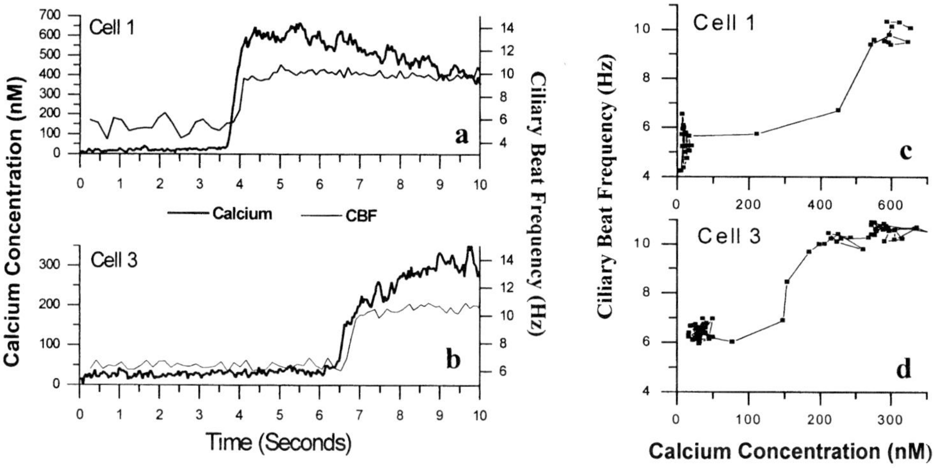

FIGURE 7 (a and b) The change in $[Ca^{2+}]_i$ and ciliary beat frequency in two different cells (shown in Fig. 6c) with respect to time during the propagation of an intercellular Ca^{2+} wave at 25°C. In both cells, the beat frequency has rapidly increased following the $[Ca^{2+}]_i$ increase. However, in contrast to the $[Ca^{2+}]_i$, the ciliary beat frequency stabilized at a maximal beat frequency even though the $[Ca^{2+}]_i$ showed both a further increase and subsequent decrease. (c and d) Scatter plots reveal of the relationship between ciliary beat frequency and $[Ca^{2+}]_i$ for two cells (left). Ciliary beat frequency is independent of the $[Ca^{2+}]_i$ below 150–200 nm. At $[Ca^{2+}]_i$ above 200 nm, the beat frequency rapidly increases to achieve a maximal rate at 300–500 nm. Scatter points are joined by lines to indicate the temporal relationship of each point.

---

at 25°C. (d) Changes in the gray level of the phase-contrast image (left scale and waveform) and beat frequency (right scale, black line) with respect to time during the passage of the intercellular Ca^{2+} wave through 3 cells. At 25°C, the basal beat frequency is slower (note difference in time scales between b and d). As the Ca^{2+} wave passes the base of the cilia of each cell, the beat frequency increases (arrows) to a maximal rate within 2–4 beat cycles. The maximal beat frequency rate is also slower at 25°C. (All) A phase-contrast image of the cells is shown at the top of each image panel. The time at which each image was recorded is indicated in each image. The white lines represent cell boundaries. The magnitude of the changes in $[Ca^{2+}]_i$ is indicated by the color scales. The positions at which $[Ca^{2+}]_i$ and ciliary beat frequency were analyzed to generate the graphs (right) are indicated by the large and small white squares, respectively. Beat frequencies are calculated for each beat cycle and plotted at the end of each beat cycle. (Also see color plate.)

up to 650 nM (Fig. 7a and b). The maximal beat frequency achieved at 37°C was higher than that achieved at 25°C. Similarly, the maximal beat frequency was achieved within three to six beat cycles at 37°C and within two to four cycles at 25°C (Fig. 6b and d). The beat frequency only began to return to prestimulus levels after the $[Ca^{2+}]_i$ had been reduced to below ~350 nM. These results indicate (1) that increases in ciliary beat frequency have a threshold $[Ca^{2+}]_i$ for activation and a maximal frequency response to transient Ca^{2+} increases and (2) that the changes in $[Ca^{2+}]_i$ act locally, perhaps at the ciliary base, to regulate beat frequency.

Ca^{2+} OSCILLATIONS INDUCED BY ATP

Airway epithelial cells possess purinergic receptors and may use extracelluar ATP as a signaling molecule to regulate beat frequency and mucus secretion (Evans and Sanderson, 1999a; Hansen et al., 1993; Homolya et al., 1999; Korngreen et al., 1998; Lazarowski et al., 1997; Woodruff et al., 1999). In addition, Ca^{2+} waves in other cell types appear, in part, to be mediated by extracellular ATP (Guthrie et al., 1999). When extracellular ATP was applied to cultures of airway ciliated epithelial cells, the cells displayed a dynamic Ca^{2+} response that consisted of two phases (Fig. 8a). ATP initially induced a large increase in $[Ca^{2+}]_i$. After this initial response and as the $[Ca^{2+}]_i$ returned to resting concentration, a series of Ca^{2+} oscillations were observed (Fig. 8b) (Evans and Sanderson, 1999a). These Ca^{2+} oscillations occur repetitively in single cells (approximately every 5–60 seconds) and frequently appear as intracellular Ca^{2+} waves that sweep across the individual cell. However, unlike mechanically stimulated Ca^{2+} waves, these intracellular Ca^{2+} waves do not propagate to adjacent cells. The failure of ATP-induced Ca^{2+} oscillations to propagate suggests that the epithelial cells may have lost their ability to communicate. However, if a single cell is mechanically stimulated after the application of ATP and during the display of Ca^{2+} oscillations, an intercellular Ca^{2+} wave is initiated and spreads through many adjacent cells. After the passage of the wave, the cells continue to oscillate (Evans and Sanderson, 1999a).

SENSITIVITY AND SPECIFICITY OF Ca^{2+} OSCILLATIONS INDUCED BY ATP AND UTP

Airway epithelial cells in culture displayed a wide range of sensitivity to extracellular ATP. This sensitivity is based on the Ca^{2+} response of the cells, and this can be classified into three groups: low, intermediate, and high sensitivity. In low-sensitivity cells, 5 μM ATP invoked only a transient increase in $[Ca^{2+}]_i$. However, increasing doses of ATP, up to 100 μM, can evoke Ca^{2+} oscillations. The frequency of the oscillations increases with ATP concentration. In high-sensitivity cells, a similar range of Ca^{2+} responses were observed, but in these cells, the

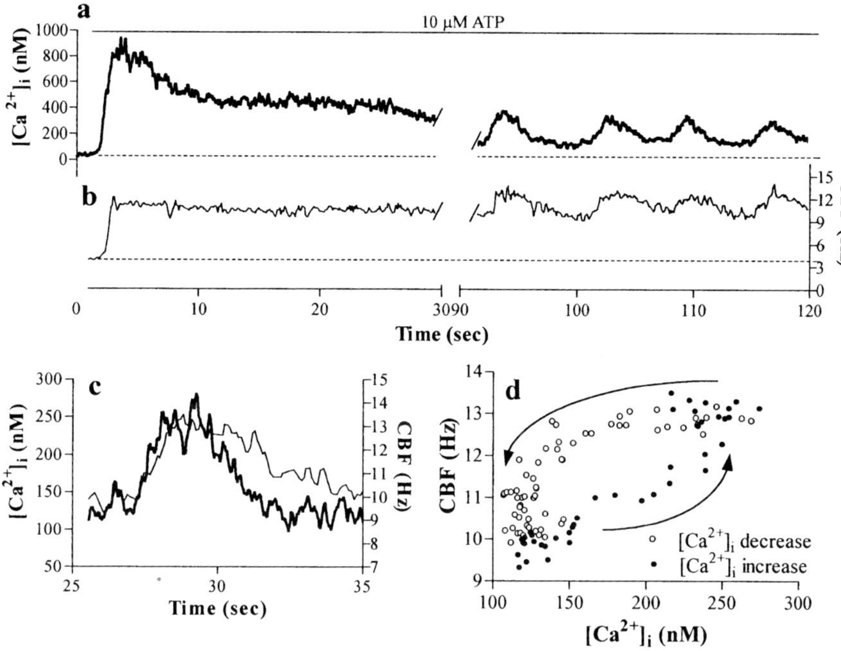

FIGURE 8 The relationship between ATP-induced Ca^{2+} oscillations and changes in ciliary beat frequency. (a) The prolonged exposure of airway cells to 10 μM ATP induces an initial large increase in $[Ca^{2+}]_i$ (thick line), which is followed by a decline in $[Ca^{2+}]_i$ and the induction of Ca^{2+} oscillations from a near basal level of $[Ca^{2+}]_i$. (b) The initial increase in $[Ca^{2+}]_i$ is accompanied by an increase in ciliary beat frequency to a higher stable rate. As the $[Ca^{2+}]_i$ declines to near basal rates the ciliary beat frequency remains elevated. In addition, the ciliary beat frequency is further elevated by each Ca^{2+} oscillation even though the magnitude of the Ca^{2+} oscillation is substantially smaller than the initial elevation induced by ATP. (c) The changes in beat frequency (thin line) associated with a Ca^{2+} oscillation (thick line) observed in (a). The increase in beat frequency closely follows the increase in $[Ca^{2+}]_i$. However, the decline in beat frequency significantly lags behind the decrease in $[Ca^{2+}]_i$. (d) The relationship between ciliary beat frequency and $[Ca^{2+}]_i$ during the Ca^{2+} oscillation illustrated in (c). Closed circles indicate data points measured during the increase (arrow) in $[Ca^{2+}]_i$ and the open circles represent data points measured during the decline (arrow) in $[Ca^{2+}]_i$. The increase in beat frequency is relatively steady with the increase in $[Ca^{2+}]_i$. However, the beat frequency remains relatively stable during the decline in $[Ca^{2+}]_i$ until the $[Ca^{2+}]_i$ falls below 150 nm, at which point the beat frequency begins to decline more rapidly.

basic response of a single transient increase in $[Ca^{2+}]_i$ is evoked by 0.1 μM ATP, and the fastest Ca^{2+} oscillations were evoked by 1 μM ATP. ATP doses of >5 μM induced a sustained increase in $[Ca^{2+}]_i$. Airway epithelial cells showed similar Ca^{2+} responses to extracellular UTP. However, for all cells examined, individual cells were more sensitive to UTP than ATP. In a low-sensitivity cell, UTP, but not ATP at similar concentrations, induced a single transient increase in $[Ca^{2+}]_i$. In an intermediate-sensitivity cell, UTP induced a single Ca^{2+} oscillation while ATP induced only a transient increase in $[Ca^{2+}]_i$. In a high-sensitivity cell, UTP induced Ca^{2+} oscillations with a frequency greater than those induced by ATP. These results suggest that the Ca^{2+} responses induced in airway epithelial cells by ATP and UTP are mediated via a P2U receptor (Homolya et al., 1999).

MECHANISMS UNDERLYING ATP-INDUCED Ca^{2+} OSCILLATIONS

ATP-induced Ca^{2+} oscillations in epithelial cells occur in the absence of extracellular Ca^{2+}, although these oscillations would run down following prolonged exposure of the cells to Ca^{2+}-free solutions. The washing of the cells with the Ca^{2+}-free solutions alone did not induce Ca^{2+} oscillations. The addition of thapsigargin (1 μM), an inhibitor of the SERCA pumps of the endoplasmic reticulum, inhibited ATP-induced Ca^{2+} oscillations. As the Ca^{2+} oscillations were inhibited, the basal Ca^{2+} level became elevated, a result consistent with the gradual emptying of Ca^{2+} stores. The addition of the PLC inhibitor U73122 (10 μM) also inhibited the ATP-induced Ca^{2+} oscillations. By contrast, the inactive sister compound, U73343, had no effect on the oscillations. Together these results indicate that that ATP-induced Ca^{2+} oscillations require continued PLC activity for the production of IP_3 and the repetitive filling and release of Ca^{2+} from intracellular Ca^{2+} stores via the IP_3 receptor (Evans and Sanderson, 1999a).

BEAT FREQUENCY AND Ca^{2+} OSCILLATIONS

The addition of extracellular ATP to ciliated epithelial cells results in an initial large increase in Ca^{2+}, and after about 30 seconds the Ca^{2+} begins to be restored to resting levels, followed by the generation of Ca^{2+} oscillations (Fig. 8). During the initial Ca^{2+} increase, the ciliary beat frequency increased rapidly and reached a maximum rate, even though the $[Ca^{2+}]_i$ continued to increase. This response is similar to that generated by a Ca^{2+} wave. However, it is also important to note that only a small decrease in the ciliary beat frequency occurred as the $[Ca^{2+}]_i$ returned to a near-basal level. Furthermore, the subsequent Ca^{2+} oscillations were able to elevate the beat frequency to rates higher than that stimulated by the initial rise in $[Ca^{2+}]_i$ despite the fact that the associated $[Ca^{2+}]_i$ increases were considerably less than the initial ATP-induced Ca^{2+} increase (Fig. 8a and b).

HYSTERESIS: $[Ca^{2+}]_i$ VERSUS CILIARY BEAT FREQUENCY

By separately analyzing increases and decreases in ciliary beat frequency with respect to $[Ca^{2+}]_i$, it became evident that changes in beat frequency were not a linear function of $[Ca^{2+}]_i$ (Fig. 8c and d). Although the increases in $[Ca^{2+}]_i$ and beat frequency were well coupled, decreases in beat frequency often lagged behind or showed hysteresis with respect to decreases in $[Ca^{2+}]_i$. During the initial ATP-induced Ca^{2+} increase, ciliary beat frequency rises to a maximum, but during the decrease in $[Ca^{2+}]_i$, the beat frequency remains constant. Data pooled from multiple Ca^{2+} oscillations indicated that the decrease in beat frequency did not occur until a lower $[Ca^{2+}]_i$ was reached. The relationship between Ca^{2+} and beat frequency is more clearly seen by analyzing a single Ca^{2+} oscillation (Fig. 8c and d). As $[Ca^{2+}]_i$ increases, beat frequency increases. However, as the $[Ca^{2+}]_i$ decreases, the decrease in beat frequency lags behind or shows hysteresis.

FREQUENCY MODULATION FOR THE REGULATION OF CILIARY ACTIVITY

A common scheme for interpreting regulatory signals is analog modulation (AM). In this scheme, the magnitude of response is generally proportional to, and in synchrony with, the magnitude of the regulatory signal. A second scheme of regulation is frequency modulation (FM) (Berridge, 1997; Putney, 1998). In this scheme, the magnitude of the response is proportional to the frequency of equal-sized regulatory signals. The possibility that ciliary beat frequency is FM regulated was suggested by two observations. First, repetitive Ca^{2+} oscillations that elevated the $[Ca^{2+}]_i$ ~300 mM are capable of maintaining a beat frequency at a rate higher than that inducible by a much larger transient increase in $[Ca^{2+}]_i$. Second, there is a hysteresis associated with the decrease in beat frequency. Experimental data for FM regulation of ciliary activity were also found by comparing the beat frequencies of cilia in cells that experienced Ca^{2+} oscillations at two different frequencies (period ~8 seconds or 5.5 seconds). A similar maximal beat frequency was induced by each Ca^{2+} oscillation, although at different periods and with slightly different peaks $[Ca^{2+}]_i$. However, the minimal beat frequency attained during the Ca^{2+} oscillations was lower for the slower oscillation, even though this occurred at similar minimal $[Ca^{2+}]_i$ (Table 1).

An explanation for these results is that the deactivation of beat frequency occurs more slowly than the activation of beat frequency, a process that is required for FM regulation. Data show that activation of beat frequency is a Ca^{2+}-dependent process, but, because the deactivation process is slower, the regulation of beat frequency by Ca^{2+} would appear to require an intermediary. A common way of regulating biological process is through phosphorylation, and Ca^{2+} may activate a kinase, via calmodulin, to phosphorylate the axoneme (Fig. 9) (De

TABLE 1 Change in Beat Frequency Associated with Ca^{2+} Oscillations of Different Frequencies

Oscillation period (sec)	Maximum frequency (Hz)	Maximum $[Ca^{2+}]_i$ (nM)	Minimum frequency (Hz)	Minimum $[Ca^{2+}]_i$ (nM)
5.3 ± 0.3^a	12.1 ± 0.1^b	346 ± 6^b	10 ± 0.1^a	228 ± 3^b
8.3 ± 0.5^a	12.1 ± 0.2^b	417 ± 3^b	8.8 ± 0.1^a	219 ± 4^b

[a] Significantly different.
[b] Not significantly different.

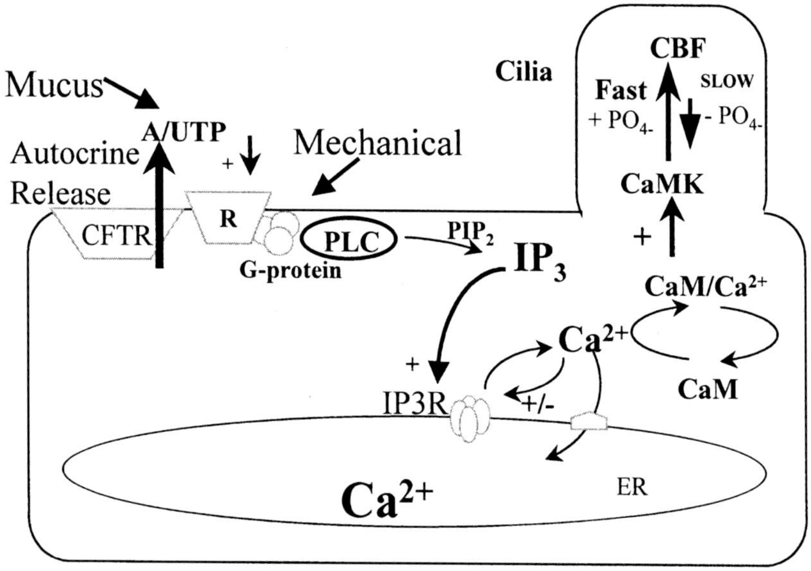

FIGURE 9 A model proposed for the mechanisms driving Ca^{2+} oscillations and beat frequency changes in airway epithelial cells. Agonist activation leads to PLC activity and the production of IP_3. IP_3 initiates the release of internal Ca^{2+} and the establishment of Ca^{2+} oscillations. Elevations in $[Ca^{2+}]_i$ lead to the activation of Cam-kinase II and the phosphorylation of axonemal proteins to elevate beat frequency. Dephosphorylation at a slower rate decreases the beat frequency of the cilia.

Koninck and Schulman, 1998). A phosphorylation event would help to explain the near-maximal responses to relatively low Ca^{2+} increases. Dephosphorylation would be required to reverse the increase in beat frequency. This process may be Ca^{2+}-independent and may occur at a steady but slower rate than phosphorylation. As a result, the time between the Ca^{2+} peaks of oscillations will determine the amount by which beat frequency declines. In other words, oscillations with a shorter time period resulted in a more stable beat frequency. At present we have no data regarding the site of phosphorylation. However, it appears that the Ca^{2+} must gain access to the axoneme via the base of the cilium, and this region may be the primary target, but this does not exclude the possibility that sites along the axoneme are not phosphorylated (Fig. 9) (Tamm and Terasaki, 1994; Wang and Satir, 1998; Yang et al., 2000).

SUMMARY

We suggest that recording techniques with high temporal and spatial resolution are required for quantifying ciliary beat frequency and $[Ca^{2+}]_i$. With these techniques, we have found that Ca^{2+}-induced increases in ciliary beat frequency occur when the $[Ca^{2+}]_i$ is above a threshold of about 150 nM. These increases in beat frequency occur very rapidly and closely follow the increase in $[Ca^{2+}]_i$ up to a steady rate induced by ~300 nM. However, decreases in ciliary beat frequency occur more slowly and significantly lag behind decreases in $[Ca^{2+}]_i$. Although extracellular ATP induces large and rapid increases in $[Ca^{2+}]_i$ and ciliary beat frequency, a sustained increase in $[Ca^{2+}]_i$ is not required to maintain an elevated ciliary beat frequency. By contrast, higher and more stable elevations in ciliary beat frequency were induced by higher frequency Ca^{2+} oscillations with lower peak $[Ca^{2+}]_i$. We suggest that airway ciliary beat frequency is regulated by frequency-modulated Ca^{2+} signaling and that ATP released by epithelial cells may locally regulate airway ciliary activity.

ACKNOWLEDGMENTS

We would like to thank Mathias Salathe for the organization of the international meeting in Italy and the invitation to present our work, and Dr. Peter Satir for organizing the session on airway ciliary activity. This work was supported by NIH grant HL49288 to MJS.

REFERENCES

MJ Berridge. The AM and FM of calcium signaling. Nature 386:759–760, 1997.
P De Koninck, H Schulman. Sensitivity of CaM kinase II to the frequency of Ca^{2+} oscillations. Science 279:227–230, 1998.

ER Dirksen, JA Felix, MJ Sanderson. Preparation of explant and organ cultures and single cells from airway epithelium. Methods Cell Biol 47:65–74, 1995.

JH Evans, MJ Sanderson. Intracellular calcium oscillations induced by ATP in airway epithelial cells. Am J Physiol 277:L30–41, 1999a.

JH Evans, MJ Sanderson. Intracellular calcium oscillations regulate ciliary beat frequency of airway epithelial cells. Cell Calcium 26:103–110, 1999b.

PB Guthrie, J Knappenberger, M Segal, MV Bennett, AC Charles, SB Kater. ATP released from astrocytes mediates glial calcium waves. J Neurosci 19:520–528, 1999.

M Hansen, S Boitano, ER Dirksen, MJ Sanderson. Intercellular calcium signaling induced by extracellular adenosine 5′-triphosphate and mechanical stimulation in airway epithelial cells. J Cell Sci 106:995–1004, 1993.

L Homolya, WC Watt, ER Lazarowski, BH Koller, RC Boucher. Nucleotide-regulated calcium signaling in lung fibroblasts and epithelial cells from normal and P2Y(2) receptor (−/−) mice. J Biol Chem 274:26454–26460, 1999.

S Inoue, KR Spring. Video Microscopy. New York: Plenum Press, 1997.

A Korngreen, Z Priel. Simultaneous measurement of ciliary beating and intracellular calcium. Biophys J 67:377–380, 1994.

A Korngreen, W Ma, Z Priel, SD Silberberg. Extracellular ATP directly gates a cation-selective channel in rabbit airway ciliated epithelial cells. J Physiol (Lond) 508: 703–720, 1998.

AB Lansley, MJ Sanderson. Regulation of airway ciliary activity by Ca^{2+}: simultaneous measurement of beat frequency and intracellular Ca^{2+}. Biophys J 77:629–638, 1999.

AB Lansley, MJ Sanderson, ER Dirksen. Control of the beat cycle of respiratory tract cilia by Ca^{2+} and cAMP. Am J Physiol 263:L232–242, 1992.

ER Lazarowski, L Homolya, RC Boucher, TK Harden. Direct demonstration of mechanically induced release of cellular UTP and its implication for uridine nucleotide receptor activation. J Biol Chem 272:24348–24354, 1997.

L Leybaert, J Sneyd, MJ Sanderson. A simple method for high temporal resolution calcium imaging with dual excitation dyes. Biophys J 75:2025–2029, 1998.

H Mao, LB Wong. Fluorescence and laser photon counting: measurements of epithelial $[Ca^{2+}]_i$ or $[Na+]_i$ with ciliary beat frequency. Ann Biomed Eng 26:666–678, 1998.

JW Putney, Jr. Calcium signaling: up, down, up, down...what's the point? Science 279: 191–192, 1998.

KP Roos, JM Parker. A low cost two-dimensional digital image acquisition subsystem for high speed microscopic motion detection. Proc Soc Photo-Opt Instrum Eng 1205:134–141, 1990.

M Salathe, MM Pratt, A Wanner. Protein kinase C-dependent phosphorylation of a ciliary membrane protein and inhibition of ciliary beating. J Cell Sci 106:1211–1220, 1993.

MJ Sanderson. Mechanisms controlling airway ciliary activity. In: DF Rogers, MI Lethem, eds. Airway Mucus: Basic Mechanisms and Clinical Perspectives. Basel: Birkhauser, 1997:91–116.

MJ Sanderson. High-speed digital microscopy. Methods 21:325–334, 2000.

MJ Sanderson, ER Dirksen. A versatile and quantitative computer-assisted photoelectronic technique used for the analysis of ciliary beat cycles. Cell Motil 5:267–292, 1985.

MJ Sanderson, ER Dirksen. Mechanosensitivity of cultured ciliated cells from the mammalian respiratory tract: implications for the regulation of mucociliary transport. Proc Natl Acad Sci USA 83:7302–7306, 1986.

MJ Sanderson, ER Dirksen. Quantification of ciliary beat frequency and metachrony by high-speed digital video. Methods Cell Biol 47:289–297, 1995.

MJ Sanderson, MA Sleigh. Ciliary activity of cultured rabbit tracheal epithelium: beat pattern and metachrony. J Cell Sci. 47:331–347, 1981.

MJ Sanderson, AC Charles, ER Dirksen. Mechanical stimulation and intercellular communication increases intracellular Ca^{2+} in epithelial cells. Cell Reg 1:585–596, 1990.

SL Tamm, M Terasaki. Visualization of calcium transients controlling orientation of ciliary beat. J Cell Biol 125:1127–1135, 1994.

P Verdugo, BV Raess, M Villalon. The role of calmodulin in the regulation of ciliary movement in mammalian epithelial cilia. J Submicrosc Cytol 15:95–96, 1983.

H Wang, P Satir. The 29 kDa light chain that regulates axonemal dynein activity binds to cytoplasmic dyneins. Cell Motil Cytoskeleton 39:1–8, 1998.

A Wanner, M Salathe, TG O'Riordan. Mucociliary clearance in the airways. Am J Respir Crit Care Med 154:1868–1902, 1996.

ML Woodruff, VV Chaban, CM Worley, ER Dirksen. PKC role in mechanically induced Ca^{2+} waves and ATP-induced Ca^{2+} oscillations in airway epithelial cells. Am J Physiol 276:L669–678, 1999.

P Yang, L Fox, RJ Colbran, WS Sale. Protein phosphatases PP1 and PP2A are located in distinct positions in the *Chlamydomonas* flagellar axoneme. J Cell Sci 113:91–102, 2000.

5

Modeling the Response of Mammalian Ciliary Beating to Changes in Cytoplasmic Calcium

Matthias Salathe and
Richard J. Bookman
University of Miami School of Medicine
Miami, Florida

INTRODUCTION

Mammalian ciliary beat frequency (CBF) is regulated by the intracellular calcium concentration. It has been shown by us and others, using a variety of methods, that CBF follows the direction of cytoplasmic calcium concentration ($[Ca^{2+}]_i$) changes (1–8). At the molecular level, however, it remains unclear how changes in $[Ca^{2+}]_i$ are transduced into changes in CBF. Recent advances in measuring $[Ca^{2+}]_i$ and CBF simultaneously, a technique first described by Korngreen and Priel (4), and the introduction of high-resolution, simultaneous measurements, where accurate kinetic relationships between these two signals can be assessed (9,10), have advanced our understanding of how Ca^{2+} regulates CBF. We have measured CBF and $[Ca^{2+}]_i$ (indicated by fura-2) at room temperature in response to activation of the G-protein coupled M3 muscarinic receptor by 10 μM acetylcholine (ACh). When CBF was estimated by a Fourier transform method with a CBF time resolution of <100 ms (10), the difference in the onset of the CBF increase compared to $[Ca^{2+}]_i$ was 70 ± 30 ms (mean ± SEM; n = 20 cilia).

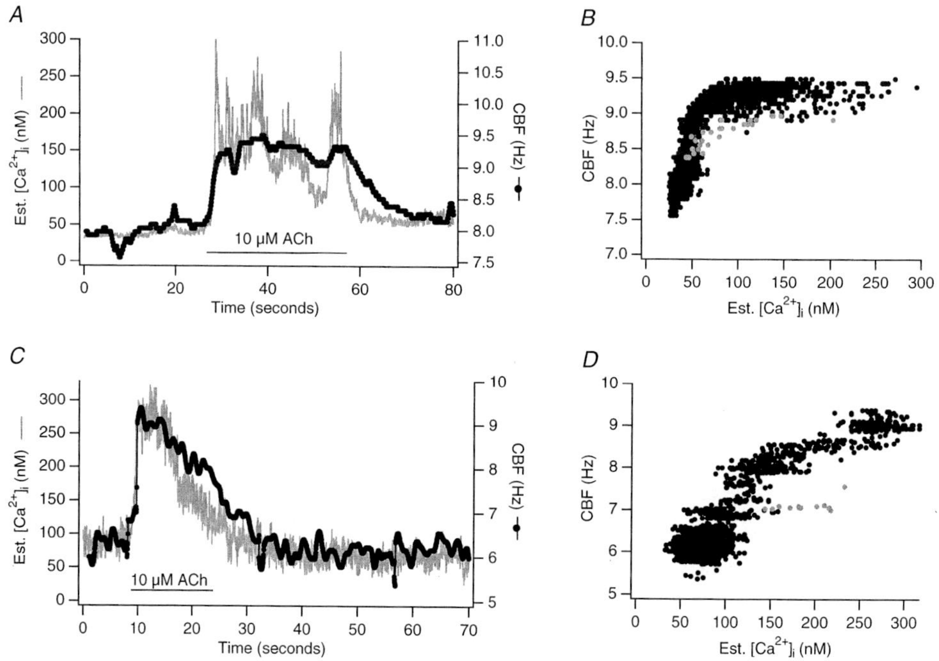

FIGURE 1 Two examples of high-resolution Fourier transform estimates of CBF are shown, correctly aligned with the estimated $[Ca^{2+}]_i$ signal to visualize the coupling between the two signals. A and C display the data on the time axis, whereas panels B and D represent the corresponding pair of $[Ca^{2+}]_i$ and CBF data plotted against each other. The trajectories in grey represent data points during the fastest change in both signals in response to 10 μM ACh. Panel A depicts an oscillatory response of Ca^{2+} and CBF to ACh as often seen in early culture (7). These measurements revealed that during the rise of both signals, no significant delay of the CBF signal, relative to the change in the $[Ca^{2+}]_i$ signal, could be measured: the point of half maximal increase in the CBF signal occurred 70 ± 30 ms (mean ± SEM; $n = 20$ cells) after the $[Ca^{2+}]_i$ increase, an amount that was within the CBF time resolution. However, after washout of ACh, the relaxation of CBF towards baseline trailed $[Ca^{2+}]_i$ by 8 ± 3 seconds (mean ± SEM, $n = 20$ cells, median 3 seconds). These differences are represented by the hysteresis of the CBF vs. $[Ca^{2+}]_i$ data (grey vs. black points). (From Ref. 10.)

During the slower return to baseline, a lag of 8 ± 3.2 s was observed, indicative of hysteresis. These data are in agreement with the ones presented by Sanderson et al. in this book (Fig. 1). While calmodulin inhibitors (calmidazolium and W-7; each $n = 5$) decreased baseline CBF by an average of 1.1 ± 0.1 Hz, they did not alter the kinetic relationship between $[Ca^{2+}]_i$ and CBF. Similarly, phosphatase inhibitors (okadaic acid and cyclosporin A; each $n = 5$) changed neither baseline CBF nor the kinetic coupling between $[Ca^{2+}]_i$ and CBF. These data suggested that the timing of Ca^{2+} action on CBF in ovine airway epithelial cells is unlikely to be determined by phosphorylation reactions involving calmodulin or kinase/phosphatase reactions. Based on these observations, we developed a model for calcium action on ciliary beating, which is presented here.

MODELING THE COUPLING OF $[Ca^{2+}]_i$ AND CBF

Three basic models of Ca^{2+}/CBF coupling were considered. A linear coupling model was rejected by the data (Fig. 1). A simple cooperative binding model was also rejected since a Hill plot (not shown) did not reveal a linear relation between $[Ca^{2+}]_i$ and CBF. A more complex model that takes the action of dynein arms into account was therefore needed. For a discussion on current understanding of the principles of dynein motor activity during ciliary beating, see Chapter 2 in this book.

Our model of the effect of $[Ca^{2+}]_i$ on CBF was based on three assumptions:

1. Ciliary dynein arms move microtubules with either a "slow" or "fast" duty cycle, in accordance with experimental microtubule sliding results using *Chlamydomonas* inner dynein arms (11).
2. Dynein ATPase activity is shifted between fast and slow modes by a change in $[Ca^{2+}]_i$.
3. N_{total} complete and sequential dynein ATPase cycles are necessary to complete one ciliary stroke, in accordance with the ciliary motility model by Holwill et al. (12,13).

Thus,

$$CBF = 1/(Ntotal \times T_{dynein\ cycle}) \quad (1)$$

where $T_{dynein\ cycle}$ is the time (in seconds) that the dynein arm actively interacts with the microtubule (thereby moving it) and not necessarily the full duration of a dynein arm cycle (which is roughly 33 ms in vitro; see Chapter 2). Using published data on dynein conformational changes, we estimated the minimum number of sequential dynein arm actions necessary for one single ciliary stroke as follows (12,13). Two outer doublets separate by ~0.1 μm at the end of a full ciliary bend, and a single dynein arm cycle can move a microtubule 4–16 nm (8 nm currently being favored; see Chapter 2). Thus, anywhere between 12 and

50 *sequential* dynein arm movements are necessary to complete a single ciliary stroke (Fig. 2). Using the currently favored 8 nm, a total of 25 sequential dynein arm movements is necessary to complete a single ciliary cycle consisting of both an effective and recovery stroke. Therefore, Eq. (1) can be rewritten as:

$$CBF = 1/(25*T_{\text{dyneina cycles}}) \quad (2)$$

or

$$CBF = 1/(25*(f_{\text{fast}}*T_{\text{fast}} + f_{\text{slow}}*T_{\text{slow}})) \quad (3)$$

where f_{fast} is the fraction of the dynein arms that are operating in fast mode, f_{slow} is the fraction of the dynein arms that are operating in slow mode (and $f_{\text{slow}} = 1 - f_{\text{fast}}$), T_{fast} is the time (in seconds) required for an active dynein cycle (interacting with the microtubule) in fast mode, and T_{slow} is the active dynein cycle time (in seconds) in slow mode.

First, we wanted to illustrate the calcium dependence of the fraction of slow (no calcium bound) vs. fast dynein arms (all binding sites with bound calcium) using the Hill equation assuming, for this example, a total K_d for all cooperative binding sites of 150 nM (Fig. 2C). CBF at low $[Ca^{2+}]$ is determined by the slow dynein cycle time. With increasing $[Ca^{2+}]$, the fraction of fast dynein will increase and therefore speed up CBF. The traces shown in the graph simulate the relationship between $[Ca^{2+}]_i$ and the fraction of fast dynein arms assuming 1 (no cooperative binding) to 2–5 cooperative Ca^{2+}-binding sites, respectively. With this use of the Hill equation to calculate the fast and slow dynein arm fraction's dependence on $[Ca^{2+}]$, CBF can be estimated.

Figure 2D shows a simulation for which the slow dynein arm duration was chosen to result in a ''resting'' CBF of 7 Hz (based on our data from recordings at 20°C), whereas the switch to all fast dynein arms was chosen to result in a maximal CBF of 12 Hz. This frequency range was only used for the purpose of simulation, and these specific values are not a general feature (or limit) of the model. The fraction of fast vs. slow was taken from the simulated Hill equation as mentioned above (Fig. 2C). Such graphs now start to resemble recorded data, especially when used with three or more cooperative Ca^{2+}-binding sites.

Finally, we fit this model to recorded data as shown in Figure 3. The slow and fast dynein cycle duration, and the K_d were free parameters of the fit, thereby allowing basal and maximal frequencies to assume any value. These fits suggested that models with four cooperative Ca^{2+}-binding sites best fit the recorded data (in a least-squares sense). Using the fit parameters for four cooperative Ca^{2+}-binding sites, the average K_d was 50 nM, the slow dynein cycle duration 9 ms, and the fast dynein cycle duration 4.5 ms ($n = 9$ cells). These nine experiments were also combined into a single data set and again fitted with the model (Fig. 3B). Again, using four cooperative Ca^{2+}-binding sites resulted in the best fit.

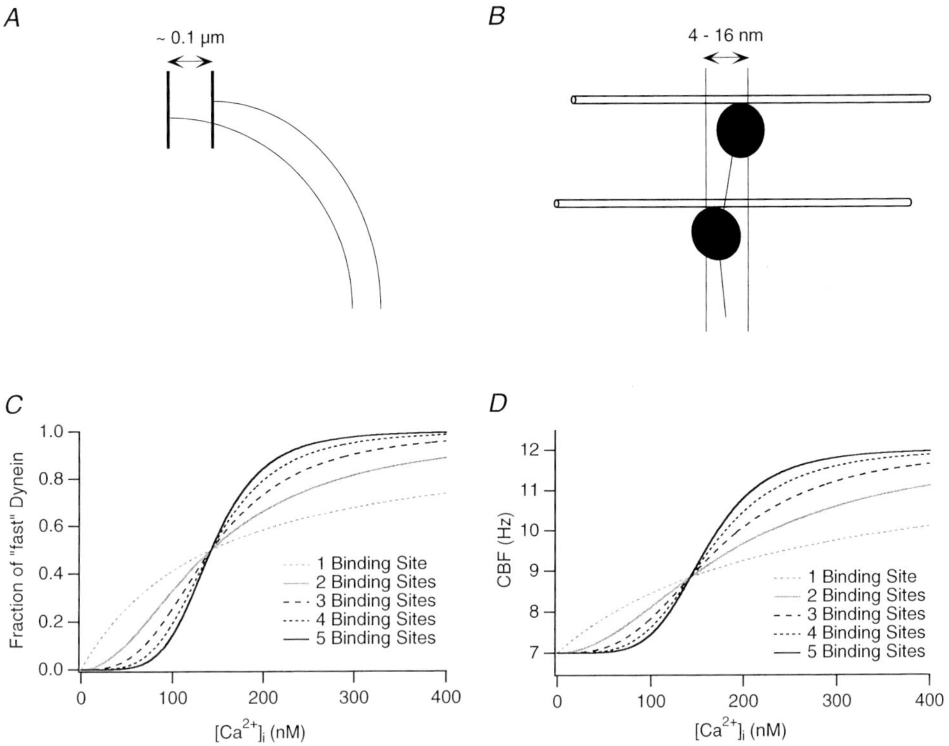

FIGURE 2 Using published data on dynein conformational changes, we estimated the minimum number of sequential dynein arm actions necessary for one single ciliary stroke. Two outer doublets are separated—0.1 μm at the end of ciliary bending, as indicated in A, and a single stroke of one dynein arm can move a microtubule 4–16 nm (B). Thus, between 12 and 50 sequential dynein arm movements are necessary to complete a single ciliary stroke (for all calculations, 25 were used). (C) The fraction of fast vs. slow dynein arms was computed using the Hill equation and is simulated here with a total K_d of 150 nM. The traces shown in the graph simulate the relationship between $[Ca^{2+}]_i$ and the fractional occupancy of Ca^{2+} from 1 to 5 cooperative Ca^{2+}-binding sites. (D) Using the Hill equation to calculate the fast and slow dynein arm fraction's dependence on $[Ca^{2+}]_i$, we simulated CBF traces, arbitrarily choosing a baseline frequency of 7 Hz and a maximal frequency of 12 Hz (however, this is not the limit of the model in general). (From Ref. 10.)

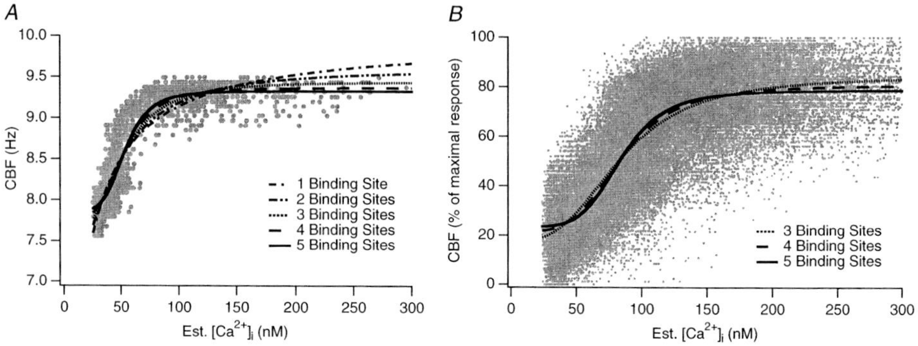

Figure 3 The CBF model was fit to actual data using the slow and fast dynein motion duration and the total K_d as parameters of the fit (thereby not limiting CBF of the model to any specific range). (A) A model with four cooperative Ca^{2+}-binding sites results in a reasonable fit (as judged by chi squares). (B) Data points from nine experiments are plotted and the model parameters are fit to the combined data set. Again, using four cooperative Ca^{2+}-binding sites resulted in the best fit. The average K_d was 50 nM, the slow dynein arm duration 9 ms, the fast dynein arm duration 4.5 ms. (From Ref. 10.)

DISCUSSION

Analyzing the data, it became rapidly clear that there was no simple, linear relationship between CBF and $[Ca^{2+}]_i$. This nonlinear relation has also been described by Sanderson et al. (9; see also Chapter 4) as well as Korngreen and Priel (5). Since it is clear that dynein arm movements provide the mechanism for ciliary bending, we started to explore a model that takes dynein arm action into account. Making three simple, literature-based assumptions, the model we developed can account for the relationship between CBF and Ca^{2+}. It should be noted here that the model does not require Ca^{2+} to bind directly to dynein. It does require that the ciliary Ca^{2+}-binding protein can communicate its Ca^{2+}-binding state to the dynein ATPase so rapidly that it is not rate limiting in this situation. In principle, even phosphorylation reactions could be part of this model as long as they are not the rate-limiting step. However, the model cannot account for the hysteresis of the response, and we decided not to complicate matters until we have further experimental data to make an informed modeling choice. There are at least two possibilities to account for the hysteresis, with little published data to favor either one. First, the off-rate of Ca^{2+} from the calcium-binding target could be much slower than the on-rate. Second, the dynein arm might have a built-in memory function, in which a particular conformation would persist long after the Ca^{2+} had unbound. There is some precedent for this in the recently described results for myosin (14): faster dynein arm rates might thereby be maintained for a period of time even after Ca^{2+} has been released from the Ca^{2+}-binding sites. Both of these explanations suggest enhancements of the model that would likely provide better fits.

The fit of the model to measured data revealed several interesting findings. First, the average value of the total K_d (a variable in the fit) was 50 nM, a reasonable value for a Ca^{2+}-response system. Second, the fast dynein cycle is about twice as fast as the slow dynein cycle (duration of slow dynein action 9 ms, duration of fast dynein action 4.5 ms in experiments measured at room temperature). As already mentioned above, this time does not reflect the whole dynein arm cycle (which is roughly 33 ms in vitro; see Chapter 2), but the duty phase of dynein interacting with the microtubule. Third, fast dynein moved microtubules over a distance of 100 nm within ~56 ms, a duration within the measured time (50–100 ms) it takes isolated single dynein molecules to move a microtubule the same distance in vitro (15). However, this is slightly slower than the measured activity of isolated outer dynein arms on glass slides in vitro where the velocity was assessed as 4–9 μm/s (16). However, faster frequencies are expected at physiological temperatures (our measurements were done at 20°C), and a frequency of 25 Hz (reducing T_{fast} to 1.6 ms) would correspond to a sliding velocity of ~4 μm/s. Therefore, this discrepancy is not a concern for the model.

In summary, we presented a simple model for the CBF regulation by Ca^{2+}. This model is not perfect but has already been proven effective to assess shifts in the Ca^{2+}/CBF coupling that may occur under different experimental conditions (17).

ACKNOWLEDGMENTS

We wish to thank our colleagues Drs. Adam Wanner and Gregory E. Conner for helpful discussions, Dr. William M. Abraham for his invaluable support, and Sara Donoghue for her technical support. Supported by grants from the NIH (HL-55341 and HL-60644), American Lung Association of Florida, and the Howard Hughes Medical Institute.

REFERENCES

1. PG Girard, JR Kennedy. Calcium regulation of ciliary activity in rabbit tracheal explants and outgrowth. Eur J Cell Biol 40:203–209, 1986.
2. P Verdugo. Calcium-dependent hormonal stimulation of ciliary activity. Nature 283: 764–765, 1980.
3. M Villalon, TR Hinds, P Verdugo. Stimulus-response coupling in mammalian ciliated cells. Demonstration of two mechanisms of control for cytosolic [Ca^{2+}]. Biophys J 56:1255–1258, 1989.

4. A Korngreen, Z Priel. Simultaneous measurement of ciliary beating and intracellular calcium. Biophys J 67:377–380, 1994.
5. A Korngreen, Z Priel. Purinergic stimulation of rabbit ciliated airway epithelia: control by multiple calcium sources. J Physiol (Lond) 497:53–66, 1996.
6. AB Lansley, MJ Sanderson, ER Dirksen. Control of the beat cycle of respiratory tract cilia by Ca^{2+} and cAMP. Am J Physiol 263:L232–242, 1992.
7. M Salathe, RJ Bookman. Coupling of $[Ca^{2+}]_i$ and ciliary beating in cultured tracheal epithelial cells. J Cell Sci 108:431–440, 1995.
8. G Di Benedetto, CJ Magnus, PTA Gray, A Mehta. Calcium regulation of ciliary beat frequency in human respiratory epithelium in vitro. J Physiol (Lond) 439: 103–113, 1991.
9. AB Lansley, MJ Sanderson. Regulation of airway ciliary activity by Ca^{2+}: simultaneous measurement of beat frequency and intracellular Ca^{2+}. Biophys J 77:629–638, 1999.
10. M Salathe, RJ Bookman. Mode of Ca^{2+} action on ciliary beat frequency in single ovine airway epithelial cells. J Physiol (Lond) 520:851–865, 1999.
11. G Habermacher, WS Sale. Regulation of flagellar dynein by phosphorylation of a 138-Kd inner arm dynein intermediate chain. J Cell Biol 136:167–176, 1997.
12. ME Holwill, P Satir. Physical model of axonemal splitting. Cell Motil Cytoskeleton 27:287–298, 1994.
13. MEJ Holwill, GF Foster, T Hamasaki, P Satir. Biophysical aspects and modelling of ciliary motility. Cell Motil Cytoskeleton 32:114–120, 1995.
14. A Ishijima, H Kojima, T Funatsu, et al. Simultaneous observation of individual ATPase and mechanical events by a single myosin molecule during interaction with actin. Cell 92:161–171, 1998.
15. C Shingyoji, H Higuchi, M Yoshimura, E Katayama, T Yanagida. Dynein arms are oscillating force generators. Nature 393:711–714, 1998.
16. WS Sale, LA Fox. Isolated beta-heavy chain subunit of dynein translocates microtubules in vitro. J Cell Biol 107:1793–1797, 1988.
17. M Salathe, T Lieb, RJ Bookman. Do cAMP and Ca^{2+} signals in cilia converge on outer dynein arms? Am J Respir Crit Care Med 161:A147, 2000.

6

Enhancement of CBF Is Dominantly Controlled by PKG and/or PKA

Alex Braiman, Natalya Uzlaner, and Zvi Priel
Ben-Gurion University of the Negev
Beer-Sheva, Israel

INTRODUCTION

Cilia are capable of transporting large and heavy objects at quite high velocities. This gigantic efficiency is achieved due to cooperative ciliary beating at high frequency. The rate of transport is dictated by the energy transferred by the tip of the cilium to the overlying blanket of mucus, during its effective stroke (the stroke in the direction of transport), multiplied by the number of cilia that are simultaneously involved in the pushing process (the number of cilia that beat synchronously per given area). The energy transferred by a cilium is proportional to the square of the ciliary beat frequency (CBF). The number of cilia involved in the pushing process is dictated by the characteristics of the metachronal wave (1). Based on energetic calculations, it is obvious that the efficiency of ciliary transport requires quite high beating frequencies. However, a high frequency of ciliary beating results in large energy expenditure. Therefore, under "normal" conditions cilia beat either at a relatively low frequency or may even be at rest, dramatically increasing their activity in response to a variety of receptor-mediated stimuli. For example, ciliary cells possess purinergic P1 and P2 (2–5), cholinergic (1,6,7), and adrenergic (8–10) receptors.

It is widely accepted that changes in intracellular levels of either calcium ions (4,8,11–17), cAMP (8,9,13,18,19), or cGMP (8,20–22) lead to CBF enhancement. Although these second messengers are well known and widely investigated modulators of ciliary activity, it is still unclear whether they act independently, each of them representing a separate pathway of ciliary stimulation, or whether their action is cooperative. Moreover, the pathways of ciliary stimulation, represented by these second messengers, co-exist in the same cell types. For example, in rabbit airway epithelium, ciliary activity is regulated by $[Ca^{2+}]_i$ (3,23,24), cAMP (13,19), and cGMP (24–26). Thus, a question can be raised about the physiological significance of this apparent multitude of ways to enhance ciliary activity in the same cell. This work aims to address these questions.

MATERIALS AND METHODS

Tissue Culture

Experiments were carried out on monolayer tissue cultures grown from rabbit trachea using the procedure described previously (3). Briefly, adult white rabbits (at least 1.5 kg in mass) were killed according to the guidelines laid down by the animal welfare committee of Ben-Gurion University by gradual exposure to carbon dioxide followed by exsanguination. Care was taken to slowly increase the gas flow over several minutes to prevent any visual signs of distress. Tracheas were removed, and the ciliary epithelium was peeled off the cartilage rings and cut into small pieces. Two or three pieces of epithelium were placed on a glass coverslip and were incubated in RPMI-1640 growth medium supplemented with 7.5% fetal calf serum, 20 units/mL penicillin, 2.5 units/mL nystatin, and 20 μg/mL streptomycin at 37°C with 5% CO_2. Prior to use, the glass coverslips were sterilized with ethanol, placed in plastic petri dishes (Nunk, Roskilde, Denmark), and incubated with a small amount of medium for 24 hours. The medium was replaced every 2 days. Measurements were performed on 8- to 20-day-old tissue cultures; this was the time period for which the epithelial monolayers were big enough to be used in the experiments.

Chemicals and Solutions

Fetal calf serum, RPMI-1640 culture medium, and antibiotics were from Biological Industries (Bet-Haemek, Israel). Fura-2/AM was from either Molecular Probes (Eugene, OR) or from Teflabs (Austin, TX). The dye was stored in solid form at −20°C, and fresh solutions were made before each experiment in dry dimethyl sulfoxide (DMSO). 6-Anilino-5,8-quinolinedione (LY-83583) was from Calbiochem (LaJolla, CA). All other materials were obtained from Sigma Chemical Co. (St. Louis, MO). The standard Ringer solution was composed of (mM): 150 NaCl, 2.5 KCl, 1.5 $CaCl_2$, 1.5 $MgCl_2$, 5 D-glucose, 5 Hepes. The pH of all the solutions was set to 7.4 with NaOH and HCl. LY-83583 was dissolved in CH_3OH/C_2H_5OH (1:4). Stock solutions of ATP were prepared in 0.05 M Hepes buffer (pH = 7.4) and diluted into the Ringer solutions immediately prior to use. All other chemicals were dissolved in water. The final concentration of the alcohol was not more than 0.5%.

Simultaneous Measurement of $[Ca^{2+}]_i$ and Ciliary Beating

We used the method of simultaneous measurement of $[Ca^{2+}]_i$ and ciliary beating extensively described previously (12). Briefly, $[Ca^{2+}]_i$ was measured with the fluorescent indicator fura-2. The dye-loaded cells were epi-illuminated with light from a 150 W xenon lamp (Oriel Corp., Stamford, CT) filtered by 340 and 380 nm interference filters (Oriel) mounted on a four-position rotating filter wheel. The fluorescence, emitted at 510 nm, was detected by a photon-counting photomultiplier (H3460-53, Hamamatsu, Japan). The 340/380 fluorescence ratio averaged over a period of 1 second was stored in a computer (Pentium®). CBF was measured by trans-illuminating the same cells with light at 600 nm (so as not to interfere with the fura-2 fluorescence at 510 nm). The light scattering from the beating cilia created amplitude modulations of the 600 nm light that were detected by a photomultiplier (R2014, Hamamatsu, Japan). This method allows measurement of an average $[Ca^{2+}]_i$ from one ciliary cell at a sampling rate of 50 Hz, while the CBF is measured from an area of 4.9 μm^2, which is a small portion of a cell, at a sampling rate of 360 Hz.

The $[Ca^{2+}]_i$ was calculated using a calibration curve that correlates fura-2 fluorescence ratio to calcium concentration. The calibration curve was constructed from measurements of the fluorescence ratio obtained from calibration solutions composed of 115 mM KCl, 20 mM NaCl, 5 mM $MgCl_2$, 5 mM D-glucose, 5 mM HEPES, 10 mM EGTA, 1 μM K_2 fura-2, and $CaCl_2$ at a range of concentrations (27). The $[Ca^{2+}]_i$ concentration was calculated directly from the calibration curve using a table look-up algorithm. The frequency response was calibrated by oscillating lamp in the range of 2–40 Hz and vibrating needle between 2 and 40 Hz. A linear response was revealed in these dynamic ranges.

Experimental Procedure and Data Presentation

Cells were preloaded with fura-2 by incubating the tissue in growth medium containing 8.5 μM fura-2/AM for 60 minutes at 37°C in a rotating water bath. The cells were then rinsed with Ringer solution and maintained at room temperature for 30 minutes to allow full deesterification and equilibration of the dye in the cytoplasm. Only evenly fluorescent cells that did not display bright spots and had a steady CBF were used. In each experiment, basal CBF (Fo) and basal $[Ca^{2+}]_i$ ($[Ca^{2+}]_{io}$) were measured for 2–5 minutes in the appropriate solution.

These were taken as reference values. Only cells that had basal CBF in the range of 9–12 Hz and basal $[Ca^{2+}]_i$ in the range of 100–250 nM were used. Then the test chemical was introduced by exchanging the bath solution, and the frequency (F) and $[Ca^{2+}]_i$ were monitored in the same ciliary area for 10–15 minutes. Each experimental condition was tested on naive cultures to avoid possible bias arising from previous manipulations. Beat frequencies are presented as the fold increase in CBF (F/Fo). Changes in $[Ca^{2+}]_i$ are presented as the difference between the observed $[Ca^{2+}]_i$ and the basal $[Ca^{2+}]_i$ ($[Ca^{2+}]_i - [Ca^{2+}]_{io}$). Each experiment was carried out on 3–15 tissue cultures from at least two or three animals. Results were averaged and are displayed as mean ± SE, with *n* (number of experiments) shown in parentheses.

RESULTS AND DISCUSSION

Cyclic-GMP and/or cAMP in CBF Enhancement

There are numerous studies showing that cAMP and/or cGMP can stimulate ciliary beating (8,9,13,18–20,25). This effect seems to be general, ranging from *Paramecium* to mammals. However, the magnitude of the ciliary stimulation induced by these agents is relatively small. Addition of 100 μM db-cGMP alone to untreated cells produced a negligible effect on CBF (24). The elevation in the db-cGMP concentration to as high as 1 mM resulted in a moderate increase in CBF, which never exceeded 30%. Similar small elevations in CBF ranging from 10 to 50% in response to an increase in either NO or cGMP levels have been observed in human airway epithelium (20–22). Thus, the cGMP pathway alone can probably elevate CBF in a calcium-independent manner, although the magnitude of this elevation is relatively small.

Cyclic-AMP is another well-known modulator of ciliary activity (8,11, 18,19). Cyclic-AMP–dependent protein kinase (PKA) has been shown to phosphorylate specific axonemal targets that increase the forward swimming speed in *Paramecium* (28,29). We have recently reported that in tissue cultures from frog esophagus, the increase in cAMP concentration induced CBF enhancement, even when a rise in $[Ca^{2+}]_i$ was deterred (18). It should be emphasized that, just as in the case of cGMP, the magnitude of the CBF enhancement, produced by the cAMP elevation in a calcium-independent manner was modest (50–70%).

To examine the effect of an increase in cAMP levels on the ciliary activity in cultured rabbit trachea, we treated the cells with forskolin, a powerful activator of adenylyl cyclase (AC). Addition of 25 μM forskolin produced a negligible effect on $[Ca^{2+}]_i$ and caused moderate enhancement of CBF (Fig. 1). The average changes in intracellular calcium concentration ($\Delta[Ca^{2+}]_i$) and CBF (F/F_o) in response to 25 μM forskolin were $\Delta[Ca^{2+}]_i = 34 \pm 5$ nM ($n = 10$) and $F/F_o =$

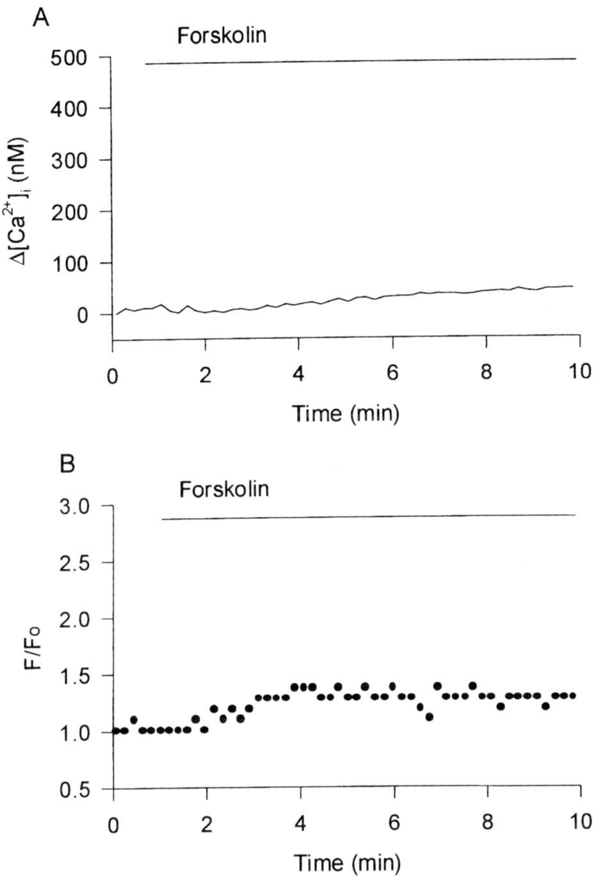

FIGURE 1 The effect of cAMP on CBF. Time course of the effect of 25 μM forskolin, a powerful activator of adenylyl cyclase, on $[Ca^{2+}]_i$(A) and CBF (B). Forskolin had a negligible effect on $[Ca^{2+}]_i$ and caused a moderate enhancement in CBF. (Courtesy of Springer-Verlag, New York.)

1.3 ± 0.1 ($n = 10$). Thus, these results conform to the results obtained in the frog esophagus (18), once again implying a general trend. It is possible to conclude that either cAMP or cGMP can enhance CBF in a calcium-independent manner. This occurs probably through phosphorylation of axonemal proteins by the related protein kinases, as has been demonstrated by Hamasaki et al. (28,29) and Salathe et al. (30). However, it is important to emphasize that such a moderate enhancement of CBF is unlikely to account for the high transport rates.

The Role of Calcium in CBF Enhancement

It is generally accepted that intracellular calcium plays a pivotal role in regulation of ciliary activity. Indeed, a rise in $[Ca^{2+}]_i$ correlates with a rise in CBF (3,14, 15,18,23,31,32). Furthermore, an elevation in $[Ca^{2+}]_i$ is essential for initiation of a response to many receptor-mediated stimuli. Thus, depletion of intracellular calcium stores, using calcium ionophores or the sarcoplasmic/endoplasmic Ca^{2+}-ATPase inhibitor thapsigargin, abolishes ciliary stimulation induced by extracellular ATP or by acetylcholine (3,15,16,24). It is presently not clear, however, whether Ca^{2+} exerts its effect by directly interacting with axonemal proteins or by serving a link in a cascade of events, eventually leading to CBF enhancement. It is possible to address these questions by simultaneously measuring $[Ca^{2+}]_i$ and CBF from the same cell.

Calmodulin is a ubiquitous Ca^{2+}-binding protein present in all animal and plant cells. Calmodulin appears to be an intracellular Ca^{2+} receptor, which, upon binding of Ca^{2+}, participates in a majority of the Ca^{2+}-regulated processes (33–35). Therefore, it is reasonable to suggest that Ca^{2+} delivers its stimulatory effect on cilia, at least partially, via the Ca^{2+}-calmodulin complex. However, the available data on the involvement of calmodulin in regulation of ciliary activity are contradictory. For example, application of a calmodulin inhibitor blocked the increase in CBF normally produced by ionomycin in human respiratory epithelium (11). However, application of a calmodulin inhibitor was ineffective in blocking the CBF rise induced by ACh in cultured ovine trachea (7). Therefore, we examined the possible involvement of calmodulin in CBF stimulation in rabbit airway epithelium. Figure 2A and B shows a normal response of ciliary cells from rabbit trachea to extracellular ATP. Figure 2C and D shows the response obtained in the presence of the calmodulin inhibitor W-7 (100 μM). The calmodulin inhibitor produced a moderate attenuation of the initial rise in $[Ca^{2+}]_I$ induced by ATP. However, despite a significant rise in $[Ca^{2+}]_i$, the elevation in CBF was almost completely abolished. The average responses to 100 μM ATP under control conditions were $\Delta[Ca^{2+}]_i = 373 \pm 7$ nM ($n = 40$) and $F/F_o = 2.2 \pm 0.1$ ($n = 40$). In contrast, the average responses to 100 μM ATP in the presence of 100 μM W-7 were $\Delta[Ca^{2+}]_i = 255 \pm 30$ nM ($n = 10$) and $F/F_o = 1.3 \pm 0.1$ ($n = 10$). This effect of a calmodulin inhibitor is not restricted to ATP or to the rabbit trachea. Similar inhibition of CBF enhancement inflicted by calmodulin inhibitors (TFP or W-7) was obtained when $[Ca^{2+}]_i$ was elevated artificially by application of the calcium ionophore ionomycin in both rabbit trachea and frog esophagus (unpublished results). Furthermore, an essentially identical effect of the calmodulin inhibitor W-7 was observed in frog esophagus stimulated by acetylcholine (36).

These results strongly suggest that the main role of Ca^{2+} in the process of CBF enhancement is to activate calmodulin, which plays an important role in the process of ciliary stimulation. Furthermore, since the effect of the calmodulin

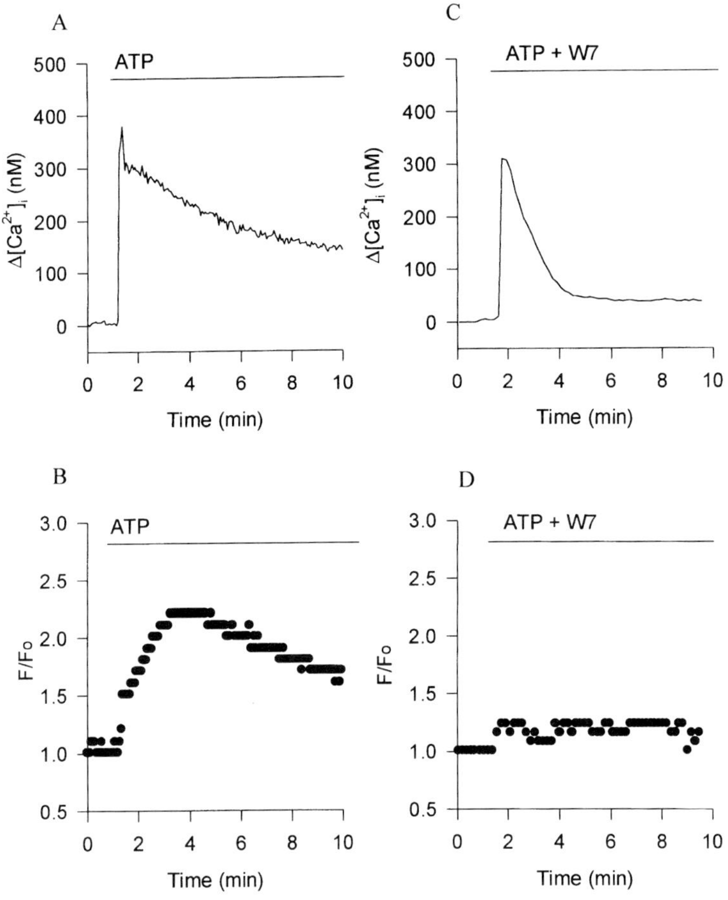

FIGURE 2 Involvement of calmodulin in ciliary activity. Time course of the effect of extracellular ATP (100 μM) on $[Ca^{2+}]_i$ (A, C) and CBF (B, D) in the presence (C, D) and the absence (A, B) of W-7 (100 μM), an inhibitor of calmodulin. W-7 decoupled the rise in $[Ca^{2+}]_i$ from CBF enhancement induced by ATP. (Courtesy of Springer-Verlag, New York).

inhibitors is limited neither to a single type of stimulation nor to a single type of ciliary tissue, it is reasonable to suggest that the critical role of calmodulin in the process of CBF enhancement is a general phenomenon.

Synergistic Action of PKG and/or PKA and Elevated $[Ca^{2+}]_i$

As was mentioned above, the Ca^{2+}-calmodulin complex can influence the activity of many cellular enzymes. Activation of the nitric oxide synthase (NOS) is of special interest in ciliary cells, since it initiates another important pathway of ciliary stimulation—the cGMP pathway. The increased production of nitric oxide (NO) leads to activation of guanylyl cyclase (GC). This results in an increase in a cGMP level and activation of the cGMP-dependent protein kinase (PKG). We have recently reported that inhibition of either NOS, or GC, or PKG blocks the CBF enhancement normally evoked by ATP, even though the $[Ca^{2+}]_i$ elevation is not significantly impaired by these procedures (24). In other words, the effect produced by these manipulations is analogous to the effect produced by the calmodulin inhibitors. The essence of this effect is uncoupling between the rise in $[Ca^{2+}]_i$ and the rise in CBF. These results suggest that at least one of the important functions performed by the Ca^{2+}-calmodulin complex in ciliary stimulation is activation of the cGMP pathway. Moreover, it appears that any manipulation that ultimately prevents activation of PKG nullifies the stimulatory effect of a high $[Ca^{2+}]_i$ on CBF.

On the other hand, direct activation of PKG without a rise in $[Ca^{2+}]_i$ leads only to small or moderate CBF enhancement. Therefore, we have hypothesized that activation of PKG in the presence of high $[Ca^{2+}]_i$ is required to achieve strong CBF enhancement. To verify this hypothesis, the following experiment has been performed in rabbit airway epithelium. First, the NO pathway was inhibited either by blocking NOS or by blocking GC to decouple between the rise in $[Ca^{2+}]_i$ and CBF stimulation. Then, PKG was directly activated by exposing the cells to db-cGMP. At 100 μM concentration of db-cGMP, there was no observable effect on $[Ca^{2+}]_i$ or CBF. However, when $[Ca^{2+}]_i$ was directly elevated by exposing the treated cells to ionomycin or to extracellular ATP, a robust enhancement in CBF was observed (Fig. 3).

In addition to the NO pathway, Ca^{2+}-calmodulin complex can stimulate AC, leading to activation of PKA. The activated PKA facilitates calcium mobilization from intracellular stores (18) and, together with high $[Ca^{2+}]_i$, enhances ciliary beating. This pathway has been recently revealed in our laboratory in tissue cultures from frog esophagus and palate (unpublished results).

Taken together, these results suggest that the elevation in $[Ca^{2+}]_i$ is neither necessary nor sufficient a condition for CBF enhancement. It is not a sufficient condition for CBF enhancement because blockage of PKG activation renders the stimulatory effect of $[Ca^{2+}]_i$ void. It is not a necessary condition because an increase in the levels of the cyclic nucleotides can induce enhancement of CBF in a calcium-independent manner. This enhancement is presumably achieved through a direct phosphorylation of axonemal proteins by PKA and/or PKG. It is important to emphasize, however, that although the ciliary activation induced in this fashion can be sustained for a long time (18), its magnitude is relatively small.

Apparently, phosphorylation(s) performed by protein kinases is/are a necessary condition for CBF enhancement, while a calcium elevation is not. On the

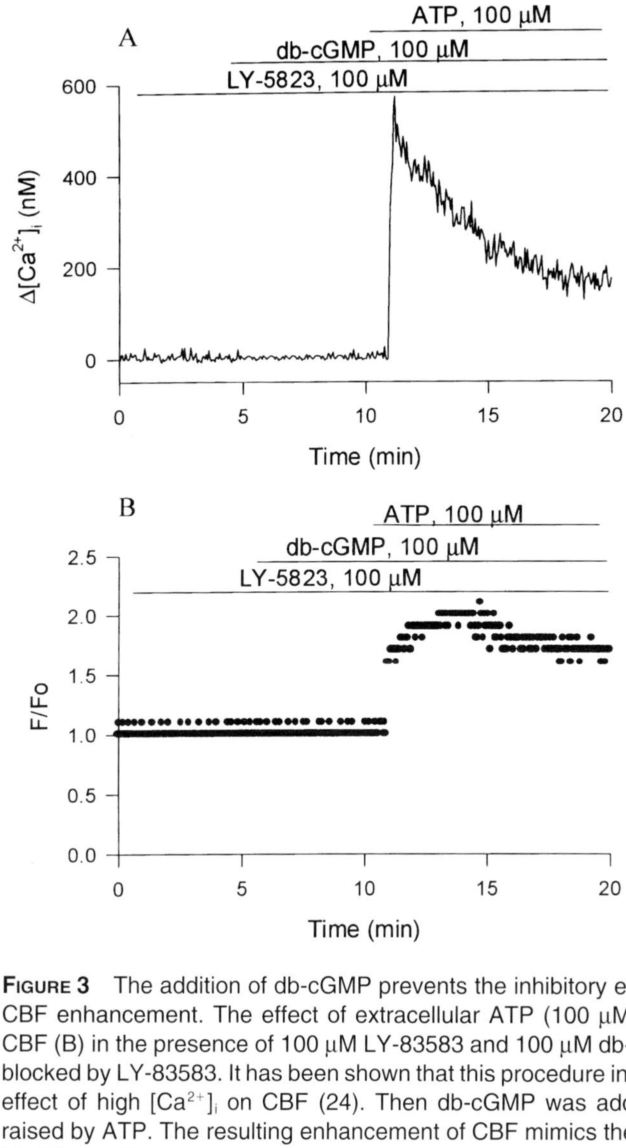

FIGURE 3 The addition of db-cGMP prevents the inhibitory effect of LY-83583 on CBF enhancement. The effect of extracellular ATP (100 μM) on $[Ca^{2+}]_i$ (A) and CBF (B) in the presence of 100 μM LY-83583 and 100 μM db-cGMP. The GC was blocked by LY-83583. It has been shown that this procedure inhibits the stimulatory effect of high $[Ca^{2+}]_i$ on CBF (24). Then db-cGMP was added and $[Ca^{2+}]_i$ was raised by ATP. The resulting enhancement of CBF mimics the usual effect of high $[Ca^{2+}]_i$ on ciliary beating.

other hand, impeding $[Ca^{2+}]_i$ elevation blocks ciliary response to the stimuli, which are conveyed by receptors coupled to phospholipase C (3,15,16,24). An elevation in $[Ca^{2+}]_i$ cannot be a single stimulatory effect of these receptors, since such an elevation alone is not sufficient for the CBF enhancement. Therefore, it appears that calmodulin plays a key role in the process of ciliary stimulation providing a link between the $[Ca^{2+}]_i$ elevation and the protein kinases (Fig. 4). Following the receptor-induced $[Ca^{2+}]_i$ elevation, the Ca^{2+}-calmodulin complex stimulates the production of cyclic nucleotides, leading to PKA- and/or PKG-dependent phosphorylations. Moreover, these phosphorylations are not only necessary for CBF enhancement but can sustain ciliary activity at a moderately enhanced level during the later stages of the stimulation, when $[Ca^{2+}]_i$ subsides to its resting level (36).

To summarize, the $[Ca^{2+}]_i$ elevation alone is insufficient to induce CBF enhancement. The cyclic nucleotide-dependent phosphorylations can produce a moderate enhancement of CBF in a calcium-independent manner. However, the combined effect of the calcium elevation and of the cyclic nucleotide-dependent kinases activated by the Ca^{2+}-calmodulin complex, manifests itself in robust and sustained CBF enhancement. Such a strong CBF enhancement together with the

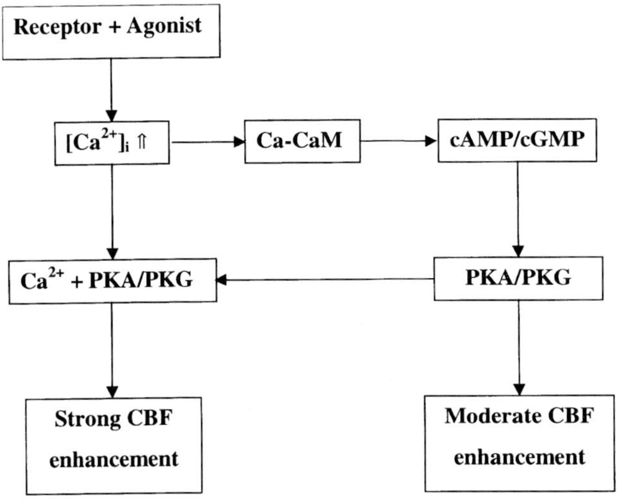

FIGURE 4 Schematic diagram of molecular events in the process of ciliary stimulation. The role of Ca^{2+} is twofold: (a) activation of protein kinases PKA/PKG through Ca^{2+}-calmodulin complex; (b) synergistic action with protein kinases PKA/PKG to strongly enhance CBF. (Courtesy of Springer-Verlag, New York).

accompanied changes in metachronal wave characteristics (1) may explain the amazing ability of cilia to transport heavy objects at a high speed.

ACKNOWLEDGMENTS

This work was partially supported by a grant from the Israeli Science Foundation. A. Braiman gratefully acknowledges the fellowship support of the Kreitman Foundation.

REFERENCES

1. L Gheber, Z Priel. Metachronal activity of cultured mucociliary epithelium under normal and stimulated conditions. Cell Motil Cytoskeleton 28:333–345, 1994.
2. L Gheber, Z Priel, C Aflalo, V Shohsan-Barmatz. Extracellular ATP binding proteins as potential receptors in mucociliary epithelium: characterization using [32P]3′-O-(4-benzoyl)benzoyl ATP, a photoaffinity label. J Memb Biol 147:83–93, 1995.
3. A Korngreen, Z Priel. Purinergic stimulation of rabbit ciliated airway epithelia: control by multiple calcium sources. J Physiol 497:53–66, 1996.
4. M Villalon, TR Hinds, P Verdugo. Stimulus-response coupling in mammalian ciliated cells. Demonstration of two mechanisms of control for cytosolic [Ca^{2+}]. Biophys J 56:1255–1258, 1989.
5. LB Wong, DB Yeates. Luminal purinergic regulatory mechanism of tracheal ciliary beat frequency. Am J Respir Cell Mol Biol 7:447–454, 1992.
6. E Aiello, J Kennedy, C Hernandez. Stimulation of frog ciliated cells in culture by acetylcholine and substance P. Comp Biochem Physiol 99:497–506, 1991.
7. M Salathe, RJ Bookman. Mode of Ca^{2+} action on ciliary beat frequency in single ovine airway epithelial cells. J Physiol 520:851–865, 1999.
8. B Yang, RJ Schlosser, TV McCaffry. Dual signal transduction mechanisms modulate ciliary beat frequency in upper airway epithelium. Am J Physiol 270:L745–L751, 1996.
9. P Verdugo, NT Johnson, PY Tam. β-Adrenergic stimulation of respiratory ciliary activity. J Appl Physiol 48:868–871, 1980.
10. I Maruyama. Conflicting effects of noradrenalin on ciliary movement of frog palatine mucosa. Eur J Pharmacol 97:239–245, 1984.
11. G Di Benedetto, CJ Magnus, PTA Gray, A Mehta. Calcium regulation of ciliary beat frequency in human respiratory epithelium in vitro. J Physiol 439:103–113, 1991.
12. A Korngreen, Z Priel. Simultaneous measurement of ciliary beating and intracellular calcium. Biophys J 67:377–380, 1994.
13. AB Lansley, MJ Sanderson, ER Dirksen. Control of the beat cycle of respiratory tract cilia by Ca^{2+} and cAMP. Am J Physiol 263:L232–L242, 1992.
14. R Levin, A Braiman, Z Priel. Protein kinase C induced calcium influx and sustained enhancement of ciliary beating by extracellular ATP. Cell Calcium 21:103–113, 1997.

15. M Salathe, RJ Bookman. Coupling of $[Ca^{2+}]_i$ and ciliary beating in cultured tracheal epithelial cells. J Cell Sci 108:431–440, 1995.
16. A Tarasiuk, M Bar-Shimon, L Gheber, A Korngreen, Y Grossman, Z Priel. Extracellular ATP induces hyperpolarization and motility stimulation of ciliary cells. Biophys J 68:1163–1169, 1995.
17. P Verdugo. Ca^{2+}-dependent hormonal stimulation of ciliary activity. Nature 283: 764–765, 1980.
18. A Braiman, O Zagoory, Z Priel. PKA induces Ca^{2+} release and enhances ciliary beat frequency in a Ca^{2+}-dependent and -independent manner. Am J Physiol 275: C790–C797, 1998.
19. J Tamaoki, M Kondo, T Takizawa. Effect of cAMP on ciliary function in rabbit tracheal epithelial cells. J Appl Physiol 66:1035–1039, 1989.
20. CA Geary, CW Davis, AM Paradiso, RC Boucher. Role of CNP in human airways: cGMP-mediated stimulation of ciliary beat frequency. Am J Physiol 268:L1021–L1028, 1995.
21. T Runer, S Lindberg. Effects of nitric oxide on blood flow and mucociliary activity in the human nose. Ann Otol Rhinol Laryngol 107:40–46, 1998.
22. T Runer, A Cervin, S Lindberg, R Uddman. Nitric oxide is a regulator of mucociliary activity in the upper respiratory tract. Oto Head Neck Surg 119:278–287, 1998.
23. AB Lansley, MJ Sanderson. Regulation of airway activity by Ca^{2+}: simultaneous measurement of beat frequency and intracellular Ca^{2+}. Biochem J 77:629–638, 1999.
24. N Uzlaner, Z Priel. Interplay between the NO pathway and elevated $[Ca^{2+}]_i$ enhances ciliary activity in rabbit trachea. J Physiol 516:179–190, 1999.
25. J Tamaoki, A Chiyotani, M Kondo, H Takemura, K Konno. Role of NO generation in β-adrenoceptor-mediated stimulation of rabbit airway ciliary motility. Am J Physiol 268:C1342–C1347, 1995.
26. J Tamaoki, A Sakai, M Kondo, H Takemura, K Konno. Role of nitric oxide in tachykinin-induced increase in potential difference of rabbit tracheal mucosa. J Physiol 488:115–122, 1995.
27. G Grynkiewicz, M Poenie, RY Tsien. A new generation of Ca^{2+} indicators with improved fluorescence properties. J Biol Chem 260:3440–3450, 1985.
28. T Hamasaki, K Barkalow, J Richmond, P Satir. cAMP-stimulated phosphorylation of an axonemal polypeptide that copurifies with the 22S dynein arm regulates microtubule translocation velocity and swimming speed in paramecium. Proc Natl Acad Sci USA 88:7918–7922, 1991.
29. T Hamasaki, K Barkalow, P Satir. Regulation of ciliary beat frequency by a dynein light chain. Cell Motil Cytoskeleton 32:121–124, 1995.
30. M Salathe, MM Pratt, A Wanner. Cyclic AMP-dependent phosphorylation of a 26 kD axonemal protein ovine cilia isolated from small tissue pieces. Am J Respir Cell Mol Biol 9:306–314, 1993.
31. H Mao, LB Wong. Fluorescence and laser photon counting: measurements of epithelial $[Ca^{2+}]_i$ or $[Na^+]_i$ with ciliary beat frequency. Ann Biomed Eng 26:666–678, 1998.
32. W Ma, A Korngreen, N Uzlaner, Z Priel, SD Silberberg. Extracellular sodium regulates airway ciliary motility by inhibiting P2X receptor. Nature 400:894–897, 1999.

33. AC Nairn, MR Picciotto. Calcium/calmodulin-dependent protein kinases. Semin Cancer Biol 5:295–303, 1994.
34. RD Hinrichsen. Calcium and calmodulin in the control of cellular behavior and motility. Biochim Biophys Acta 1155:277–293, 1993.
35. FA Antoni. Calcium regulation of adenylate cyclase. Relevance for endocrine control. Trends Endocrin Metab 8:7–14, 1997.
36. O Zagoory, A Braiman, L Gheber, Z Priel. The role of calcium and calmodulin in ciliary stimulation induced by acetylcholine. Am J Physiol 280:C100–C109, 2001.

7

Modulation of Ciliary Motility by Na^+

Shai D. Silberberg, Alon Korngreen,*
Weiyuan Ma, Natalya Uzlaner,
and Zvi Priel
Ben-Gurion University of the Negev
Beer-Sheva, Israel

INTRODUCTION

Extracellular Na^+ is unlikely to be an ion-channel regulator since its physiological concentration in the serum and interstitial fluid is high and maintained within a narrow range. In a number of confined bodily compartments, however, the extracellular fluid is functionally segregated from the serum. One such compartment is the fluid that lines the lumen of the airways. If the concentration of Na^+ in this confined bodily compartment can fluctuate appreciably, then Na^+ could potentially function as a physiological regulator. Using rabbit airway epithelium in vitro as an experimental model, we have recently uncovered that extracellular Na^+ can regulate airway ciliary motility in a dose-dependent manner by inhibiting an ATP-gated ion channel ($P2X_{cilia}$ channel). The effect of Na^+ could be overcome by raising the concentration of ATP, suggesting an allosteric mechanism of regulation. The following sections outline the major aspects of mucociliary activation

**Current affiliation*: Max-Planck Institut für medizinische Forschung, Heidelberg, Germany.

and regulation pertinent to this discovery as well as a concise account of these recent findings and a brief discussion of their potential implications.

THE ROLE OF EXTRACELLULAR ATP IN THE REGULATION OF CILIARY MOTILITY

The mucociliary system is responsible for maintaining the airways clean of inhaled particles and pathogens. This important task is performed by the beating of cilia and the consequent movement of mucus from the lungs to the upper airways (1,2). Ciliary activity is strongly regulated by hormones and neurotransmitters, which provide fine control over the efficiency of mucus transport (1). In various mucociliary systems, including human airways, extracellular ATP potently stimulates ciliary activity (3–7). Cell-surface receptors for purines and pyrimidines are classified into two major types (8): G-protein coupled receptors (P2Y receptors) and ligand-gated ion channels (P2X receptors) (9–11). Most P2Y receptors are coupled to phospholipase C, leading to the formation of inositol 1, 4, 5-trisphosphate ($InsP_3$) and, thus, to the mobilization of Ca^{2+} from internal stores. The P2Y receptor in rabbit airway ciliated cells was recently shown to have a similar sensitivity to extracellular ATP and extracellular UTP and to activate the same signal transduction pathway when activated by either extracellular ATP or extracellular UTP (12). P2X receptors are ligand-gated cation channels that are permeable to Ca^{2+} (9–11). It is evident, therefore, that the activation of either P2X or P2Y receptors would elevate the cytoplasmic concentration Ca^{2+} ($[Ca^{2+}]_i$).

THE ROLE OF Ca^{2+} IN EXTRACELLULAR ATP-INDUCED MUCOCILIARY ACTIVATION

It is well established that Ca^{2+} is an important mediator in mucociliary activation. In many mucociliary systems, including human airway, stimulation of ciliary activity by various agonists is correlated with a rise in $[Ca^{2+}]_i$ (4,6,13–17). The dynamic effects of extracellular ATP on ciliary beating frequency (CBF) and $[Ca^{2+}]_i$ were simultaneously measured in rabbit airway epithelium (4). Extracellular ATP induces a rapid rise in $[Ca^{2+}]_i$ and in CBF. In the continuous presence of extracellular ATP, $[Ca^{2+}]_i$ declined over several minutes to a lower elevated $[Ca^{2+}]_i$ plateau, while CBF remained at a high level during the ten minutes of exposure to extracellular ATP. In contrast, in low extracellular Ca^{2+}, extracellular ATP induced only a transient rise in both $[Ca^{2+}]_i$ and CBF. A similar time course of $[Ca^{2+}]_i$ in response to extracellular ATP, and a similar dependence on extracellular Ca^{2+}, was observed also in human airway epithelium (18). Taken together, these findings show that in extracellular solutions containing 140 mM Na^+, the initial rapid rise in both $[Ca^{2+}]_i$ and in CBF induced by extracellular ATP results primarily from the mobilization of Ca^{2+} from internal stores, while the sustained elevated ciliary activity requires Ca^{2+} influx. There is accumulating evidence that a pathway for Ca^{2+} influx that has not yet been identified is most likely activated by the P2Y pathway. In addition, as outlined below, extracellular ATP can directly activate a Ca^{2+}-permeable ion channel (P2X receptor).

EXTRACELLULAR ATP AND Ca^{2+} INFLUX IN RABBIT AIRWAY CILIATED CELLS

In an attempt to identify the pathway for Ca^{2+} influx activated by extracellular ATP, we developed a procedure to dissociate rabbit airway epithelium to obtain single viable ciliated cells (19). The patch clamp technique (20) was then used to control the composition of the intracellular environment while simultaneously recording membrane currents activated by various nucleotides. In these whole-cell experiments, the intracellular (pipette) solution contained only a physiological concentration of salt, Mg-ATP, and GTP. Rabbit airway ciliated cells were found to express an ATP-gated cation-selective channel ($P2X_{cilia}$ channel) that is strongly attenuated by extracellular divalent cations (19). Extracellular ATP (but not extracellular UTP or ADP) activated the $P2X_{cilia}$ channels, which then remained activated for several minutes in the presence of extracellular ATP. The conductance was permeable to monovalent and divalent cations with approximate relative permeability for $P_{Ba}:P_{Cs}:P_{TEA}$ of 4:1:0.1. Permeability to Cl^- was negligible. The permeability of $P2X_{cilia}$ channels to Ca^{2+} was estimated to be ninefold greater than the permeability to Cs^+, indicating that Ca^{2+} is more permeant than Ba^{2+} (21). As Ca^{2+} plays a pivotal role in ciliary function, and $P2X_{cilia}$ channels remain activated for at least 30 minutes with extracellular ATP, $P2X_{cilia}$ channels might be involved in maintaining ciliary activation during prolonged exposures to extracellular ATP. Indeed, the following section shows that activation of $P2X_{cilia}$ channels contributes to sustained enhancement of ciliary motility by extracellular ATP.

EXTRACELLULAR Na^+ AND CILIARY MOTILITY

Recently we discovered that extracellular Na^+ (Na^+_O) inhibits $P2X_{cilia}$ channels and thereby attenuates extracellular ATP-induced ciliary motility (22). The major observations can be summarized as follows: (1) Na^+_O directly and specifically inhibits $P2X_{cilia}$ channels, (2) inhibition by Na^+_O of an extracellular ATP-activated conductance is seen in airway ciliated cells from a variety of mammalian species including rabbit, rat, pig, and human, and (3) the degree of inhibition by Na^+_O depends on interplay between the effective concentrations of extracellular ATP and Na^+_O.

In order to investigate the contribution of $P2X_{cilia}$ channels to ciliary function and consequently the possible contribution of Na^+_O to mucociliary clearance, we performed experiments in which the P2Y-mediated pathway was suppressed by depleting the intracellular Ca^{2+} stores with thapsigargin and by inhibiting phospholipase C with U-73122. Ciliary activity and $[Ca^{2+}]_i$ were simultaneously measured from these cells in extracellular solutions containing either low (10 mM) or high (140 mM) Na^+. In control cells (with functional P2Y receptors), in high Na^+_O, extracellular ATP (100 μM) increased $[Ca^{2+}]_i$ and CBF (Fig. 1A). In contrast, in high Na^+_O, after suppressing the P2Y pathway, extracellular ATP (100 μM) failed to elevate $[Ca^{2+}]_i$ or to enhance ciliary activity (Fig. 1B). This result clearly shows that the P2Y pathway contributes to sustained ciliary activity. In addition, these results showed that the experimental procedure effectively suppressed ciliary activation via the P2Y pathway and thus provided a means to investigate the possible contribution of the $P2X_{cilia}$ channels to ciliary activation by extracellular ATP.

Contrary to the inability of extracellular ATP to enhance CBF after suppressing the P2Y pathway when Na^+_O is high, in low Na^+_O, extracellular ATP elevated $[Ca^{2+}]_i$ and enhanced CBF (Fig. 2A). To show that the effects of extracellular ATP in low Na^+_O are mediated via $P2X_{cilia}$ channels, two additional sets of experiments were performed. First, extracellular Ca^{2+} was lowered from 0.5 mM to 0.5 μM, demonstrating that the ATP-induced rise in $[Ca^{2+}]_i$ and in CBF requires extracellular Ca^{2+} (Fig. 2B). Second, the effect of extracellular UTP on

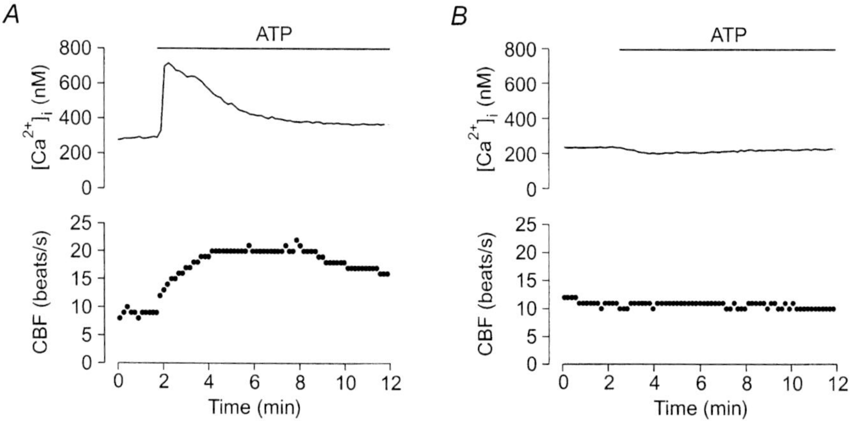

FIGURE 1 Extracellular ATP does not enhance ciliary activity in high Na^+_O when the P2Y pathway is suppressed. Simultaneous measurements of $[Ca^{2+}]_i$ (top panels) and CBF (bottom panels) in high (140 mM) Na^+_O, in response to extracellular ATP (100μM), from a control cell (A) and from a cell in which the P2Y pathway was suppressed by depleting the intracellular Ca^{2+} stores with thapsigargin and by inhibiting phospholipase C with U-73122 (B).

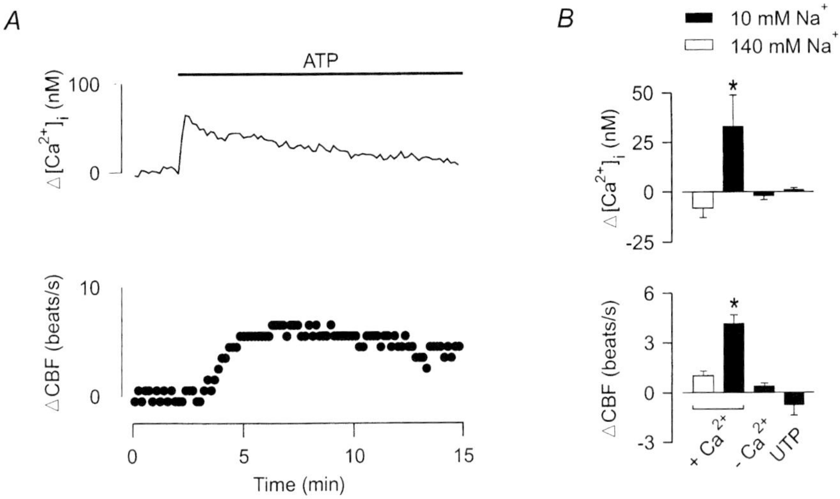

FIGURE 2 Extracellular ATP can enhance ciliary activity when the P2Y pathway is suppressed if Na^+_O is low. (A) Simultaneous measurements of $[Ca^{2+}]_i$ (top panel) and CBF (bottom panel) in low (10 mM) Na^+_O, from a cell in which the P2Y pathway was suppressed. Extracellular ATP (100 μM) enhanced $[Ca^{2+}]_i$ and CBF. (B) A summary of the change in $[Ca^{2+}]_i$ (top panel) and CBF (bottom panel) induced by either extracellular ATP (100 μM) or extracellular UTP (100 μM) on cells in which the P2Y pathway was suppressed. In high (140 mM) Na^+_O, extracellular ATP was without effect. Extracellular ATP increased both $[Ca^{2+}]_i$ and CBF in low Na^+_O when the extracellular solution contained 0.5 mM $CaCl_2(+Ca^{2+})$, but not when the extracellular solution contained 0.5 μM $CaCl_2(-Ca^{2+})$. Extracellular UTP was without effect (UTP). (From Ref. 22.)

$[Ca^{2+}]_i$ and CBF in low Na^+_O was examined. Since extracellular UTP does not activate $P2X_{cilia}$ channels, we predicted that extracellular UTP would not elevate $[Ca^{2+}]_i$ or stimulate the cilia. Indeed, extracellular UTP (100 μM) was without effect (Fig. 2B). Taken together, these results show that Ca^{2+} influx through $P2X_{cilia}$ channels can increase ciliary activity and that Na^+ inhibits both Ca^{2+} influx through $P2X_{cilia}$ channels and the resulting ciliary activity.

THE ROLE OF $P2X_{cilia}$ CHANNELS IN MUCOCILIARY FUNCTION

Little is known on the concentrations of extracellular ATP in the airways, and thus it is presently not possible to estimate the extent to which $P2X_{cilia}$ channels are physiologically activated. Nevertheless, comparing the effects of Na^+_O on

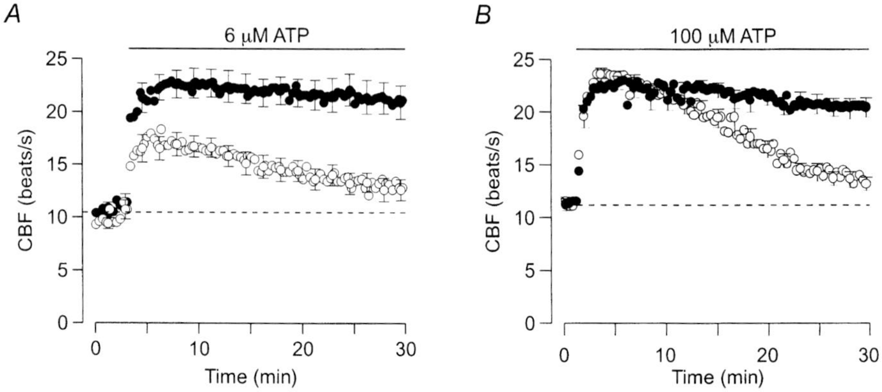

FIGURE 3 The relative contributions of the P2Y pathway and the $P2X_{cilia}$ pathway to ciliary activation by extracellular ATP. Average CBF recorded in 140 mM (empty circles) or 10 mM (full circles) extracellular NaCl in response to 6 μM extracellular ATP (A), or 100 μM extracellular ATP (B). Control cells (i.e., the P2Y pathway is functional). Every 10th error bar is presented. (B from Ref. 22.)

ciliary activation induced by 100 μM extracellular ATP to the effects of Na^+_O on ciliary activation induced by 6 μM extracellular ATP (Fig. 3) provides insight as to the relative contributions of the P2Y pathway and the $P2X_{cilia}$ pathway to ciliary activation. Assuming that ATP^{4-} is the active agonist (19), then under the experimental conditions used, 100 μM and 6 μM extracellular ATP are equivalent to approximately 95 and 6 nM of ATP^{4-}, respectively.

The initial rise in CBF induced by 100 μM extracellular ATP is the same in low and high Na^+_O (Fig. 3B), indicating that robust activation of the P2Y pathway can maximally stimulate the cilia. However, when the $P2X_{cilia}$ pathway is inhibited (i.e., in high Na^+_O), ciliary activation induced by extracellular ATP is transient. Thus, the $P2X_{cilia}$ pathway contributes to sustaining the cilia at a high level of activation. In contrast to the ability of 100 μM ATP to maximally stimulate the cilia in both high and low Na^+_O, the initial rise in ciliary activation induced by 6 μM extracellular ATP is partially inhibited by Na^+_O (Fig. 3A). This suggests that the $P2X_{cilia}$ pathway can also contribute to the initial rise in ciliary activation induced by extracellular ATP when the concentration of extracellular ATP is relatively low. Thus, modulation of $P2X_{cilia}$ channels by Na^+_O might be a physiological mechanism for regulating ciliary function.

A POSSIBLE ROLE FOR Na^+_O IN AIRWAY PHYSIOLOGY

The physiological relevance of the findings described above may depend on the fundamental question of whether $P2X_{cilia}$ channels are exposed to the airway surface fluid. Using the patch clamp technique, we have detected $P2X_{cilia}$ channels in the basolateral membrane of isolated ciliary cells, suggesting that $P2X_{cilia}$ channels are exposed to the interstitial fluid in which the concentration of Na^+ is high. However, it is well established that dissociated epithelial cells lose their polarity and membrane proteins redistribute in the membrane. Thus, the patch-clamp data do not undermine the possibility that in intact epithelium the $P2X_{cilia}$ channels are restricted to the apical membrane of the ciliary cells. Without specific antibodies raised against $P2X_{cilia}$ channels, this question cannot easily be addressed since it is presently not possible to record single-channel currents from the membrane of the cilium.

Another question of key importance is what are the physiological concentrations of ATP and Na^+ in the airways. In the proximal airways, the Na^+ content of the airway surface fluid (ASF) has been reported to be lower than in plasma (23–26) and has been reported to be either unchanged (24–27) or elevated (23,28) in CF. Little information is available, however, on the composition of the ASF in the distal airways (29,30). If Na^+ in the distal airways is elevated in disorders such as CF, then one of the early events leading to impaired mucociliary clearance in CF may be a reduced ability of extracellular ATP to regulate ciliary motility. Since human β-defensin, the peptide antibiotic expressed by the surface epithelia in lung, is also modulated by Na^+ (31–33), modulation of protein function by Na^+_O may be a fundamental physiological mechanism of airway function.

While the above results suggest that $P2X_{cilia}$ channels contribute to the regulation of ciliary motility, the physiological role of Na^+ as a modulator of mucociliary clearance remains to be determined. Preventing the inhibition of $P2X_{cilia}$ channels by Na^+ may, nevertheless, provide a novel route for improving mucociliary clearance.

ACKNOWLEDGMENTS

This work was supported by the Israel Science Foundation, founded by the Israel Academy of Sciences and Humanities–Dorot Science Foundation.

REFERENCES

1. A Wanner, M Salathé, TG O'Riordan. Mucociliary clearance in the airways. Am J Respir Crit Care Med 154:1868–1902, 1996.
2. H Matsui, SH Randell, SW Peretti, CW Davis, RC Boucher. Coordinated clearance of periciliary liquid and mucus from airway surfaces. J Clin Invest 102:1125–1131, 1998.
3. CA Geary, CW Davis, AM Paradiso, RC Boucher. Role of CNP in human airways: cGMP-mediated stimulation of ciliary beat frequency. Am J Physiol 268:L1021–L1028, 1995.
4. A Korngreen, Z Priel. Purinergic stimulation of rabbit ciliated airway epithelia: control by multiple Ca^{2+} sources. J Physiol (Lond) 497:53–66, 1996.

5. D Ovadyahu, D Eshel, Z Priel. Intensification of ciliary motility by extracellular ATP. Biorheology 25:489–501, 1988.
6. M Villalón, TR Hinds, P Verdugo. Stimulus-response coupling in mammalian ciliated cells. Demonstration of two mechanisms of control for cytosolic [Ca^{2+}]. Biophys J 56:1255–1258, 1989.
7. LB Wong, DB Yeates. Luminal purinergic regulatory mechanism of tracheal ciliary beat frequency. Am J Respir Cell Mol Biol 127:447–454, 1992.
8. M Abbracchio, G Burnstock. Purinoceptors: Are there families of P2X and P2Y purinoceptors? Pharmacol Ther 64:445–475, 1994.
9. SS Bhagwat, M Williams. P2 purine and pyrimidine receptors: emerging superfamilies of G-protein-coupled and ligand-gated ion channel receptors. Eur J Med Chem 32:183–193, 1997.
10. RA North, EA Barnard. Nucleotide receptors. Curr Op Neurobiol 7:346–357, 1997.
11. V Ralevic, G Burnstock. Receptors for purines and pyrimidines. Pharmacol Rev 50: 413–492, 1998.
12. N Uzlaner, Z Priel. Interplay between the NO pathway and elevated $[Ca^{2+}]_i$ enhances ciliary activity in rabbit trachea. J Physiol (Lond) 516:179–190, 1999.
13. ER Dirksen, MJ Sanderson. Regulation of ciliary activity in the mammalian respiratory tract. Biorheology 27:533–545, 1989.
14. AB Lansley, MJ Sanderson, ER Dirksen. Control of the beat cycle of respiratory tract cilia by Ca^{2+} and cAMP. Am J Physiol 263:L232–L242, 1992.
15. R Levin, A Braiman, Z Priel. Protein kinase C induced Ca^{2+} influx and sustained enhancement of ciliary beating by extracellular ATP. Cell Calcium 21:103–113, 1997.
16. M Salathe, RJ Bookman. Coupling of $[Ca^{2+}]_i$ and ciliary beating in cultured tracheal epithelial cells. J Cell Sci 108:431–440, 1995.
17. AB Lansley, MJ Sanderson. Regulation of airway ciliary activity by Ca^{2+}: simultaneous measurement of beat frequency and intracellular Ca^{2+}. Biophys J 77:629–638, 1999.
18. AM Paradiso, SJ Mason, ER Lazarowski, RC Boucher. Membrane-restricted regulation of Ca^{2+} release and influx in polarized epithelia. Nature 377:643–646, 1995.
19. A Korngreen, W Ma, Z Priel, SD Silberberg. Extracellular ATP directly gates a cation-selective channel in rabbit airway ciliated epithelial cells. J Physiol (Lond) 508:703–720, 1998.
20. B Sakmann, E Neher. Single Channel Recording. 2d ed. New York: Plenum Press, 1995.
21. W Ma, Z Priel, SD Silberberg. Pore properties of $P2X_{cilia}$ channels. Biophys J 78: A357, 2000.
22. W Ma, A Korngreen, N Uzlaner, Z Priel, SD Silberberg. Extracellular sodium regulates airway ciliary motility by inhibiting a P2X receptor. Nature 400:894–897, 1999.
23. L Joris, I Dab, PM Quinton. Elemental composition of human airway surface fluid in healthy and diseased airways. Am Rev Respir Dis 148:1633–1637, 1993.
24. MR Knowles, JM Robinson, RE Wood, CA Pue, WM Mentz, GC Wager, JT Gatzy, RC Boucher. Ion composition of airway surface liquid of patients with cystic fibrosis

as compared with normal and disease-control subjects. J Clin Invest 100:2588–2595, 1997.
25. J Hull, W Skinner, C Robertson, P Phelan. Elemental content of airway surface liquid from infants with cystic fibrosis. Am J Respir Crit Care Med 157:10–14, 1998.
26. S Baconnais, R Tirouvanziam, JM Zahm, S de Bentzmann, B Peault, G Balossier, E Puchelle. Ion composition and rheology of airway liquid from cystic fibrosis fetal tracheal xenografts. Am J Respir Cell Mol Biol 20:605–611, 1999.
27. H Matsui, BR Grubb, R Tarran, SH Randell, JT Gatzy, CW Davis, RC Boucher. Evidence for periciliary liquid layer depletion, not abnormal ion composition, in the pathogenesis of cystic fibrosis airway disease. Cell 95:1005–1015, 1998.
28. J Zabner, JJ Smith, PH Karp, JH Widdicombe, MJ Welsh. Loss of CFTR chloride channels alters salt absorption by cystic fibrosis airway epithelia in vitro. Mol Cell 2:397–403, 1998.
29. JJ Wine. The genesis of cystic fibrosis lung disease. J Clin Invest 103:309–312, 1999.
30. RC Boucher. Molecular insights into the physiology of the 'thin film' of airway surface liquid. J Physiol (Lond) 516:631–638, 1999.
31. JJ Smith, SM Travis, EP Greenberg, MJ Welsh. Cystic fibrosis airway epithelia fail to kill bacteria because of abnormal airway surface fluid. Cell 85:229–236, 1996.
32. MJ Goldman, GM Anderson, ED Stolzenberg, UP Kari, M Zasloff, LM Wilson. Human β-defensin-1 is a salt-sensitive antibiotic in lung that is inactivated in cystic fibrosis. Cell 88:553–560, 1997.
33. R Bals, X Wang, Z Wu, T Freeman, V Bafna, M Zasloff, JM Wilson. Human β-defensin 2 is a salt-sensitive peptide antibiotic expressed in human lung. J Clin Invest 102:874–880, 1998.

8

Presentation of Ciliary Beat Frequency Results to Our Peers

X. Mwimbi, R. Muimo, and Anil Mehta
Ninewells Hospital and Medical School
Dundee, United Kingdom

The interactions between different cell signaling pathways governing ciliary beat frequency (CBF) are as complex as the interpretation of the resultant CBF profiles, leading to many arguments about the best analytical methods. Here we are not concerned with those arguments but will concentrate on the alternative modes of presentation of the resultant data. For example, when CBF is studied in ciliated cells brush-biopsied from the human inferior turbinate using our previously described protocol (1), they do not all have the same setpoint of CBF. Thus, even if we accept that the in vitro CBF acts as a surrogate marker for the integrated output from such signaling pathways, then the wide range of intrabiopsy CBF values found in freshly isolated cells incubated in medium 199 (Fig. 1) suggests intercell differences in the regulation of the CBF response. For example, note how individual cells differ so substantially in the set point of their intrinsic CBF. Is this difference a true reflection of cell biology or a methodological artefact due to cell injury? The individual patient data sets in Figure 1 represent the *averaged* CBF responses from at least two cell borders (typically three cell borders) per patient recorded co-temperaneously from the microscopic field of view in an aliquot of each normal subject's nasal brushing. The average of the curves shown,

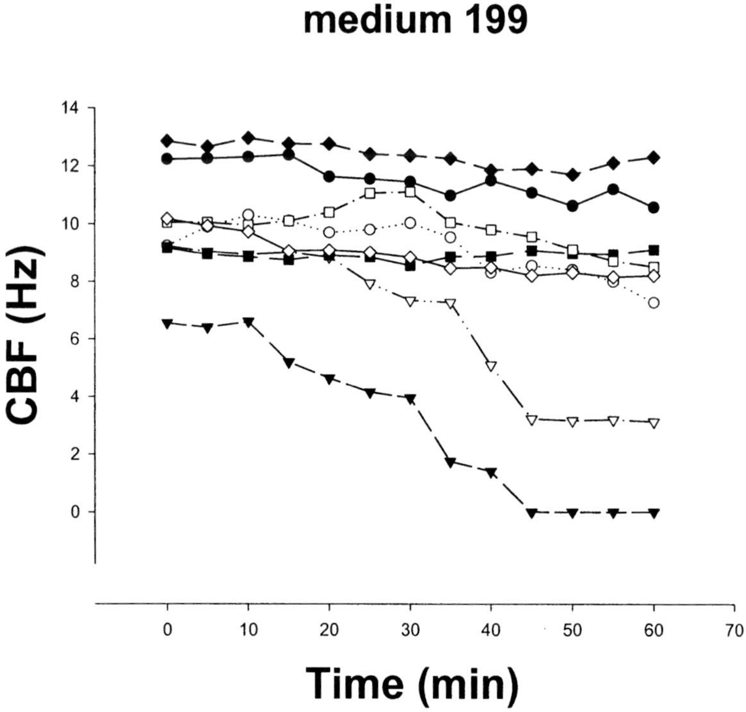

Figure 1 Nasal brushings from eight different normal healthy subjects. Each line shows the averaged CBF response from at least three cell borders per subject. Note the decay of CBF in two subjects. (See Fig. 2 for the underlying data).

as recommended by conventional statistical advice, provides measurements that are truly independent of one another, and typically, our publications (1–3) have reported such averaged data from approximately 10 different patients (over 30 cells per study). However, if the hypothesis that the spectrum of observed CBF is a true manifestation of different cell biology has any credence, then the averaged response calculated from Figure 1 may be misleading. Figure 2 shows two different groups of underlying raw data: first from those apparently normal subjects whose cilia had an *unstable* CBF in one or more cells (upper panel) or, second, from the more typical subject whose CBF showed both internal consistency and stability between different cell borders (lower panel). Which is the "true" response? Methodological differences cannot explain this dichotomy because post-harvest, the CBF was recorded identically for the two groups from cell borders which were immediately placed in medium 199 on ice, transferred to the perfu-

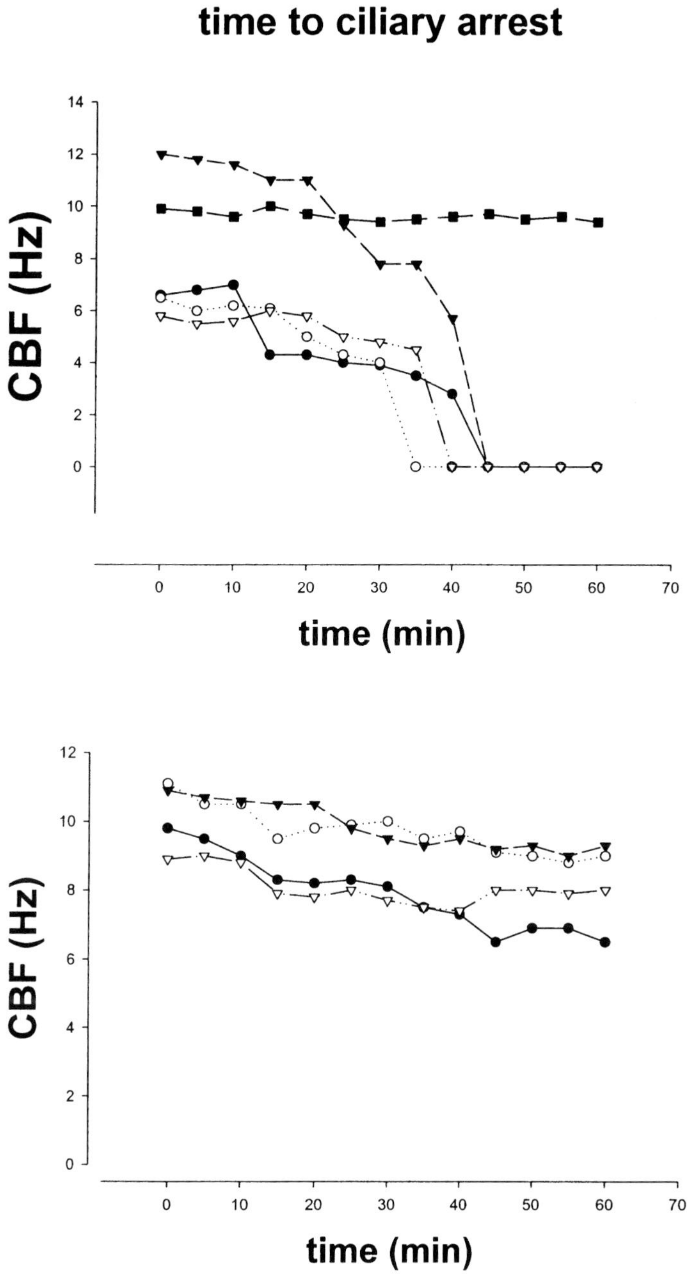

Figure 2 Individual patient CBF responses showing the degree of instability of CBF in different cells. Upper graph: note how one cell continues to beat at 10 Hz, whereas neighbors decay to zero with variable timing. In the lower graph, all cells continue to beat in a consistent manner. Survival curve analysis is appropriate for the cells showing a decay in frequency, but this is not appropriate for the second set.

sion chamber, and the chamber temperature restored to the normal nasal temperature of 32°C, all within 60 minutes. Despite such standardization, the spectrum of CBF is wide. We have previously not studied cells beating at around 6 Hz (at 32°C) assuming that they are in some way damaged, thus explaining a decay in their CBF that is so extreme that ciliary arrest occurs in some cases. However, this application of such a notion must then also include a cell beating at approximately 12 Hz (having been stable enough for inclusion at the beginning of the study), which nevertheless also shows the same decay to ciliary arrest phenomenon. Are such cells dead? Figure 3 shows that this conclusion is also premature because reperfusion with medium transiently restores CBF in some cells, whereas for others, CBF reverts to predecay values (or higher; not shown). The simplest conclusion is that such cells are "resting" since reperfusion does not merely refresh the chamber with new energy from the medium (or wash out toxins) because throughout the phase of ciliary arrest adjacent cells within the same field of view continue to beat normally (see Fig. 2). Could ciliary arrest reflect an adaptive response to the high energy requirements induced by ciliary beat? We

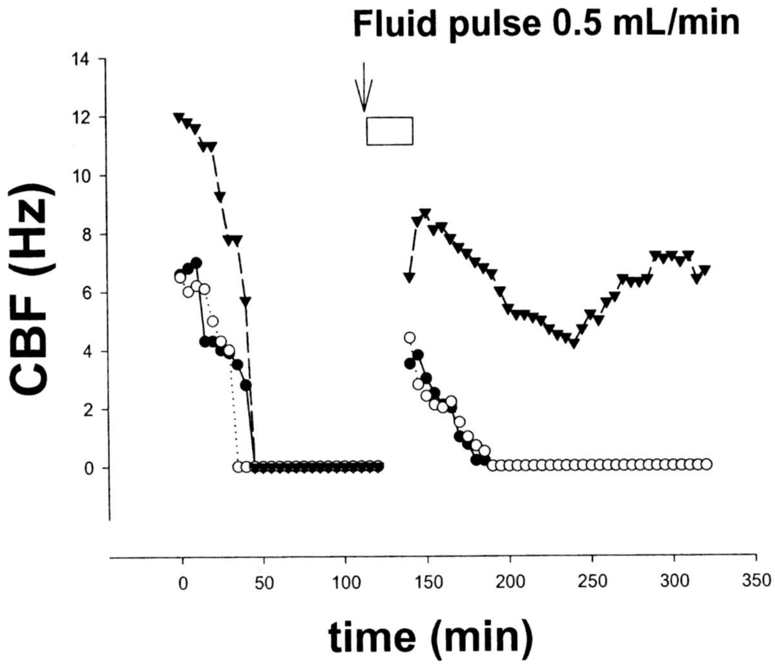

FIGURE 3 CBF falls to zero but is revived when cells are mechanically stimulated by flow.

do not know the answers to such questions at present but obtained a clue from another set of studies. We were investigating the effects of pM concentrations of the phorbol ester PMA. Figure 4 shows a composite pattern of results. Note that in the left panel, which shows raw frequencies against time, CBF does not decay to zero in any of the cells. Is this a chance (false-negative) observation, or does pM phorbol ester promote the stability of CBF? The right panel shows comparative box plots of the difference in CBF (ΔCBF) between the beginning and end of the study for each cell. Note that (1) a few cells increase their CBF, but only when PMA is present, (2) the median decline (solid line intra-box) is less than with medium alone, and (3) the mean CBF (dotted line) lies on different sides of the median for the two groups, suggesting a different distribution of CBF. Conventionally, pM PMA might be thought to be too low for a biological effect, and our preliminary data suggest that this notion may have to be revised for intact cells.

In the above examples, the study temperature was chosen to reflect the normal nasal value of 32°C. The temperature used for CBF studies varies between investigators—indeed, it has been our experience over the last decade that many referees have asked why we report our results at the "nonphysiological" 20°C

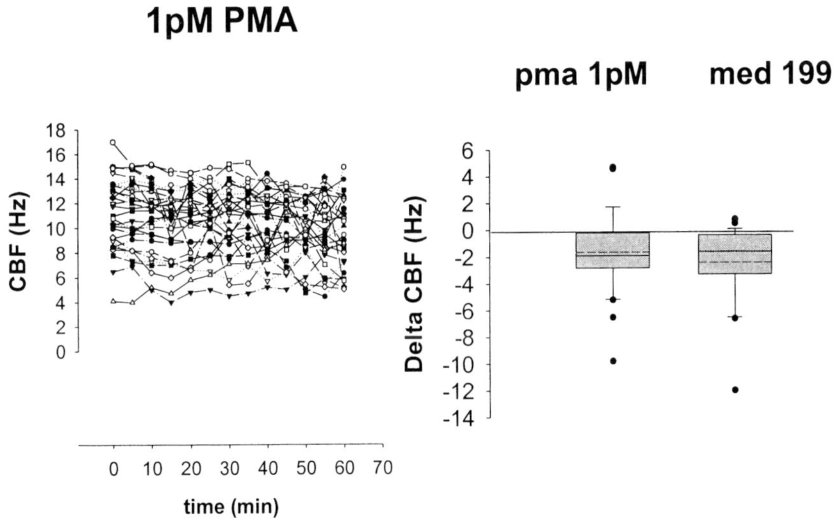

Figure 4 The pico-pharmacology of cilia: Individual cell CBF was recorded in the presence of 1 pM PMA at room temperature. Note that no cells decay to zero. Box plots (dotted line shows the mean; solid line shows median) show that the difference in CBF (ΔCBF) between 0 and 60 minutes differs dependent on the presence of PMA. PMA appears to promote ciliary stability.

(1–3). We have argued that the nasal temperature in the antarctic deep winter will be substantially different from that in the tropical rain forest. Thus, the nasal epithelium will need to adapt to climate-induced temperature change, thus making the study of CBF relevant whatever the temperature. We (and others) have noted that CBF is more stable at room temperature compared to 32°C (see Fig. 5), which probably explains why so many studies are undertaken at the 'non' physiological temperature. Our recent work (3) suggests that at 20°C the stability of CBF is partially dependent on the immediate history of the cell—specifically the flow-stress imparted on that cell by fluid impinging on the cell surface. When freshly biopsied human nasal epithelial cells were subjected to a flow rate of 0.5 mL/min, we found that there was a precipitate decline in CBF of ~15% within 30 seconds of the applied flow followed by an attempt at recovery over the next 30 minutes to preflow baseline. Figure 5 shows that this decline in ciliary response is also present at 32°C and may therefore play an important regulatory role. Our data at 20°C (3) suggest that PKC activation prior to the flow stress eliminates this flow-induced decline in CBF. We do not know what happens at other temperatures.

The observations reported here lead to many questions. First, does the control of cell calcium differ in cells that show a decay to arrest phenomenon in

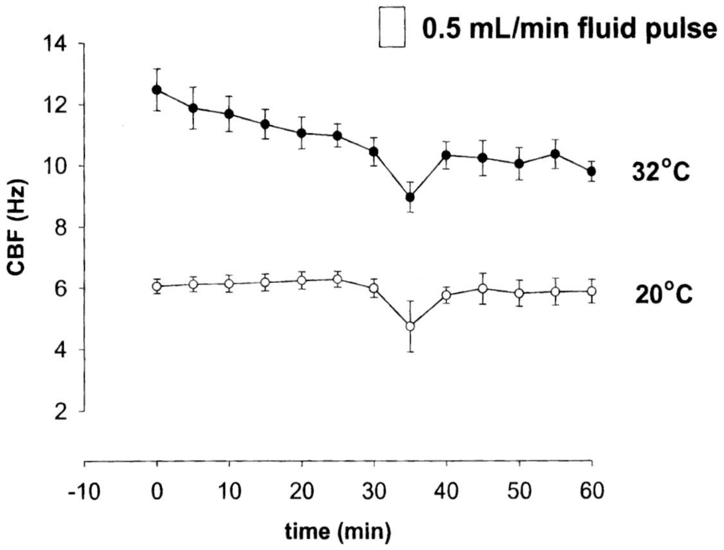

FIGURE 5 Note the ciliary beat frequency declines faster at 32°C compared to 20°C. The dip in CBF when cilia are subjected to a fluid pulse occurs at both temperatures but is more marked at the higher temperature.

respect of their CBF? Second, do phorbol esters control the temporal profile of CBF? Third, if they do, then do they act via cell calcium or not (4)? However, when we provide our peers with answers to such questions, we will have to present our data in a transparent manner. Only then will we take peer review to a new dimension. In 1690, John Locke at the age of 57, having published virtually nothing in his academic life, wrote in the third book of his classic essay on human understanding, that total ignorance was a ''want of ideas'' and partial ignorance was a want of ''connexions between ideas.'' Three hundred and ten years later we remain partially ignorant about the subtleties of CBF control (5,6). Perhaps there should be a MCC website (CBF.org) where researchers can place the underlying raw data from their publications for all to analyze remotely. Only then will Locke rest easy knowing that we are all getting connected.

ACKNOWLEDGMENTS

XM and RM are supported by the Wellcome Trust and the Anonymous Trust. XM holds an International Research Training Fellowship.

REFERENCES

1. RP Smith, R Shellard, DP Dhillon, J Winter, A Mehta. Asymmetric interactions between phosphorylation pathways regulating ciliary beat frequency in human nasal respiratory epithelium in vitro. J Physiol 496:883–889, 1996.
2. G Di Benedetto, C Magnus, P Gray, A Mehta. Calcium regulation of ciliary activity in human respiratory epithelium in vitro. J Physiol 439:103–113, 1991.
3. X Mwimbi, R Muimo, A Mehta. Protein kinase C regulates the flow rate dependent decline in human nasal ciliary beat frequency (CBF) in vitro. J Aerosol Med 13(3): 273–300, 2000.
4. B Wong, LC Park, DB Yeates. Neuropeptide Y (NPY) inhibits ciliary beat frequency (CBF) in cultured human ciliated cells via protein kinase C (PKC) dependent mechanism. Am J Respir Crit. Care Med 155:A432, 1997.
5. M Salathe, MM Pratt, A Wanner. Protein kinase C-dependent phosphorylation of a ciliary membrane protein and inhibition of ciliary beating. J Cell Sci 106:1211–1220, 1993.
6. P Satir, K Barkalow, T Hamasaki. The control of ciliary beat frequency. Trends Cell Biol 3:409–412, 1993.

9

Identifying the Genes for Primary Ciliary Dyskinesia and Kartagener Syndrome

Michał Witt and Ewa Rutkiewicz
Institute of Human Genetics
Poznań, Poland

Yue-Fen Wang, Cui-e Sun, Diego F. Wyszynski,* and Scott R. Diehl
National Institutes of Health
Bethesda, Maryland

Jacek Pawlik and Jerzy Żebrak
Institute of Tuberculosis and Lung Diseases
Rabka, Poland

Kartagener syndrome (KS) is characterized by a classic triad of symptoms: *situs inversus*, bronchiectasis, and chronic sinusitis. It is considered a subtype of primary ciliary dyskinesia (PCD), formerly known as immotile cilia syndrome (ICS) (1). It is *situs inversus*, a reversal of the usual left-right asymmetry of the abdominal and thoracic internal organ locations, that distinguishes KS from other forms of PCD. All forms of PCD are characterized by dysmotility or immotility of cilia in airway epithelial cells, spermatozoa, and other ciliated cells of the body (1–

**Current affiliation*: Boston University School of Medicine, Boston, Massachusetts.

4). Most of the cases of ciliary dysmotility or immotility reported to date are associated with visible ultrastructural defects of cilia, predominantly by a total or partial absence of dynein arms (70–80% of all electron microscopy–detectable defects), defects of radial spokes, nexin links, and general axonemal disorganization with microtubular transposition. Ciliary discoordination can also be caused by random ciliary orientation (5). Thus far in our studies, we have found cilia ultrastructural defects in all PCD-affected subjects tested. We found lack of both dynein arms in 73% of PCD-affected individuals. In a couple of these cases, some of the cilia tested also had defects of radial spokes or lacked only the inner or outer dynein arms. In 27% of PCD cases, all had missing outer dynein arms in most or all cilia tested, and 21% had missing inner dynein arms in most or all cilia tested. In a number of cases various ultrastructural defects coexisted in one individual. Other abnormalities, such as defects of radial spokes, were observed in 4% of our PCD patients.

One explanation for the relationship between axonemal ultrastructural defects and *situs inversus* may be that normal ciliary function plays a role in organ orientation, whereas organ orientation in PCD is a random event because of dysfunctional cilia in early embryonic development. Recent findings in mouse models support this mechanism (6,7). An alternative hypothesis for the association between PCD and *situs inversus* is that the same gene mutation in PCD-affected subjects causes both defective cilia and organ laterality through independent molecular pathways, rather than the cilia defect itself directly causing the loss of left-right asymmetry. This mechanism could involve either pleiotropic effects of a single KS gene or action of a relatively common variation at a *situs inversus* modifier gene or genes linked or unlinked to the KS gene, but expressed only when the KS gene mutation is present (i.e., gene interaction). A pair of monozygotic PCD twins were reported to be discordant in their organ laterality, with one diagnosed as KS, the other as PCD without *situs inversus* (8). This observation is most consistent with the view that the KS gene mutation itself causes a randomization of the direction of left-right asymmetry and argues against the influence of genetic variation at ''modifier'' genes affecting the outcome. Caution should be exercised, however, against generalizing too broadly to all cases of KS and PCD from this one observation.

Clinical consequences of PCD cover a wide spectrum, mainly affecting lower and upper airways and the male reproductive system. The most frequent symptoms of PCD detected in our study group of 83 PCD patients are shown in Table 1. The prognosis in PCD is good, but morbidity can be considerable if patients are not correctly managed (9). Lesions detectable by bronchography were found mainly in medial and lower left lobes. It should be recognized that ascertainment bias attributable to our primary focus of recruitment through pulmonary problems may have affected the frequency of these symptoms in our subjects.

TABLE 1 Clinical Manifestations of PCD in 83 Patients Studied in Poland

Symptoms	Prevalence (%)
Recurrent/chronic bronchitis	100
Recurrent/chronic sinusitis	95.1
Recurrent/chronic otitis media	83.3
Situs inversus viscerum totalis	67.3
Bronchiectasis	57.9
Nasal polyps	10.3

Our recruitment through pulmonary problems may have enriched our samples for cases having more severe respiratory problems and perhaps also enriched for cases with a major gene etiology leading to multiple cases occurring more often among close relatives. It should be recognized, therefore, that the population-wide frequency of PCD and KS symptoms may be somewhat different from that observed in our study subjects. For example, a study of 13 individuals with *situs inversus* but without serious respiratory complaints found dynein arm deficiencies in cilia in four subjects, with abnormal ciliary activity in three of these four (10). This suggests that ciliary dysfunction may be common in subjects with *situs inversus*, even in the absence of any obvious respiratory problems. Other recent reports confirm the importance of objective methods of diagnosis including careful evaluation of ciliary activity and ultrastructure (11–13).

Estimates of incidence of PCD range from 1/16,000 to 1/60,000 live births (2). Since about half of PCD cases display *situs inversus* (although this might be an underestimation of true embryonic occurrence, since L-R asymmetry defects may often be lethal because of cardiac abnormalities), incidence of KS is estimated to range between 1/32,000 and 1/120,000 live births. However, the possibly frequent occurrence of PCD without respiratory complaints in *situs inversus* cases and other circumstances noted above suggest an incidence of PCD as high as 1/12,500 (10,13). Inheritance in most cases appears to be via a single major locus autosomal recessive mode of transmission (1,2,4), although pedigrees showing autosomal dominant or X-linked modes of inheritance have also been reported (14). It has been suggested that primary ciliary dyskinesia represents a cluster of organelle disorders involving cilia, analogous to mitochondrial, lysosomal, or peroxysomal diseases (1). Nearly 200 different polypeptides have been identified within the ciliary axoneme of lower organisms; at least the same number of proteins can be expected in axonemes of humans (15). It is rather unlikely that mutations of as many as 200 different genes coding for various ciliary proteins cause the same or similar pathological consequences of ciliary dysfunction.

If this were true, one might expect incidence of PCD to be much higher than actually occurs (1). It is possible that many ciliary protein gene mutations may be lethal even if heterozygous, while mutations in other genes may not affect cilia function at all, perhaps due to redundancy of functions of gene families such as dyneins. In any case, the low incidence of KS and PCD suggests that perhaps mutations in a small number of genes may underlie these diseases. This potentially reduced level of locus heterogeneity provides cautious optimism for success of gene-mapping strategies.

A number of animal models of PCD have been identified, including examples in the mouse (16), rat (17), dog (18,19), and pig (20). Some of these helped focus attention on particular regions of the human genome that contain homologs of genes involved in animal models of PCD, thus creating a list of candidate genes for humans.

Several human chromosomal regions can be considered candidates for PCD gene locations. Chromosome 7q33-q34 is syntenic to a fragment of mouse chromosome 6 containing the *hop* mutation (previously named *hpy*). In a homozygous state, this mutation causes a dynein defect in cilia and flagella similar to that seen in PCD (21). The gene for a β heavy chain of the outer dynein arm maps to 7p15 region (22) and clones containing sequences homologous to dynein gene family map to 7q21-q22 and to 7p21. The case of a rare chromosomal anomaly called uniparental disomy of chromosome 7 strengthened the candidacy of this genomic region for the PCD locus (23). The patient expressed complete *situs inversus*, motionless respiratory cilia, and cystic fibrosis (homozygous for the delta F508 mutation of the CFTR gene located on human chromosome 7). His mother was not a carrier of the CF mutation and both chromosomes 7 were inherited from his father (a documented CF carrier), and false maternity was excluded.

Following up on this report, our group reported linkage data obtained from 30 PCD families recruited in Poland that strongly excluded a KS gene (at least for a high proportion of our families) for all or most of chromosome 7 with a multipoint LOD score < -2.0 (24). Our PCD families were subclassified either as Kartagener syndrome (KS) families (if at least one PCD-affected member was diagnosed as having KS, i.e., exhibiting *situs inversus*) or as ciliary dysfunction only (CDO) families (if none of the members affected with PCD exhibited *situs inversus*). For instance, a mutation affecting only 9 + 2 cilia, but not nodal 9 + 0 cilia, would result in ciliary dyskinesia but not in *situs inversus*. We analyzed KS and CDO families separately because of the possibility that different molecular pathologies could underlie these subtypes of PCD. Such a hypothesis is consistent with the fact that mice carrying the *hop* mutation that maps to a region syntenic to human 7q33-q34 exhibit CDO but not *situs inversus*. We performed linkage analyses with 17 microsatellite markers spanning chromosome 7. The average interval between markers was 10.8 cM and the analysis was performed using fluorescence-based, semiautomated DNA sizing technology. Our 23 KS families included 25 affected individuals, and our 7 CDO families included 9 affected individuals. Our results of pairwise and multipoint LOD scores provided a weak suggestion of linkage in only the CDO subset of our PCD families. The highest multipoint LOD score of 1.41 for these families occurred precisely at the position where the gene for the β-heavy chain of the outer dynein arm is located. Additional data will be needed to confirm this suggestion of linkage.

The *HLA* region of chromosome 6p contains the β-tubulin gene (*TUBB*) (25) and was suggested to be linked to PCD in one study (26). However, the motilin gene (*MLN*) that also resides in this region was not supported as a candidate for involvement in PCD etiology in another study (27), and recombination was also observed in another family (28). Another candidate gene located in the HLA region is the kinesin-like 2 gene (*KNSL2*) (29). Kinesins are involved in ciliary beating (7), and this gene is located at the centromeric end of the major histocompatibility complex. Four single base substitutions were detected in the *HSET* gene in two PCD families studied (29).

The echinoderm microtubule-associated protein-like gene (*EMAPL*) is located on chromosome 14q32. Mutations in this gene have been shown to cause Usher syndrome type 1A, and these patients sometimes exhibit ultrastructural defects in the axonemes of their respiratory cilia similar to PCD (30,31). Human chromosome 14qter is homologous to mouse chromosome 12 that contains the *iv* (*inversus viscerum*) mutation. In homozygous form, *iv* leads to a loss of the normal direction of left-right asymmetry of the heart and viscera, with 50% of homozygous mice exhibiting *situs inversus*, and 50% exhibiting the normal *situs solitus* pattern of left-right asymmetry (32). Human chromosome 14qter contains an axonemal dynein heavy-chain gene (33). It has recently been shown that mutations of this gene (*left/right dynein, lrd*) in mice is the cause of the *iv* and *legless* (*lgl*) mutants, which exhibit randomization of L-R asymmetry (34) through effects on cilia of the embryonic ventral node (6,7).

Another mouse mutant, *inv* (*inversion* of embryonic turning), is the only mutation that causes almost 100% of homozygous embryos to develop *situs inversus* (35). These embryos do not survive and they have cardiovascular, renal, and pancreatic defects. However, cilia show no structural or activity abnormalities. The gene causing the *inv* mutation has been cloned, and the protein, called inversin, has ankyrin-like repeats and potential nuclear localization signals (36). This gene reverses the left-right pattern of expression of the *modal*, *lefty-1*, and *lefty-2* genes, suggesting that inversin acts upstream of these TGF-β–like genes that are also involved in left-right asymmetry (7). Human inversin is located at chromosome 9q31.

It has been shown that kinesin superfamily proteins KIF3A and KIF3B determine L-R asymmetry by participating in the assembly of motile cilia that produce the leftward nodal flow of extraembryonic fluid in the mouse embryo (7,37). These genes are located on human chromosomes 5q31 (*KIF3A*) and 20 (*KIF3B*) (38).

Randomization of L-R asymmetry due to complete absence of cilia was observed in mice homozygous for mutation of the hepatocyte nuclear factor/forkhead homolog (HFH)-4 gene (39). The human homolog has been mapped to chromosome 17q23-q25 (38).

Several recent successes in human genetic studies of PCD have identified one gene and two chromosomal regions of the human genome that are involved in the pathogenesis of KS and PCD:

Chromosome 9p13-p21—An axonemal dynein intermediate chain gene (*DNAI1*), related to a *Chlamydomonas reinhardtii* gene *IC78*, has mutations associated with absence of outer dynein arms (40). Two loss-of-function mutations of this gene have been identified in a patient with PCD characterized by immotile cilia lacking outer dynein arms. However, linkage between this gene and PCD was excluded in five other families, demonstrating locus heterogeneity.

Chromosome 19q13.3—A genome scan in five PCD families of Arabic origin (four with reported consanguinity) revealed a maximum multipoint LOD score of 4.4 (41). Linkage was not found for this region in two of these families, again confirming locus heterogeneity.

Chromosome 15q24—We recruited 51 KS families (one or more PCD-affected members having *situs inversus*) with 168 individuals providing blood for genotype analyses, including 51 PCD-affected. We also recruited and separately analyzed 19 CDO families (none of the PCD-affected members having *situs inversus*) with 52 individuals genotyped, including 19 PCD affected. After obtaining suggestive evidence of linkage in a preliminary, partial genome scan using a subset of these subjects, we conducted a high-density scan of chromosome 15 using 66 microsatellite markers. In our KS families, we obtained a maximum pairwise LOD score of 4.23 for marker D15S154 (42). This was supported by further multipoint and nonparametric analyses and localizes a KS gene to a 10 cM region of chromosome 15q. Linkage analyses in our CDO families, by contrast, were negative for chromosome 15, thus demonstrating locus heterogeneity dependent on the presence or absence of *situs inversus* among PCD-affected cases. Another recently reported genome scan also implicated chromosome 15q, 11, and other chromosomal regions, but none of these had more suggestive levels of statistical support (43).

In summary, data reported thus far have identified one dynein gene and several chromosomal regions of the human genome for PCD. It is clear that PCD is genetically heterogeneous, with different genes involved in different families, depending, in part, on whether *situs inversus* is present (i.e., KS versus CDO subtypes of PCD). The successes obtained thus far already confirm that PCD is not commonly caused by hundreds of different genes, even though cilia themselves have so many structural and assembly proteins and gene regulatory signals. Future priorities are to identify the genes that reside in the mapped regions and to identify new genes by combinations of linkage and linkage disequilibrium gene mappings and by mutation screening of appropriate candidate genes. With the imminent completion of the human genome sequence (44), many genes affecting cilia and left-right asymmetry such as dyneins, nexins, kinesins, myosins, and others (45–49) are very rapidly accumulating in genomic databases. This tremendous wealth of new knowledge promises to greatly accelerate progress towards the goal of identifying the genes that cause KS and PCD.

ACKNOWLEDGMENT

This research was supported by project PAN-NIH-97-310 of the Maria Skłodowska-Curie Fund II.

REFERENCES

1. V McKusick, ed. Kartagener syndrome (244400) and Immotile Cilia Syndrome (242650). In: Online Mendelian Inheritance of Man, 2000. *http://www.ncbi.nlm.nih.gov/omim/*
2. BA Afzelius, B Mossberg. Immotile-cilia syndrome (primary ciliary dyskinesia), including Kartagener syndrome. In: C Scriver, AL Beaudet, W Sly, D Valle, eds. The Metabolic and Molecular Bases of Inherited Diseases, 7th ed. New York: McGraw-Hill, 1995, pp 3943–3954.
3. DV Schidlow. Primary ciliary dyskinesia (the immotile cilia syndrome). Ann Allergy 73:457–470, 1994.
4. BA Afzelius. Genetics and pulmonary medicine. 6. Immotile cilia syndrome: past, present, and prospects for the future. Thorax 53:894–897, 1998.
5. J Rutland, RU de Iongh. Random ciliary orientation: a cause of respiratory tract disease. N Engl J Med 323:1681–1684, 1990.
6. How to get a heart in the right place. Science 285:23, 1999.
7. DM Supp, S Potter, M Brueckner. Molecular motors: the driving force behind mammalian left-right development. Trends Cell Biol 10:41–45, 2000.
8. PG Noone, D Bali, JL Carson, A Sannuti, CL Gipson, LE Ostrowski, PA Bromberg, RC Boucher, MR Knowles. Discordant organ laterality in monozygotic twins with primary ciliary dyskinesia. Am J Med Genet 15:155–160, 1999.
9. A Bush, P Cole, M Hariri, I Mackay, G Phillips, C O'Callaghan, R Wilson, JO Warner. Primary ciliary dyskinesia: diagnosis and standards of care. Eur Respir J 12:982–988, 1998.

10. AA Kroon, JM Heij, WA Kuijper, AJ Veerman, S van der Baan. Function and morphology of respiratory cilia in situs inversus. Clin Otolaryngol 16:294–297, 1991.
11. F Santamaria, MM de Santi, G Grillo, P Sarnelli, M Caterino, L Greco. Ciliary motility at light microscopy: a screening technique for cilary defects. Acta Paediatr 88:853–857, 1999.
12. D Holzmann, PM Ott, H Felix. Diagnostic approach to primary ciliary dyskinesia: a review. Eur J Pediatr 159:95–98, 2000.
13. TN Teknos, R Metson, T Chasse, G Balercia, GR Dickersin. New developments in the diagnosis of Kartagener's syndrome. Otolaryngol Head Neck Surg 116:68–74, 1997.
14. D Narayan, SN Krishnan, M Upender, TS Ravikumar, MJ Mahoney, TF Dolan Jr, AS Teebi, GG Haddad. Unusual inheritance of primary ciliary dyskinesia (Kartagener syndrome). J Med Genet 31:493–496, 1994.
15. DJL Luck, B Huang, G Piperno. Genetic and biochemical analysis of the eukaryotic flagellum. Soc Exp Biol Symp 35:399–407, 1982.
16. JHD Bryan, RL Hughes, TJ Bates. Brain development in hydrocephalic-polydactyl, a recessive pleiotropic mutant in the mouse. Virchows Arch Pathol Anat Histol 374: 205–214, 1977.
17. C Torikata, C Kijimoto, M Koto. Ultrastructure of respiratory cilia of WIV-Hyd male rats. An animal model for human immotile cilia syndrome. Am J Pathol 138: 341–347, 1991.
18. DF Edwards, CS Patton, DA Bemis. Immotile cilia syndrome in three dogs from a litter. J Am Vet Med Assoc 183:667–672, 1983.
19. PJ Watson, ME Herrtege, D Sargan. Primary ciliary dyskinesia in Newfoundland dogs. Vet Rec 143:484, 1998.
20. F Roperto, O Galati, P Rossacco. Immotile cilia syndrome in pigs. A model for human disease. Am J Pathol 143:634–647, 1993.
21. MA Handel. Allelism of *hop* and *hpy*. Mouse News Lett 72:124, 1985.
22. K Kastury, WE Taylor, M Gutierrez, L Ramirez, PJ Coucke, P Van Hauwe, G Van Camp, S Bhasin. Chromosomal mapping of two members of the human dynein gene family to chromosome regions 7p15 and 11q13 near the deafness loci DFNA 5 and DFNA 11. Genomics 44:362–364, 1997.
23. Y Pan, CD McCaskill, KH Thompson, J Hicks, B Casey, LG Shaffer, WJ Craigen. Paternal isodisomy of chromosome 7 associated with complete situs inversus and immotile cilia. Am J Hum Genet 62:1551–1555, 1998.
24. M Witt, Y-F Wang, S Wang, C Sun, J Pawlik, E Rutkiewicz, J Zebrak, SR Diehl. Exclusion of chromosome 7 for Kartagener syndrome but suggestion of linkage in families with other forms of primary ciliary dyskinesia. Am J Hum Genet 64:313–318, 1999.
25. A Volz, E Weiss, J Trowsdale, A Ziegler. Presence of an expressed β-tubulin gene (TUBB) in the HLA class I region may provide the genetic basis for HLA-linked microtubule dysfunction. Hum Genet 93:42–46, 1994.
26. E Bianchi, S Savasta, A Calligaro, G Beluffi, P Poggi, M Tinelli, E Mevio, M Martinetti. HPA haplotype segregation and ultrastructural study in familial immotile-cilia syndrome. Hum Genet 89:270–274, 1992.
27. P Gasparini, A Grifa, S Savasta, I Merlo, L Bisceglia, A Totaro, L Zelante. The

motilin gene: subregional localisation, tissue expression, DNA polymorphisms and exclusion as a candidate gene for the HLA-associated immotile cilia syndrome. Hum Genet 94:671–674, 1994.
28. P Gasparini, A Griffa, N Oggiano, E Fabrizzi, PL Giorgi. Immotile cilia syndrome: a recombinant family at HLA-linked gene locus. Am J Med Genet 49:450–451, 1994.
29. K Janitz, A Wild, S Beck, S Savasta, G Beluffi, A Ziegler, A Volz. Genomic organization of the HSET locus and the possible association of HLA-linked genes with immotile cilia syndrome (ICS). Immunogenetics 49:644–652, 1999.
30. D Bonneau, F Raymond, C Kremer, J-M Klossek, J Kaplan, F Patte. Usher syndrome type I associated with bronchiectasis and immotile nasal cilia in two brothers. J Med Genet 30:253–254, 1993.
31. JD Eudy, M Edmonds, SF Yao, CB Talmadge, PM Kelley, MD Weston, WJ Kimberling, J Sumegi. Isolation of a novel human homologue of the gene coding for echinoderm microtubule-associated protein (EMAP) from the Usher syndrome type 1a locus at 14q32. Genomics 42:104–106, 1997.
32. M Brueckner, P D'Eustachio, AL Horwich. Linkage mapping of a mouse gene, "*iv*", that controls left-right asymmetry of the heart and viscera. Proc Natl Acad Sci USA 86:5035–5038, 1989.
33. D Narayan, T Desai, A Banks, SR Patanjali, TS Ravikumar, DC Ward. Localization of the human cytoplasmic dynein heavy chain (DNECL) to 14qter by fluorescence in situ hybridization. Genomics 22:660–661, 1994.
34. DM Supp, DP Witte, SS Potter, M Brueckner. Mutation of an axonemal dynein affects left-right asymmetry in inversus viscerum mice. Nature 389:963–966, 1997.
35. T Yokoyama, N Copeland, N Jenkins, C Montgomery, F Elder, P Overbeek. Reversal of left-right asymmetry: a situs inversus mutation. Science 260:679–682, 1993.
36. D Morgan, L Turnpenny, J Goodship, W Dai, K Majumder, L Matthews, A Gardner, G Schuster, L Vien, W Harrison, FFB Elder, M Penman-Splitt, P Overbeek, T Strachan. Inversin, a novel gene in the vertebrate left-right axis pathway, is partially deleted in the inv mouse. Nature Genet 20:149–156, 1998.
37. S Nonaka, Y Tanaka, Y Okada, S Takeda, A Harada, Y Kanai, M Kido, N Hirokawa. Randomizaton of left-right asymmetry due to loss of nodal cilia generating leftward flow of extraembryonic fluid in mice lacking KIF3B motor protein. Cell 95:829–837, 1998.
38. NCBI LocusLink: http://www.ncbi.nlm.nih.gov/LocusLink/
39. J Chen, HJ Knowles, JL Hebert, BP Hackett. Mutation of the mouse hepatocyte nuclear factor/forkhead homologue 4 gene results in an absence of cilia and random left-right asymmetry. J Clin Invest 102:1077–1082, 1998.
40. G Pennarun, E Escudier, C Chapelin, AM Bridoux, V Cacheux, G Roger, A Clement, M Goossens, S Amselem, B Duriez. Loss-of-function mutations in a human gene related to Chlamydomonas reinhardti dynein IC78 result in primary ciliary dyskinesia. Am J Hum Genet 65:1508–1519, 1999.
41. M Meeks, A Walne, S Spiden, H Simpson, H Mussafi-Georgy, HD Hamam, EL Fehaid, M Cheehab, M Al-Dabbagh, S Polak-Charcon, H Blau, A O'Rawe, HM Mitchison, RM Gardiner, E Chung. A locus for primary ciliary dyskinesia maps to chromosome 19q. J Med Genet 37:241–244, 2000.

42. MP Witt, DF Wyszynski, Y-F Wang, A Miller-Chisholm, J Pawlik, MH Khoshnevisan, C Sun, S Wang, Y-J Zhang, E Rutkiewicz, J Zebrak, A Kapelerova, SR Diehl. Linkage of a gene for Kartagener syndrome to chromosome 15q. Am J Hum Genet 65:A31, 1999.
43. JL Blouin, M Meeks, U Radhakrishna, A Sainsbury, C Gehring, GD Sail, L Bartoloini, V Dombi, A O'Rawe, A Walne, E Chung, BA Afzelius, M Armengot, M Jorissen, DW Schidlow, L van Maldergem, H Walt, RM Gardiner, D Probst, PA Guerne, CD Delozier-Blanchet, SE Antonarakis. Primary ciliary dyskinesia: a genome-wide linkage analysis reveals extensive locus heterogeneity. Eur J Hum Genet 8:109–118, 2000.
44. ES Lander, RA Weinberg. Genomics: journey to the center of biology. Science 287: 1777–1782, 2000.
45. V Mermall, PL Post, MS Moosaker. Unconventional myosins in cell movement, membrane traffic, and signal transduction. Science 279:527–533, 1998.
46. N Hirokawa. Kinesin and dynein superfamily proteins and the mechanism of organelle transport. Science 279:519–526, 1998.
47. S Izraeli, LA Lowe, VL Bertness, DJ Good, DW Dorward, IR Kirsch, MR Kuehn. The SIL gene is required for mouse embryonic axial development and left-right specification. Nature 399:691–694, 1999.
48. K Tamura, S Yonei-Tamura, JC Belmonte. Molecular basis of left-right asymmetry. Dev Growth Differ 41:645–694, 1999.
49. CT Rankin, T Bunton, AM Lawler, SJ Lee. Regulation of left-right patterning in mice by growth/differentiation factor-1. Nat Genet 24:262–265, 2000.

10

Homozygosity Mapping as an Approach for Identifying Genes Involved in Primary Ciliary Dyskinesia

S. L. Spiden and H. M. Mitchison
Royal Free and University College Medical School
London, United Kingdom

INTRODUCTION

Primary ciliary dyskinesia (PCD, OMIM 242650) is an autosomal recessive disorder affecting cilia motility with an estimated incidence of 1 in 20,000 live births (1). Impaired or absent motility of respiratory tract cilia leads to recurrent sinopulmonary infections and permanent lung damage (bronchiectasis) (2). Abnormalities of the related flagella structure of sperm lead to male subfertility (3). Other ciliated tissues such as those of the female reproductive tract and the brain ependyma can also be affected and are associated with reduced fertility, ectopic pregnancy, and hydrocephalus in some patients (4,5).

Patients also exhibit abnormal left-right body axis (LRA) determination, and laterality defects are observed in 50% of PCD patients—commonly complete mirror-image reversal of internal body organs (*situs inversus*). Patients with ciliary defects and situs inversus are classified as having Kartagener syndrome (KS, OMIM 244400) (1). Studies in mice have suggested a link between movement of monocilia at the embryonic node during development and left-right pattern formation, although such an association has yet to be proven in PCD patients (6).

PCD is a genetically heterogeneous disorder, and this is reflected by the variation in ciliary ultrastructural defects and clinical presentations that have been documented in patients (7,8). The most common defect is a reduced number or complete absence of the inner and/or outer axonemal dynein arms. Defects of other ciliary structural components, such as the radial spokes and central pair microtubules, as well as elongation and incorrect orientation of the cilia have also been documented (9,10).

CANDIDATE GENES FOR PCD

Cilia have an evolutionarily conserved structure and are composed of over 250 different polypeptides (11,12). The genes encoding structural components of the ciliary axoneme are obvious candidates to underlie different ultrastructural defects seen in PCD patients. These include dynein protein components of the dynein arms (13), the radial spoke head proteins (11), and other microtubule-associated proteins, such as kinesins (14). Much of the knowledge of the structural components of the human ciliary axoneme, and the genes encoding their proteins, comes from studies of motility mutants of *Chlamydomonas reinhardtii* (15) and the subsequent identification of homologous genes in humans. To date, 11 dynein genes and 12 kinesin and kinesin-related genes have been assigned chromosomal localizations in humans (http:/www.gene.ucl.ac.uk/nomenclature).

Another source of human candidate disease genes comes from the study of animal models of PCD and of genes involved in the LRA determination pathway (16). Several rodent mutants have been investigated that display features of the PCD disease phenotype. The *hop*^{hpy} mouse mutant has defects of the ciliary axoneme in respiratory and reproductive tissue and the sperm flagellae (17,18). The *hfh4* mouse mutant has a complete absence of respiratory cilia in conjunction with randomized LRA determination (19,20). The *kif3B* (21) and *iv* mouse (22) have randomized LRA determination and immotile nodal monocilia but display normal functional respiratory cilia. The *WIC-Hyd* rat has immotile respiratory and ependymal cilia in conjunction with situs inversus and hydrocephalus (23).

APPROACHES USED TO IDENTIFY GENES INVOLVED IN PCD

Candidate Gene Approach

A candidate gene approach can be used to try and identify the genes mutated in PCD. A multitude of putative candidate genes for PCD are distributed across the human genome, and this technique has been used successfully to identify the first known PCD gene. Sequencing of the human *DNAI1* gene located on human 9p13–p21, a homolog of the *Chlamydamonas IC78* gene, which is essential for axonemal outer dynein arm assembly (24), identified two separate loss-of-function mutations in a PCD patient with an outer dynein arm defect (25). Another candidate gene, analyzed in a collaborative effort including our laboratory, is *hfh4*, a winged-helix/forkhead transcription factor with expression restricted to ciliated cells (26). PCD patients were screened that, like the *hfh4* mouse mutant, had no respiratory cilia, or who were consistent with linkage to the region of 17q that *HFH4* mapped, but no sequence abnormalities were observed.

Positional Cloning Approach Using Genetic Linkage Analysis

Linkage analysis allows mapping of disease genes without prior knowledge of the functional components involved in the disorder. It is based on detection of the cosegregation of chromosome DNA markers with a disease trait in families. Linkage is significant if a lod score—defined as the logarithm of the odds ratio that the disease and marker loci are linked rather than unlinked—of >3 is achieved. A number of linkage studies in PCD have been reported. In a study by Witt et al. (27), linkage analysis was performed on chromosome 7, which was of interest because (1) the hop^{hpy} mouse gene maps to a region syntenic to human 7q33–q34; (2) several dynein related genes have been positioned on the p and q arms of chromosome 7; and (3) uniparental disomy of chromosome 7 has been observed in an individual with KS and cystic fibrosis (28). A weak suggestion of linkage was obtained at the β-dynein heavy-chain locus on 7p15 in PCD families with no situs inversus. Witt et al. also reported linkage of 51 KS families to chromosome 15p with a lod score of 4.23 (29). Linkage analysis at the HLA complex on chromosome 6 has been performed due to the location there of a β-tubulin gene, an axoneme structural component (30). Affected individuals from two unrelated PCD families showed an association with the HLA-DR7; DQW2 haplotype. A genome-wide linkage screen was performed in 31 nuclear PCD families by Blouin et al. (31), but this identified no single major locus, confirming locus heterogeneity. Under the assumption that 40% of the individuals were linked to a single locus, slightly positive lod scores were obtained on chromosomes 3p, 4p, 5p, 8q, 15q, 16p, 17q, and 19q.

Homozygosity Mapping as a Genetic Analysis Technique

Linkage analysis in a genetically heterogeneous disorder such as PCD, where more than one locus may be mutated in a particular ultrastructural phenotype and/or group of families from similar geographical region, can prove problematic due to the large number of families required for analysis. Identification of PCD loci by linkage analysis has been greatly enhanced by homozygosity mapping, an approach that allows recessively inherited traits to be mapped in families of consanguineous unions (usually first or second cousin marriages) (32). It is based on the premise that a child of a consanguineous marriage affected with an autosomal recessive disorder will have inherited the same ancestral disease chromosome from each parent and will therefore contain a region of the chromosome spanning the disease locus that is homozygous by descent (HBD). An affected child from a consanguineous marriage would be expected to have several regions that are HBD since 1/16 genome is HBD in a first cousin marriage (33), but these regions will vary from child to child. By searching for a region HBD that is shared among all affected individuals in a number of families, a common region of the chromosome containing a locus for the shared disease may be positioned (see Fig. 1). One affected offspring from a first cousin marriage has been calculated to yield a lod score of 1.2, and this means that just three affected individuals are needed to obtain a lod score of >3 (33). Homozygosity mapping can also be applied to nuclear pedigrees from small, inbred populations, due to the high level of inbreeding and the likelihood that a single common mutation arose in the founder population (33).

MAPPING OF A LOCUS FOR PCD TO 19q13.4 BY HOMOZYGOSITY MAPPING

Homozygosity mapping identified a locus for PCD in five Arabic families, four of which were consanguineous due to first cousin marriages. The resource consisted of 24 individuals, with 12 affected individuals, 5 of whom had *situs inversus*. Family information and linkage data are reported in Meeks et al. (34). A genome-wide screen identified a region of excess homozygosity among affected individuals of the four consanguineous families on chromosome 19q13.4 (34). Haplotype analysis is summarized in Figure 2. A region of shared homozygosity between affected individuals of Families 66/67 and 89 limits the region to between *D19S572* and *D19S218* (Fig. 2).

GENEHUNTER linkage analysis (35) was performed on the four consanguineous plus one nuclear family (072) (34), assuming recessive inheritance at α values (proportion of linked families) of 0.35, 0.65, and 1 (Fig. 3). The highest multipoint lod score allowing for heterogeneity (HLOD), of 5 at $\alpha = 0.65$, was obtained between *D19S572* and *D19S890*. These data provide statistically significant evidence for a locus for PCD on 19q13.4 and confirm locus heterogeneity.

CONCLUSIONS

The identification of a locus for PCD on chromosome 19q13.4 illustrates the power of homozygosity mapping as an approach for linkage analysis of rare, recessive diseases in consanguineous families or very inbred populations, even

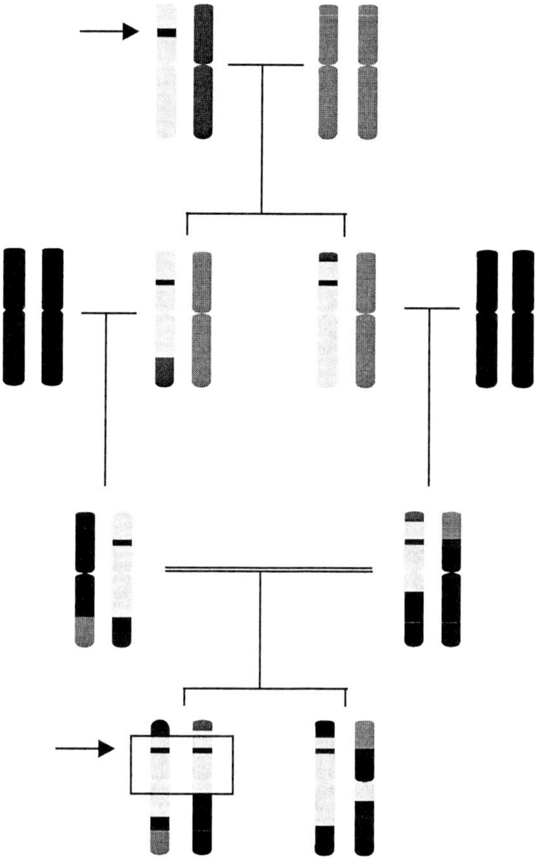

Figure 1 The principle of homozygosity mapping. The ancestral disease causing mutation, indicated by the arrow, is passed on to each of the offspring in the second and third generations. Each of the disease alleles is passed on to the affected offspring of the first cousins in the third generation, who are both carriers for the ancestral mutation. The box indicates the region of shared homozygosity surrounding both copies of the ancestrally mutated disease chromosome, inherited in the affected child in the fourth generation. The unaffected sibling has inherited a single copy of the disease chromosome and therefore does not have a homozygous region in common with its affected sib.

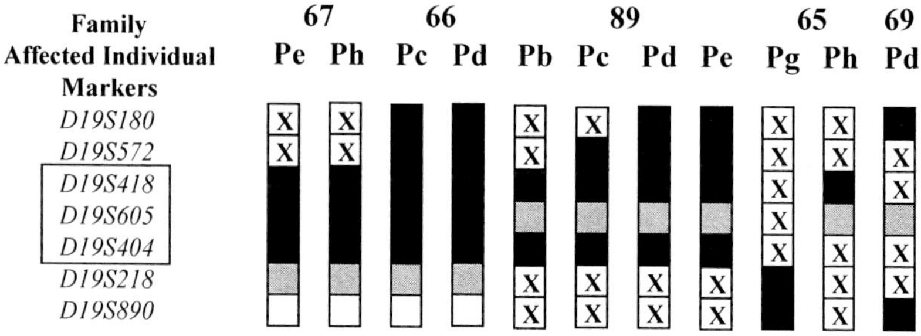

FIGURE 2 Haplotype analyses showing homozygosity present in the affected individuals of four Arabian consanguineous families on chromosome 19q13.4. Genotypes for 11 affected individuals from 4 families (66/67, 89, 65, and 69) are indicated. Microsatellite markers that are homozygous are indicated by filled boxes, black being those that are HBD (parents informative) and grey being those that are homozygous by state (HBS; parents uninformative). Markers that are heterozygous are indicated by boxes containing black crosses, and for markers that were not typed the boxes are unfilled. The region of homozygosity that is shared among all the affected individuals of families 66/67 and 89 that link to the region on 19 is indicated by the black box. Family 65 is not linked to this region because recombination events are present between the two affected individuals. Family 69 is unlikely to be linked as it is heterozygous and shares a common haplotype with an unaffected sib.

in the presence of genetic heterogeneity. Genetic loci can be identified in small family groups selected according to phenotypic trait and/or ethnic origin.

Homozygosity mapping uses relatively simple and well-established methodology and is unbiased if the genome is analyzed as a whole when searching for regions that are HBD. In contrast, candidate gene analysis is limited by current understanding of the disease pathology and the number of candidate genes that have been characterized. Conclusive mutation analysis of potential candidates can also be problematic if the full genomic structure of the gene is unavailable.

Work to isolate the 19q13.4 PCD disease gene is currently in progress. Refinement of the genetic map and subsequent transcript identification at this and other PCD loci will be greatly expediated as a result of the accumulation of genetic information from the human genome and the genomes of numerous other species (36). DNA sequences available in public databases as a result of the Human Genome Project initiative will allow more effective genetic linkage and positional candidate gene analysis. Identification of the genes involved in PCD should enhance understanding of the molecular mechanisms involved in cilia function and LRA determination as well as allowing investigation into therapeutic methods.

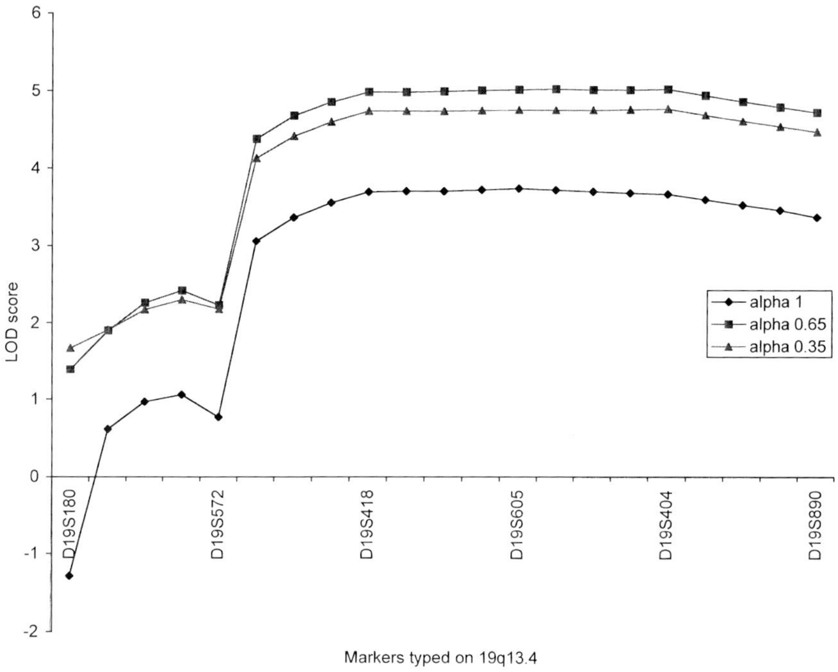

FIGURE 3 GENEHUNTER analysis showing the lod score in relation to the proportion of linked families.

ACKNOWLEDGMENTS

This work was supported by the MRC (UK) and the Welcome Trust (UK). We would like to thank Dr. E. Chung for his intellectual input in the preparation of this manuscript.

REFERENCES

1. HD Rott. Kartagener's syndrome and the syndrome of immotile cilia. Hum Genet 46:249–261, 1979.
2. BA Afzelius. A human syndrome caused by immotile cilia. Science 193:317–319, 1976.
3. G Lungarella, L Fonzi, AG Burrini. Ultrastructural abnormalities in respiratory cilia and sperm tails in a patient with Kartagener's syndrome. Ultrastruct Pathol 3:319–323, 1982.
4. M Meeks. Primary ciliary dyskinesia. Curr Paeds 8:231–236, 1998.

5. MA Greenstone, RW Jones, A Dewar, BG Neville, PJ Cole. Hydrocephalus and primary ciliary dyskinesia. Arch Dis Child 59:481–482, 1984.
6. MK Wagner, HJ Yost. The roles of nodal cilia. Curr Biol 10:R149–R151, 2000.
7. BA Afzelius. Genetical and ultrastructural aspects of the immotile-cilia syndrome. Am J Hum Genet 33:852–864, 1981.
8. A Bush, P Cole, M Hariri, I Mackay, G Phillips, C O'Callaghan, R Wilson, JO Warner. Primary ciliary dyskinesia: diagnosis and standards of care. Eur Respir J 12:982–988, 1998.
9. MA Sleigh. Kartagener's syndrome, ciliary defects and ciliary function. Eur J Respir Dis Suppl 127:157–161, 1983.
10. BA Afzelius, G Gargani, C Romano. Abnormal length of cilia as a possible cause of defective mucociliary clearance. Eur J Respir Dis 66:173–180, 1985.
11. SK Dutcher. Flagellar assembly in two hundred and fifty easy-to-follow steps. Trends Genet 11:398–404, 1995.
12. I Manton. The fine structure of plant cilia. Symposia Soc Exp Biol 306–319, 1952.
13. R Kamiya. Exploring the function of inner and outer dynein arms with *Chlamydomonas* mutants. Cell Motil Cytoskeleton 32:98–102, 1995.
14. N Hirokawa, Y Noda, Y Okada. Kinesin and dynein superfamily proteins in organelle transport and cell division. Curr Opin Cell Biol 10:60–73, 1998.
15. PA Lefebvre, CD Silflow. *Chlamydomonas*: the cell and its genomes. Genetics 151: 9–14, 1999.
16. DM Supp, M Brueckner, SS Potter. Handed asymmetry in the mouse: understanding how things go right (or left) by studying how they go wrong. Semin Cell Dev Biol 9:77–87, 1998.
17. DR Johnson, DM Hunt. Hop-sterile, a mutant gene affecting sperm tail development in the mouse. J Embryol Exp Morphol 25:223–236, 1971.
18. RT Bronson, PW Lane. Hydrocephalus with hop gait (hyh): a new mutation on chromosome 7 in the mouse. Brain Res Dev Brain Res 54:131–136, 1990.
19. J Chen, H Knowles, J Hebert, B Hackett. Mutation of the mouse hepatocyte nuclear factor/forkhead homologue 4 gene results in an absence of cilia and random left-right asymmetry. J Clin Invest 102:1077–1082, 1998.
20. GJ Pelletier, SL Brody, H Liapis, RA White, BP Hackett. A human forkhead/winged-helix transcription factor expressed in developing pulmonary and renal epithelium. Am J Physiol 274:L351–359, 1998.
21. S Nonaka, Y Tanaka, Y Okada, S Takeda, A Harada, Y Kanai, M Kido, N Hirokawa. Randomization of left-right asymmetry due to loss of nodal cilia generating leftward flow of extraembryonic fluid in mice lacking KIF3B motor protein. Cell 95:829–837, 1998.
22. DM Supp, DP Witte, SS Potter, M Brueckner. Mutation of an axonemal dynein affects left-right asymmetry in inversus viscerum mice. Nature 389:963–966, 1997.
23. C Torikata, C Kijimoto, M Koto. Ultrastructure of respiratory cilia of WIC-Hyd male rats. An animal model for human immotile cilia syndrome. Am J Pathol 138: 341–347, 1991.
24. CG Wilkerson, SM King, A Koutoulis, GJ Pazour, GB Witman. The 78,000 M(r) intermediate chain of *Chlamydomonas* outer arm dynein is a WD-repeat protein required for arm assembly. J Cell Biol 129:169–178, 1995.

25. G Pennrun, E Escudier, C Chapelin, AM Bridoux, V Cacheux, G Roger, A Clement, M Goossens, S Amselem, B Duriez. Loss-of-function mutations in a human gene related to *Chlamydomonas reinhardtii* dynein IC78 results in primary ciliary dyskinesia. Am J Hum Genet 65:1508–1519, 1999.
26. AK Maiti, L Bartoloni, HM Mitchison, et al. No deleterious mutations in the FOXJ1 (alias HFH-4) gene in patients with primary ciliary dyskinesia. Cytogenet Cell Genet 90:119–122, 2000.
27. M Witt, Y Wang, S Wang, C Sun, J Pawlik, E Rutkiewicz, J Zebrak, S Diehl. Exclusion of Chromosome 7 for Kartegeners syndrome but suggestion of linkage in families with other forms of primary ciliary dyskinesia. Am J Hum Genet 64:313–318, 1999.
28. Y Pan, C McCaskell, K Thompson, J Hicks, B Casey, L Shaffer, W Craigen. Paternal isodisomy of chromosome 7 associated with complete situs inversus and immotile cilia. Am J Hum Genet 62:1551–1555, 1998.
29. M Witt, D Wyszynski, Y Wang, A Miller-Chisholm, J Pawlik, M Khoshnevisan, C Sun, S Wang, Y-J Zhang, E Rutkiewica, J Zebrak, A Kapelerova, S Diehl. Linkage of a gene for Kartegener syndrome to chromosome 15q. Am J Hum Genet 65:A31, 1999.
30. E Bianchi, S Savasta, A Calligaro, G Beluffi, P Poggi, M Tinelli, E Mevio, M Martinetti. HLA haplotype segregation and ultrastructural study in familial immotile-cilia syndrome. Hum Genet 89:270–274, 1992.
31. J-L Blouin, U Radhakrishna, C Gehrig, GD Sail, A Sainsbury, L Bartoloni, M Meeks, V Dombi, D Probst, B Afzelius, M Armengot, E Chung, M Jorissen, D Schidlow, R Gardiner, H Walt, LV Maldergam, P Guerne, CD Blanchet, S Antonarakis. Primary Ciliary Dyskinesia: A genome-wide linkage analysis reveals extensive locus heterogeniety. Am J Hum Genet 65:A244, 1999.
32. ES Lander, D Botstein. Homozygosity mapping: a way to map human recessive traits with the DNA of inbred children. Science 236:1567–1570, 1987.
33. M Farrall. Homozygosity mapping: familiarity breeds debility. Nat Genet 5:107–108, 1993.
34. M Meeks, A Walne, S Spiden, H Simpson, H Mussaffi-Georgy, H Hamam, E Fehaid, M Cheebab, M Al-Dabbagh, H Blau, A O'Rawe, H Mitchison, R Gardiner, E Chung. A locus for primary ciliary dyskinesia maps to chromosome 19q. J Med Genet 37: 241–244, 2000.
35. A Kong, NJ Cox. Allele-sharing models: LOD scores and accurate linkage tests. Am J Hum Genet 61:1179–1188, 1997.
36. SJ O'Brien, M Menotti-Raymond, WJ Murphy, WG Nash, J Wienberg, R Stanyon, NG Copeland, NA Jenkins, JE Womack, JA Marshall Graves. The promise of comparative genomics in mammals. Science 286:458–462, 479–481, 1999.

11

Mutations in the Novel Mammalian Gene *DNAI1* Result in Primary Ciliary Dyskinesia*

Gaëlle Pennarun, Catherine Chapelin, Anne-Marie Bridoux, Valère Cacheux, Michel Goossens, Serge Amselem, and Bénédicte Duriez
INSERM U468, Hôpital Henri-Mondor, Créteil, France

Estelle Escudier
Groupe Hospitalier Pitié-Salpêtrière, Paris, France

Gilles Roger and Annick Clément
Hôpital Armand-Trousseau, Paris, France

INTRODUCTION

Functional and ultrastructural abnormalities of respiratory cilia have been described in patients with a congenital respiratory disease known as primary ciliary dyskinesia (PCD) (1,2). PCD, previously described as immotile cilia syndrome, represents a heterogeneous group of genetic disorders affecting 1 in 16,000 indi-

* This work was previously published by the University of Chicago Press in *American Journal of Human Genetics* 65(6):1508–1519, 1999.

viduals; the PCD phenotype is characterized by impaired mucociliary clearance resulting from a lack of ciliary movements believed to be responsible for chronic lung, sinus, and middle ear diseases. Approximately 50% of the patients with PCD display a *situs inversus*, thereby defining the Kartagener's syndrome. Most male patients are infertile because of nonmotile spermatozoa in relation to functional and ultrastructural abnormalities of sperm flagella (3). Studies to date, however, have failed to decipher the molecular basis of this disease. PCD is usually transmitted as an autosomal recessive trait. The various functional and ultrastructural abnormalities of respiratory cilia documented in PCD patients suggest that a genetic heterogeneity underlies this condition (4).

The main ciliary defect found in PCD is an absence of dynein arms affecting almost all cilia (3). Dyneins consist of a large family of proteins involved in many types of microtubule-dependent cell motility in both lower and higher eukaryotes. The axonemal dyneins are found in the dynein arms of the ciliary and flagellar axonemes; they are essential for ciliary and flagellar beating.

Among the several immotile strains of *Chlamydomonas reinhardtii*, the mutants carrying a defect in IC78 gene, which encodes a dynein intermediate chain, have been the subject of detailed functional, ultrastructural, and molecular studies (5). Of particular interest, the flagellar ultrastructural phenotype of these mutants is similar to the axonemal ultrastructural abnormality observed in several PCD patients (i.e., absence of outer dynein arms); we have therefore isolated a human gene related to IC78, the *DNAI1* gene (6), to test its involvement in PCD.

CLINICAL AND ULTRASTRUCTURAL PHENOTYPES OF THE PCD PATIENTS

A 9-year-old Caucasian boy (patient II-1) born to unrelated parents (family 1) presented in early childhood with chronic respiratory symptoms characterized by chronic sinusitis, serous otitis, and recurrent episodes of bronchitis associated with severe segmental atelectasis, having led to partial lobectomy. Chest radiograph showed normal cardiac and visceral *situs*. Neither his parents nor other relatives have a history of respiratory disease. At the time of a bronchoscopy, samples of trachea mucosa were obtained and processed for ciliary studies, as described (7). No ciliary beating was observed, and transmission electron microscopy showed the absence of outer dynein arms in all cilia, asserting the diagnosis of PCD (Fig. 1A). Five other unrelated consanguineous PCD families were also investigated (families 2 to 6). In two independent families (families 2 and 3), the ultrastructural phenotype was identical to that documented in patient II-1 from family 1; however, in one case (family 3), the PCD patients also displayed a *situs inversus*. In the three remaining families (families 4, 5, and 6), both outer and inner dynein arms were absent in the patients' cilia; in one case (family 6), the disease phenotype was associated with a *situs inversus*. In all these individu-

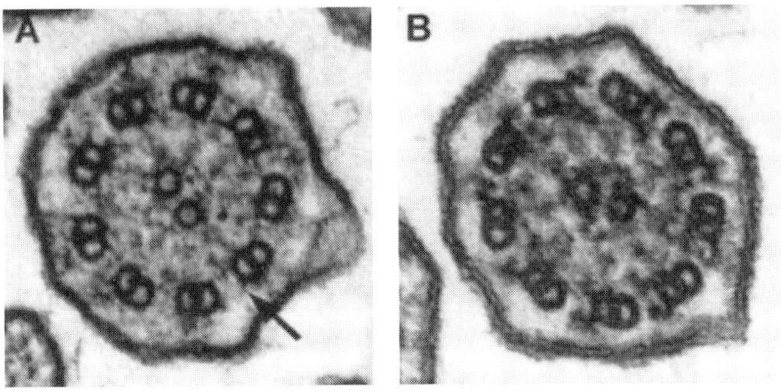

Figure 1 Electron micrograph of cross sections of respiratory cilia (magnification ×90,000). (A) The absence of outer dynein arms (arrow) is observed on all the peripheral doublets of the ciliary sections obtained from patient II-1 with PCD. (B) Normal ciliary ultrastructure from a control.

als, DNA samples were isolated from peripheral blood samples, according to standard techniques.

DNAI1 MUTATIONS LEAD TO A PCD PHENOTYPE

All the coding regions and intron-exon boundaries of the DNA sample from the patient II-1 from family 1 were amplified with *DNAI1*-specific primers and run on single-strand conformation polymorphism (SSCP) gels or sequenced directly. Two SSCP variants were identified in his genomic DNA; these variants were located in two PCR fragments spanning exon 5 (Fig. 2A) and 1 (Fig. 3A), respectively. To characterize these molecular variations, we cloned the corresponding PCR products. A 4-bp insertion, located at codon 95, was present in 10 of the 20 clones spanning exon 5 (Fig. 2B). A similar experimental approach led to the identification of a 1-bp insertion in the splice-donor site following exon 1 (Fig. 3B). The patient therefore carries two different *DNAI1* mutations, thereby demonstrating compound heterozygosity in keeping with the absence of consanguinity documented in this family.

The maternal 4-bp insertion results in a frameshift leading to premature stop codon 24 amino acids downstream. To determine the consequences of the paternally inherited splice-donor-site mutation on the processing of *DNAI1* transcripts, total RNA obtained from nasal epithelial cells of patient II-1 was reverse-transcribed. The resulting products were used as templates in a PCR assay performed with *DNAI1*-specific primers (Fig. 3C). Two molecular species were generated: one of expected size (290 bp), and a 422-bp fragment, which was observed only in the patient's sample and not in the control. Sequencing of these PCR

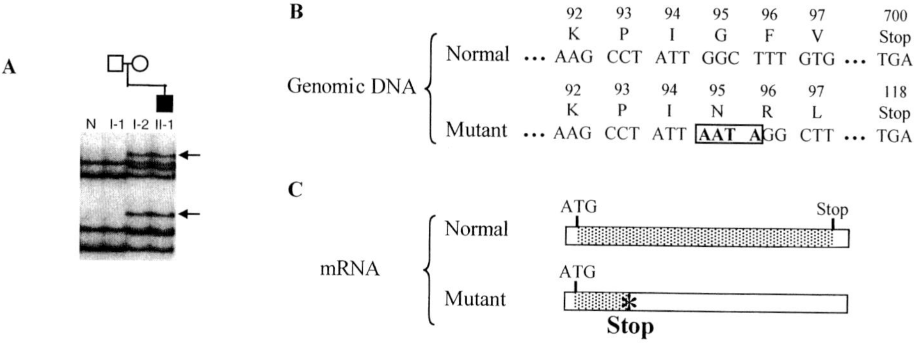

FIGURE 2 The exonic insertion identified on the maternal *DNAI1* allele of patient II-1 from family 1. (A) SSCP analysis of the PCR products of exon 5 from three members (I-1, I-2, and II-1) of this family and a control (N), showing bandshifts (arrows) in patient II-1 (filled-in black square) and his mother (I-2). (B) Nucleotide sequence of the normal (top) and mutant (bottom) *DNAI1* alleles. The insertion of 4 nucleotides in exon 5 is boxed. (C) Schematic representation of the presumptive normal (top) and mutant (bottom) mRNAs (open boxes) and translated regions (shaded gray).

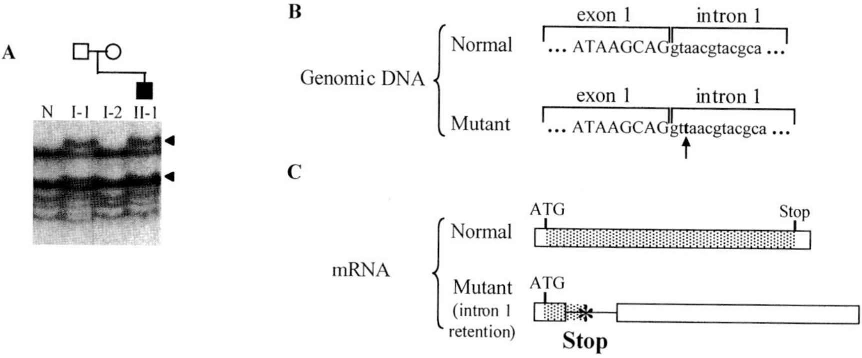

FIGURE 3 The splice mutation identified on the paternal *DNAI1* allele of patient II-1 from family 1. (A) SSCP analysis of the PCR products of exon 1 from three members of this family and a control (N), showing bandshifts (arrows) in patient II-1 (filled-in black square) and his father (I-1). (B) Nucleotide sequence of the normal (top) and mutant (bottom) *DNAI1* alleles. The 1-bp insertion (t) at nucleotide +3 of the intronic sequence following exon 1 is in bold character (arrow). The position of the exon-intron boundary is indicated above the sequence. (C) Schematic representation of the presumptive normal (top) and mutant (bottom) mRNAs (open boxes) and translated regions (shaded gray).

products (290 bp and 422 bp) revealed that the larger fragment contained a 132-bp insertion corresponding to the 5′ intronic sequence following exon 1, indicating the use of a cryptic splice-donor site in this intron at position 133 (Fig. 3D). Indeed, sequence analysis of the intronic region spanning this site revealed the existence of a perfect splice-donor-site consensus sequence at position 133 of intron 1. If translated, this abnormal *DNAI1* transcript would result in a premature stop codon at position 73. To further test whether these two mutations are responsible for the PCD phenotype, 50 unrelated control individuals were screened for these *DNAI1* variations. None of their 100 chromosomes contained such mutations.

LOCUS HETEROGENEITY IN PCD

In the course of this study, we identified two intragenic nucleotide polymorphisms: one is located at nucleotide 42 (G > C) of intron 11; the other one is a G-to-A transition at nucleotide 1003 resulting in the V3351 substitution. These two intragenic polymorphisms were used to test the involvement of *DNAI1* in the PCD phenotype identified in the five remaining families (families 2–6) in which the patients were born to consanguineous unions. The genotype at these two loci was characterized by genomic DNA sequencing. In family 2, the two affected children, who displayed an absence of outer dynein arms in all cilia, and the healthy older sister share the same *DNAI1* genotype, thereby demonstrating an exclusion of linkage between the *DNAI1* gene and the PCD phenotype. In family 5, the two affected children who displayed a PCD phenotype characterized by an absence of both outer and inner dynein arms, carried different *DNAI1* genotypes at nucleotide 42 of intron 11. In the three remaining consanguineous families (families 3, 4, and 6), the patients were found to be heterozygous at one or at the two intragenic loci (data not shown).

CONCLUSION

We postulated that the human *DNAI1* gene, which is homologous to the *Chlamydomonas* IC78 gene, was an excellent candidate sequence to investigate in patients with a PCD phenotype characterized by an absence of outer dynein arms. Indeed, *Chlamydomonas* IC78 protein belongs to outer dynein arms (8), and a lack of outer dynein arms has clearly been documented in the axonemes of *Chlamydomonas* strains in which the IC78 gene is either deleted or disrupted by a large insertion (5). Two different trans-allelic germline *DNAI1* mutations leading to frameshifts were indeed identified in one patient presenting with a PCD phenotype associated with an absence of outer dynein arms. The 4-bp insertion identified on the maternal *DNAI1* allele, which probably arose by slippage replication, is predicted to produce a frameshift introducing a premature stop codon located

24 amino acids downstream. The paternal mutation is a splice defect; if translated, the abnormal *DNAI1* transcript, which retains the first 132 nucleotides of intron 1, would result in a premature TGA stop codon at position 73. Both the maternal mutation and the paternal mutation should, therefore, generate severally truncated polypeptides lacking 85% and 95%, respectively, of the DNAI1 protein. Taken together, these data highly suggest that such truncated proteins, even if synthesized, could not play their key role in the outer dynein arm assembly. We therefore conclude that these mutations, which were absent from 100 control chromosomes, underlie the ultrastructural phenotype observed in this PCD patient. The documented immotile cilia are also consistent with the absence of outer dynein arms; this may be responsible for mucociliary impairment, which led to the severe chronic respiratory symptoms observed in this patient.

We have also identified two nucleotide polymorphisms in *DNAI1* that allowed us to demonstrate an exclusion of linkage between the *DNAI1* gene and the PCD phenotype in five other families. These linkage data, which provide the first clear-cut demonstration of a locus heterogeneity in this condition, also reveal a higher complexity level of heterogeneity than first expected, since patients with identical ultrastructural defects may have mutations in other still unidentified genes; it is tempting to speculate that these latter genes may encode functional partners of DNAI1 involved in dynein arm assembly. Given the documented clinical, ultrastructural and genetic heterogeneity in PCD, it is, therefore, not surprising that genetic analyses performed in different sets of families did not allow to establish a reproducible genetic linkage between the disease phenotype and a given chromosomal region (9–11). However, in theory this kind of approach could be fruitful if applied to the study of separate families, each being considered individually (12). Nevertheless, such genetic linkage studies have been hampered by the small size of affected families in which male patients are usually infertile. This is the reason why, as shown in the present study, approaches based on the investigation of candidate genes represent powerful alternatives. These candidate genes include the dynein gene family. This study illustrates the use of one particular *Chlamydomonas* flagellar mutant as an excellent model for axonemal abnormalities observed in PCD; as several other light (13), intermediate (14,15), and heavy chains (16–20) of dynein have been implicated in other *Chlamydomonas* flagellar mutants, we consider the corresponding genes as good candidates for other subsets of PCD and related developmental diseases.

ACKNOWLEDGMENTS

The authors wish to thank M. C. Millepied and M. Couprie (Ecole Supérieure d'Ingénieurs en Electrotechnique et Electronique (ESIEE)) for their help in determining the ciliary ultrastructural phenotype. This work was supported by grants from the Chancellerie des Universités (legs Poix), the Assistance Publique/

Hôpitaux de Paris (CRC96125), and the Université Paris XII (BQR). GP is the recipient of a fellowship from the Ministère de l'Education Nationale, de la Recherche et de la Technologie.

REFERENCES

1. BA Afzelius. A human syndrome caused by immotile cilia. Science 193:317–319, 1976.
2. CM Rossman, JB Forrest, RM Lee, MT Newhouse. The dyskinetic cilia syndrome. Ciliary motility in immotile cilia syndrome. Chest 78:580–582, 1980.
3. BA Afzelius. The immotile-cilia syndrome: a microtubule-associated defect. CRC Crit Rev Biochem 19:63–87, 1985.
4. J Chao, JA Turner, JM Sturgess. Genetic heterogeneity of dynein-deficiency in cilia from patients with respiratory disease. Am Rev Respir Dis 126:302–305, 1982.
5. CG Wilkerson, SM King, A Koutoulis, GJ Pazour, GB Witman. The 78,000 M(r) intermediate chain of *Chlamydomonas* outer arm dynein is a WD-repeat protein required for arm assembly. J Cell Biol 129:169–178, 1995.
6. G Pennarun, E Escudier, C Chapelin, AM Bridoux, V Cacheux, G Roger, A Clement, M Goossens, S Amselem, B Duriez. Loss-of-function mutations in a human gene related to *Chlamydomonas reinhardtii* dynein IC78 result in primary ciliary dyskinesia. Am J Hum Genet 65:1508–1519, 1999.
7. E Escudier, D Escalier, MC Pinchon, M Boucherat, JF Bernaudin, J Fleury-Feith. Dissimilar expression of axonemal anomalies in respiratory cilia and sperm flagella in infertile men. Am Rev Respir Dis 142:674–679, 1990.
8. SM King, CG Wilkerson, GB Witman. The Mr 78,000 intermediate chain of Chlamydomonas outer arm dynein interacts with alpha-tubulin in situ. J Biol Chem 266: 8401–8407, 1991.
9. C Chapelin, S Amselem, M Jean-Pierre, B Duriez, A Coste, E Lesprit, J Janaud, et al. Genetic analysis of primary ciliary dyskinesia in six multiplex families. Am J Respir Crit Care Dis 155:A504, 1997.
10. M Witt, Y Wang, S Wang, C Sun, J Pawlik, E Rutkiewicz, J Zebrak, SR Diehl. Exclusion of chromosome 7 for Kartagener syndrome but suggestion of linkage in families with other forms of primary ciliary dyskinesia. Am J Hum Genet 64:313–318, 1999.
11. JL Blouin, M Meeks, U Radhakrishna, A Sainsbury, C Gehring, GD Sail, L Bartoloni, V Dombi, A O'Rawe, A Walne, E Chung, BA Afzelius, M Armengot, M Jorissen, DV Schidlow, L van Maldergem, H Walt, RM Gardiner, D Probst, PA Guerne, CD Delozier-Blanchet, SE Antonarakis. Primary ciliary dyskinesia: a genome-wide linkage analysis reveals extensive locus heterogeneity. Eur J Hum Genet 8:109–118, 2000.
12. M Meeks, A Walne, S Spiden, H Simpson, H Mussaffi-Georgy, HD Hamam, EL Fehaid, M Cheehab, M Al-Dabbagh, S Polak-Charcon, H Blau, A O'Rawe, HM Mitchison, RM Gardiner, E Chung. A locus for primary ciliary dyskinesia maps to chromosome 19q. J Med Genet 37:241–244, 2000.
13. M LeDizet, G Piperno. ida4-1, ida4-2, and ida4-3 are intron splicing mutations af-

fecting the locus encoding p28, a light chain of *Chlamydomonas* axonemal inner dynein arms. Mol Biol Cell 6:713–723, 1995.
14. DR Mitchell, Y Kang. Identification of oda6 as a *Chlamydomonas* dynein mutant by rescue with the wild-type gene. J Cell Biol 113:835–842, 1991.
15. CA Perrone, P Yang, E O'Toole, WS Sale, ME Porter. The *Chlamydomonas* IDA7 locus encodes a 140-kDa dynein intermediate chain required to assemble the I1 inner arm complex. Mol Biol Cell 9:3351–3365, 1998.
16. H Sakakibara, DR Mitchell, R Kamiya. A *Chlamydomonas* outer arm dynein mutant missing the alpha heavy chain. J Cell Biol 113:615–622, 1991.
17. H Sakakibara, S Takada, SM King, GB Witman, R Kamiya. A *Chlamydomonas* outer arm dynein mutant with a truncated beta heavy chain. J Cell Biol 122:653–661, 1993.
18. ME Porter, JA Knott, LC Gardner, DR Mitchell, SK Dutcher. Mutations in the SUP-PF-1 locus of *Chlamydomonas reinhardtii* identify a regulatory domain in the beta-dynein heavy chain. J Cell Biol 126:1495–1507, 1994.
19. G Rupp, E O'Toole, LC Gardner, BF Mitchell, ME Porter. The sup-pf-2 mutations of *Chlamydomonas* alter the activity of the outer dynein arms by modification of the gamma-dynein heavy chain. J Cell Biol 135:1853–1865, 1996.
20. SH Myster, JA Knott, E O'Toole, ME Porter. The *Chlamydomonas* Dhc1 gene encodes a dynein heavy chain subunit required for assembly of the I1 inner arm complex. Mol Biol Cell 8:607–620, 1997.

12

Molecular Strategies for the Study of Human Cilia

Lawrence E. Ostrowski
University of North Carolina at Chapel Hill
Chapel Hill, North Carolina

INTRODUCTION

For many years, molecular and biochemical studies of mammalian (especially human) respiratory cilia and ciliated cells have been hampered by the relative lack of available material. In contrast, studies of simpler organisms, such as *Paramecium* and *Chlamydomonas*, which are easily grown in large quantities in a laboratory setting, have provided a substantial amount of knowledge about the structural organization and function of proteins in the cilia and flagella of these organisms. While many of the basic structural features of cilia have been highly conserved between species, a complete understanding of human respiratory cilia will require detailed studies of human-derived material. Recently, advances in the techniques used to culture human airway epithelial cells have provided researchers with the ability to cultivate increased numbers of ciliated cells. The availability of substantial amounts of high-quality starting material, combined with scientific advances in other areas, including the sequencing of the human genome and the ability to rapidly identify proteins by mass spectrometry, has made possible studies of human cilia that previously were impractical. This chapter will briefly describe our model for culturing human ciliated cells and isolating ciliary axonemes in sufficient quantities for biochemical analysis. An example of a current research project utilizing purified human axonemes will also be presented.

THE MODEL

For the majority of our studies, we use human bronchial epithelial (HBE) cells, although cultures derived from nasal epithelial cells are also suitable. Briefly, large airways are excised from excess surgical tissue obtained by Institutional Review Board–approved protocols and digested with protease as described by Bernacki et al. (1). The isolated HBE cells are washed and plated on collagen-coated tissue culture dishes (100 mm diameter) at a density of $1–2 \times 10^6$ cells per dish in modified LHC9 medium. Using these conditions, the cells expand as an undifferentiated monolayer and are approximately 75% confluent after 7 days. The cells are collected by trypsinization and plated on collagen coated permeable membranes (Transwell-COL inserts; Costar) at a density of $1–2.5 \times 10^5$ cells/cm^2 in ALI medium, a mixture of LHC9 and DMEM that results in improved mucociliary differentiation (1,2). When the cultures reach confluence, usually after about 3 days, the apical medium is removed and the cells are fed basally for the remainder of the culture period. Exposing the apical surface of cultures of airway epithelium to air has been previously shown to significantly enhance ciliated cell differentiation (3,4). During the next 3–4 weeks, the cultures progress

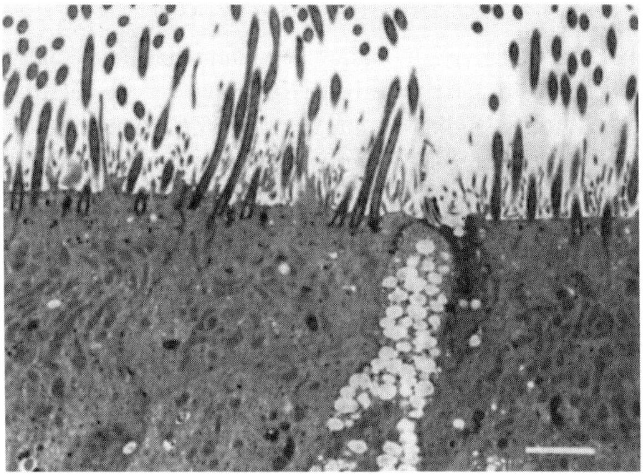

FIGURE 1 Transmission electron micrograph showing several mature ciliated cells and a secretory cell from a well-differentiated culture of human airway epithelial cells. Bar = 2.5 μM.

from a single layer of undifferentiated, basal type cells, to a multilayered columnar epithelium containing well-differentiated secretory cells and abundant ciliated cells (Fig. 1). An interesting feature of these well-differentiated cultures is the appearance of small areas where the cilia have become orientated in such a manner that mucociliary transport occurs in a circular pattern. Although clearly different from the unidirectional transport observed in vivo, the ability of these cultures to reproduce the essential features of mucociliary transport in vitro provides further evidence that these cultures provide a good model for studies of cilia and ciliated cells. In addition, cultures that spontaneously develop these areas of mucus transport provide a model for additional studies (5).

ISOLATION OF CILIARY AXONEMES

Many of the projects underway in our laboratory require the isolation of ciliary axonemes. We have found that the basic procedure developed by Hastie et al. (6) effectively removes ciliary axonemes from our HBE cultures with a minimal amount of contamination. After washing the cultures multiple times with PBS to remove mucus and any accumulated cellular debris, a small volume of buffer containing 0.1% Triton X-100 and 10 mM $CaCl_2$ is added to the surface of the cultures. The culture is gently rocked for approximately 1 minute, the supernatant is collected, and the procedure is repeated. The two washings are pooled and centrifuged at low speed (1000 × *g*) to pellet cellular debris, and the ciliary axonemes are then collected from the supernatant by centrifugation at 16,000 *g*. For most studies, the axonemes are washed by resuspension in 0.1% Triton X-100 containing buffer and centrifuged as above. As shown in Figure 2A, this

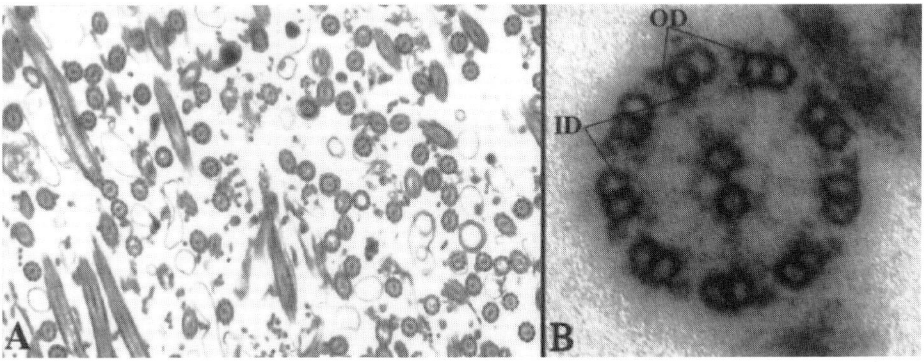

FIGURE 2 Preparation of human ciliary axonemes isolated from cultured airway epithelial cells examined by low (A) and high (B) magnification electron microscopy. In panel B, inner (ID) and outer (OD) dynein arms are clearly visible.

procedure yields a highly enriched preparation of ciliary axonemes, which when examined at higher magnification (Fig. 2B), demonstrate the characteristic structural features of axonemes, including the 9 + 2 microtubules, inner and outer dynein arms, and radial spokes. The ability to use heavily ciliated cultures of HBE cells eliminates many sources of contamination and variability that are inherent when using in vivo–derived material and certainly contributes to the purity of this axonemal preparation.

ANALYSIS OF AXONEMAL PROTEINS

One of the research interests in our laboratory is the identification of mutations that cause primary ciliary dyskinesia (PCD). Primary ciliary dyskinesia is an inherited disease in which defects in the structure of the ciliary axoneme result in impaired or absent mucociliary clearance. Individuals with PCD suffer from recurring infections in the sinuses, middle ear, and respiratory tract. The most commonly reported ciliary defect in PCD, as determined by electron microscopy, is the absence of outer dynein arms, although many other abnormalities have been reported and the disease is believed to be heterogeneous. Because the defects in ciliary axonemes frequently involve the absence of protein structures, we reasoned that a careful comparison of proteins present in purified axonemes from normal and PCD individuals might allow us to identify proteins missing in PCD patients. Proteins that are missing in the axonemes from PCD patients would be excellent candidates for further analysis to determine if the gene coding for the protein contained a causative mutation. This approach is similar to that used to identify the mutated protein in radial spoke mutants of *Chlamydomonas* (7,8).

To maximize the sensitivity of our analysis, HBE cells obtained from normal individuals or a patient with PCD were cultured in the presence of ^{35}S-labeled methionine. Ciliary axonemes were isolated as described above and equal amounts of radioactive proteins from each sample were separated by two-dimensional gel electrophoresis. Preliminary experiments, which compared several normal samples, demonstrated that the pattern of axonemal proteins is highly reproducible. However, our initial comparisons between axonemes isolated from normal and PCD cells (Fig. 3) revealed no obvious differences. Currently we are exploring the use of narrow range pH strips for the first dimension to increase the resolution of the axonemal proteins and improve the detection of missing or altered proteins. We have also begun to analyze the high molecular weight dynein heavy chains, which do not focus on 2-D gels, by SDS-PAGE. Proteins that are missing or altered in the axonemes of PCD patients will be identified by using mass spectrometry and searching of databases. In addition, a number of proteins from different regions of the 2-D gel pattern of normal axonemes will be identified by the same technique. These proteins will provide reference points so that additional samples can be easily compared.

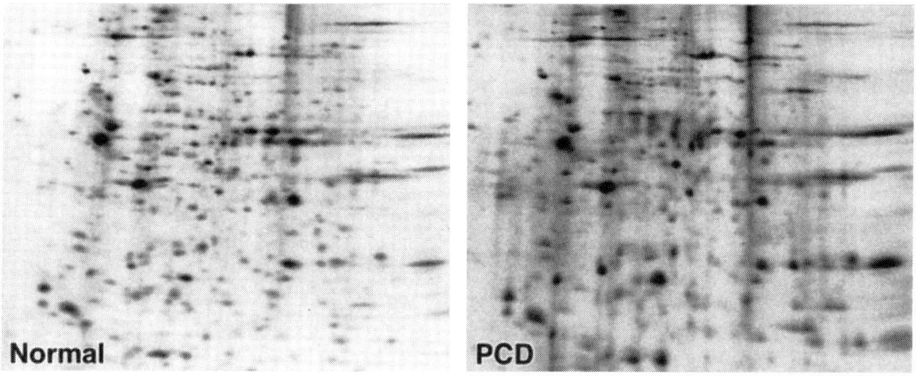

FIGURE 3 Two-dimensional gel electrophoresis of S-35–labeled axonemal proteins isolated from well-differentiated cultures of normal and PCD airway epithelial cells. Equal amounts of radioactive proteins were loaded on the two gels.

SUMMARY AND CONCLUSIONS

The ability to culture HBE cells under conditions that reproducibly allow for ciliated cell differentiation to occur has made it possible to begin to analyze human ciliated cell differentiation and function in vitro. In addition, these cultures provide sufficient quantities of ciliary axonemes for detailed biochemical analysis. For example, we are comparing the 2-D protein pattern of labeled axonemes obtained from normal and PCD cultures to identify proteins that may be mutated in PCD. The availability of purified preparations of human axonemes will allow researchers to begin to identify and characterize the molecular components of human respiratory cilia.

ACKNOWLEDGMENTS

The author would like to thank Drs. Boucher and Randell for many helpful discussions of this research and K. Burns for providing excellent electron microscopy services. This work was supported in part by NHLBI grant HL63103.

REFERENCES

1. SH Bernacki, AL Nelson, L Abdullah, et al. Mucin gene expression during differentiation of human airway epithelia in vitro. Muc4 and muc5b are strongly induced. Am J Respir Cell Mol Biol 20(4):595–604, 1999.

2. TE Gray, K Guzman, CW Davis, LH Adbullah, P Nettesheim. Mucociliary differentiation of serially passaged normal human tracheobronchial epithelial cells. Am J Respir Cell Mol Biol 14:104–112, 1996.
3. AB Clark, SH Randell, P Nettesheim, TE Gray, B Bagnell, LE Ostrowski. Regulation of ciliated cell differentiation in cultures of rate tracheal epithelial cells. Am J Respir Cell Mol Biol 12:329–338, 1995.
4. LE Ostrowski, P Nettesheim. Inhibition of ciliated cell differentiation by fluid submersion. Exp Lung Res 21:957–970, 1995.
5. H Matsui, BR Grubb, R Tarran, et al. Evidence for periciliary liquid layer depletion, not abnormal ion composition, in the pathogenesis of cystic fibrosis airways disease. Cell 95(7):1005–1015, 1998.
6. AT Hastie, DT Dicker, ST Hingley, F Kueppers, ML Higgins, G Weinbaum. Isolation of cilia from porcine tracheal epithelium and extraction of dynein arms. Cell Motil Cytosketeton 6:25–34, 1986.
7. G Piperno, B Huang, DJL Luck. Two-dimensional analysis of flagellar proteins from wild-type and paralyzed mutants of *Chlamydomonas reinhardtii*. Proc Natl Acad Sci USA 74:1600–1604, 1977.
8. D Luck, G Piperno, Z Ramanis, B Huang. Flagellar mutants of *Chlamydomonas*: studies of radial spoke-defective strains by dikaryon and revertant analysis. Proc Natl Acad Sci USA 74:3456–3460, 1977.

13

Ciliated Ependymal Cells: The Effect of *Streptococcus pneumoniae* on the Beat Frequency Response

Robert A. Hirst
Adelaide University, Adelaide, Australia

T. J. Mitchell
University of Glasgow, Glasgow, United Kingdom

P. W. Andrew
University of Leicester, Leicester, United Kingdom

C. O'Callaghan
University of Leicester and Leicester Royal Infirmary, Leicester, United Kingdom

INTRODUCTION

The ependymal layer is in direct contact with the cerebrospinal fluid within the ventricles and aqueducts of the brain. Cilia covering the ependymal cells maintain a constant flow of cerebrospinal fluid (CSF), helping to keep the ventricular and aqueductal surface clear of debris (1). Approximately 40 cilia, 8 μm in length, project from each ependymal cell and beat, continuously, at up to 40 beats per second (Hz) (2). We have established a model that allows measurement of ependymal ciliary beat frequency by high-speed video photography during incubation with bacteria or their toxins. We were particularly keen to use our model to study pathogenic modes of action of the pneumococcus. For pneumococcal meningitis, mortality rates of up to 20% remain despite effective antibiotic killing of bacteria (3), and significant problems may occur in survivors, most commonly due to sensorineural deafness (4). The ependymal cells have recently been identified as adult neural stem cells (5). Whether their loss during infection results in a reduced regenerative capacity of certain brain tissues remains speculative.

It has been postulated that the rapid release of the pneumococcal toxin, pneumolysin, on antibiotic-induced bacterial lysis may cause local tissue damage and may play a part in the development of sensorineural deafness (6). Indeed, we have shown (2) that pneumolysin causes rapid ependymal ciliary stasis at levels that may be encountered clinically during meningitis. The contribution of pneumolysin to the damage seen in pneumococcal meningitis is still under debate (7). In this chapter we will present evidence to show that pneumolysin in vitro is an important part of the mechanism of damage of ependymal ciliary function by pneumococci.

AIMS OF EXPERIMENTS

The aims of the experiments are as follows:

1. To develop a ciliated tissue culture model of the ependyma so that long-term ciliary beat frequency can be measured.
2. To test the effects of pneumolysin on ependymal cells from rat brain slices and cultured ependymal cells.
3. Examine the effects of pneumolysin-deficient and -sufficient pneumococci on the ependyma.
4. Examine the role of pneumococcal hydrogen peroxide as an ependymal toxin.

METHODS

Brain Slices

Brains were isolated from 14 to 17-day-old rats. The cerebellum was removed and mounted on a vibratome under M199 medium (ICN Laboratories, UK). The brain was sliced (250 μm) through the medulla oblongata and pons into the floor of the fourth ventricle. The slices were mounted in M199 medium prior to use.

Cell Culture

Ependymal cells were grown using a method adapted from Wiebel and colleagues (8). Newborn rat brains were removed using careful dissection. The brain regions (containing ependymal cells and ventricles) were mechanically dissociated in 2 mL of tissue culture medium. Ependymal cells were grown on a fibronectin extracellular matrix in serum-free Minimum Essential Medium (Gibco Life Technologies, Paisley, Scotland) containing penicillin, streptomycin, fungizone, bovine serum albumin, insulin, transferrin, selenium, and from day 3 onwards thrombin. The medium was replaced 3 days after seeding. The ciliated ependymal cells were used between 14–17 days in culture when cell proliferation was optimal (8).

Cilia Beat Frequency

The method was identical as previously described (2,9,10). The brain slices were placed in an incubation (37°C) chamber surrounding a light the microscope (Diphot, Nikon, UK). Beating cilia were recorded using a digital high-speed video camera (Kodak Ektapro Motion Analyser, Model 1012). Video sequences were played back at reduced frame rates in order to calculate the CBF.

Pneumococci

The strains used were a type 2 wild-type (D39) and a pneumolysin-negative version (PLN-A) (11). Bacteria were grown in brain heart infusion broth to late log phase or on 10% blood agar. In some experiments the pneumococci were lysed by exposure to penicillin.

Purified Pneumolysin

Recombinant pneumolysin was purified from *E. coli* containing the pneumococcal pneumolysin gene, as previously described (12). The bacteria were lysed by sonication and the pneumolysin purified by hydrophobic and ion exchange chromatography.

Viable Counting

Suspensions of bacteria were serially (1 in 10) diluted in water. Blood agar plates were split into four sectors, and in each sector 60 μL of culture was diluted between 10^{-3} and 10^{-6} times. When dry, the cultures were incubated overnight and the resulting colonies counted.

Hemolytic Assay

The hemolytic fraction was serially diluted (1 in 1) into phosphate buffered saline. A 2% suspension of compacted sheep's red blood cells was then added to each well. The plate was incubated for 30 minutes at 37°C and the hemolytic units (HU) calculated, as described before (10).

Hydrogen Peroxide Release and Assay

Pneumococci were grown and harvested when the OD_{500} was between 0.5 and 0.7. Following centrifugation, the pellet was resuspended in Hanks-HEPES. The bacterial suspension was incubated (37°C) for 60 minutes. One hundred μL of the suspension was sampled and centrifuged in a microcentrifuge tube. H_2O_2 was measured in the supernatant using a fluorometric assay (13).

RESULTS

We showed that the cilia lining the fourth ventricle of rat brain slices had a beat frequency identical to that observed in primary rat ependymal cells (Fig. 1). The beat frequency declined in the brain slice as it was cultured, such that by day 5 no beating cilia were evident (Fig. 1). In contrast, the primary rat ependymal cells were able to maintain a beat frequency of 40 Hz for up to 48 days in culture (Fig. 1). The addition of pneumolysin caused a dose-dependent decline in the ciliary beat frequency (CBF) in both brain slices and ependymal cells in culture. The 50% inhibitory concentrations (slice = 4.4 hemolytic units (HU)/mL; cells

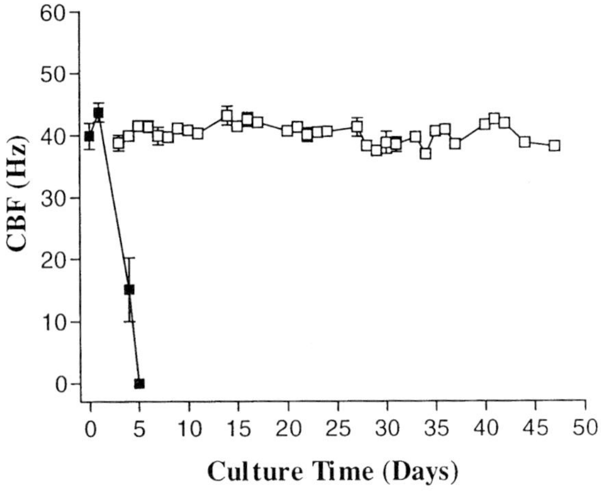

FIGURE 1 CBF measurements of ependymal cells (□) and brain slices (■) in culture. Data are mean ± SE mean of 3–5 preparations up to day 33, all subsequent data are from a single culture.

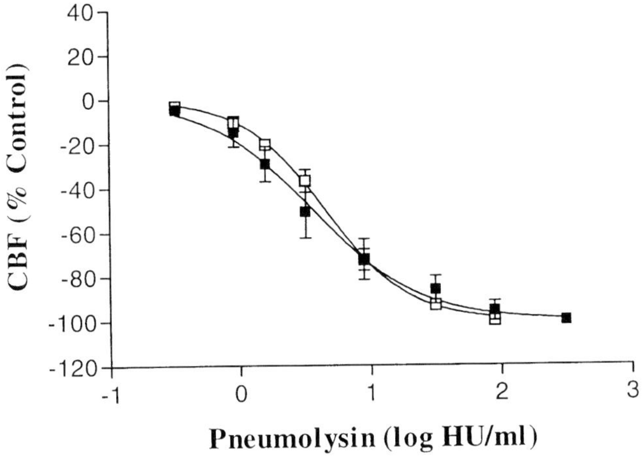

FIGURE 2 Pneumolysin inhibition of CBF in ependymal cells (■) and brain slices (□). Data are mean ± SE mean of 6 individual experiments.

= 3.7 HU/mL) were similar, indicating a similar mode of action of pneumolysin in the two models (Fig. 2).

The brain slice model was selected for a study in which we used a pneumolysin-sufficient and -deficient strain of pneumococcus in order to identify further the action of these bacteria in meningitis (10). Both D39 (wild-type) and PLN-A pneumococci caused ciliary slowing (Fig. 3). In order to investigate any potential role for pneumococcal production of H_2O_2 in this inhibition, brain slices were co-incubated with catalase (2000 EU/mL). There was a small but significant ($p < 0.05$) decrease in the CBF of D39 treated slices only at 30 and 60 minutes compared with D39 + catalase (Fig. 3A). The addition of catalase to PLN-A (Fig. 3B) prevented the reduction in ciliary beat frequency seen on exposure to PLN-A alone (Fig. 3B).

In order to examine whether an intact bacterial cell wall was a prerequisite for the inhibition of ciliary beat frequency, the pneumococci were lysed using penicillin. Lysed D39 bacteria caused rapid ciliary stasis, an effect not reversed by catalase (Fig. 4A). Lysed PLN-A did not reduce ciliary beat frequency, and catalase had no effect on ciliary beat frequency (Fig. 4B). In the presence of 100 μM, 1 mM, and 10 mM H_2O_2 there was a statistically significant ($p < 0.05$) inhibition of the CBF at 15 minutes compared with control. The inhibitory effect of each concentration of H_2O_2 on ciliary beat frequency was reversed by co-incubation with 2000 EU/mL catalase. The release of H_2O_2 from D39 and PLN-A was 81.4 ± 18 and 127 ± 58 μM, respectively (Table 1). In the presence of brain slice, there was a similar release of H_2O_2 from D39 and PLN-A pneumo-

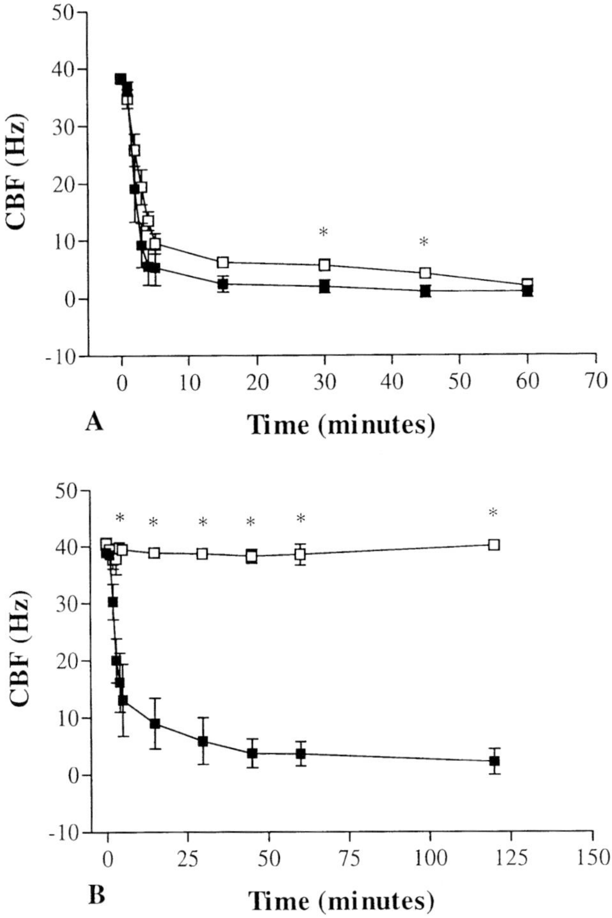

Figure 3 (A) The effect of D39 (10^8 CFU/ml) on ependymal CBF in the absence (■) or presence (□) of catalase (2000 EU/mL). *Statistically ($p < 0.05$, paired t-test) increased compared with D39. (B) The protective effect of catalase (2000 EU/mL, □) on PLN-A–induced inhibition (■) of ependymal ciliary beat frequency. *Statistically ($p < 0.05$, paired *t*-test) increased compared with PLN-A. All data are mean ± SE mean of 4 independent experiments.

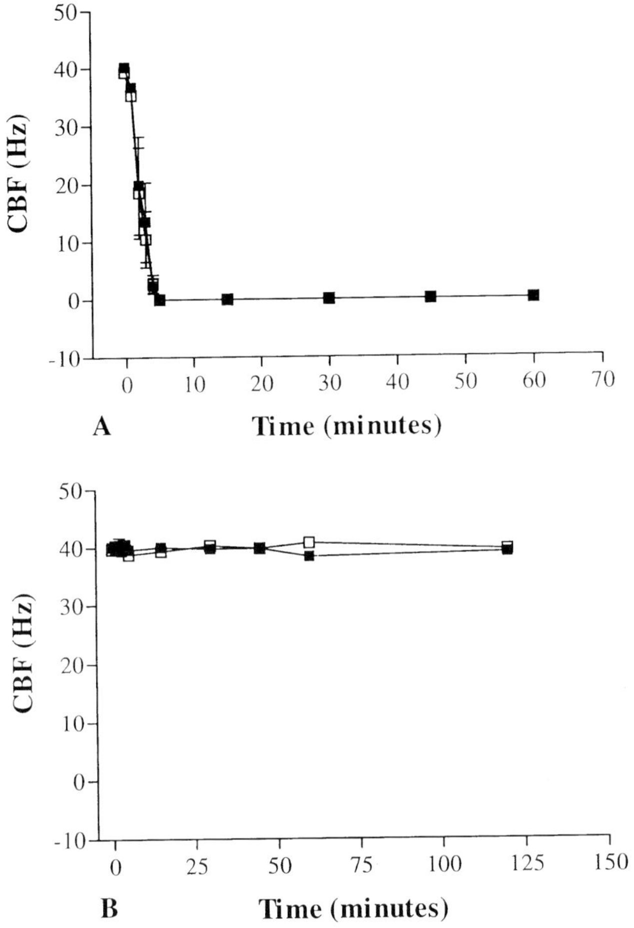

Figure 4 The effect of penicillin (1 mg/mL) lysed 10^8 CFU/mL pneumococci (D39 in Fig. 4A and PLN-A in Fig. 4B) in the presence (□) or absence (■) of catalase (2000 Eu/mL). There were no statistical differences in the data (mean ± SE mean of 4 experiments).

TABLE 1 Hydrogen Peroxide Release from a 10^8 cfu/mL of Pneumococci After 60 Minutes at 37°C

Strain/Tissue	CFU/mL	H_2O_2 concentration (μM) at 60 minutes
D39	10^8	81.4 ± 18
D39 + brain slice	10^8	73.6 ± 23
D39 + catalase (2000 EU/mL)	10^8	3.3 ± 0.5*
D39 + penicillin (1 mg/mL)	10^8 lysed	4 ± 2.2*
Brain slice	—	0 ± 0
PLN-A	10^8	127 ± 58
PLN-A + brain slice	10^8	78 ± 8
PLN-A + catalase (2000 EU/mL)	10^8	0 ± 0**
PLN-A + penicillin (1 mg/mL)	10^8 lysed	2.8 ± 1.7**

Data are mean ± SEM of 4–13 individual experiments.
* Significantly ($p < 0.05$, unpaired *t*-test) reduced compared with D39; ** significantly ($p < 0.05$) reduced compared with PLN-A.

cocci. The brain slice alone did not release measurable levels of H_2O_2 (Table 1). Co-incubation with catalase converted the H_2O_2 produced by the pneumococci (D39 and PLN-A) to H_2O (Table 1). Not surprisingly, penicillin-lysed D39 and PLN-A did not synthesize H_2O_2 (Table 1).

CONCLUSIONS

Ciliary function appears to be a good indicator of ependymal toxicity, and ciliary beat frequency measurements can be made quickly. This technique will be useful for identifying and screening of factors active in diseases of the cerebrospinal fluid. In addition, destruction of the ciliary function may have a role to play in the disease process directly, and thus determination of means of reversal of this destruction may lead to clinical interventions (9).

For long-term studies the ciliated ependymal cells in culture represent a major improvement over brain slices. In addition, ependymal cells in culture are advantageous in that the proportion of ciliated cells is increased over nonciliated cells when compared to brain slices, thus reducing the number of animals used for experiments. However, the brain slice model is ideal for acute studies of ependymal ciliary function where rapid measurements of ciliary beat frequency are required as the transverse section of ependyma can be readily visualized without the need for 14 days of prior culture. Brain slices would also be advantageous in experiments to study neuronal-ependymal cell communication due to the presence of the underlying neuronal tissue. Pneumolysin was equipotent at causing inhibition of ciliary beat frequency in both ependymal models, which suggests it has a similar mode of action in each.

Previous studies have shown that H_2O_2 disrupts the respiratory epithelium, causing ciliary stasis (14–16). H_2O_2 has also been shown to deplete epithelial ATP levels and because ciliary beating is heavily ATP dependent (17), this was suggested as the mechanism for H_2O_2-induced epithelial ciliary stasis (18). Our results demonstrate that H_2O_2 can inhibit ependymal CBF and that pneumococci release sufficient amounts of H_2O_2, to cause damage ependymal cilia, at concentrations of pneumococci that are commonly observed in patients with pneumococcal meningitis (19). Pneumolysin caused ciliostasis independently of H_2O_2 production. This means that it has a different mode of action for causing ciliary stasis compared with H_2O_2 alone.

Ependymal ciliary stasis caused by pneumolysin-negative pneumococci (PLN-A) was predominantly caused by H_2O_2 release. The subtle effects of bacterial H_2O_2 cannot be determined from our experiments due to the high levels of bacterial H_2O_2 to which the ependymal cells were exposed, but the mechanisms underlying H_2O_2 induced brain ciliary inhibition remains unclear. It is conceivable that ATP depletion, Ca^{2+}, PKC, or even membrane perturbation may be involved. H_2O_2 produced by bacteria readily diffuses across plasma membranes, where it can react rapidly to form other reactive species (20). These reactive species cause a wide array of biochemical changes in host cellular organelles. These include stimulation of diacylglycerol production and subsequent activation of protein kinase C (PKC) and the inhibition of Ca^{2+} homeostasis, all of which have been suggested as the reason for H_2O_2-induced ciliary stasis (16,21).

Leib and colleagues showed that reactive oxygen species were produced in the CSF of rats with bacterial meningitis (22). Antioxidants (23) have been shown to attenuate pathophysiological responses associated with experimental pneumococcal meningitis, and therefore, the released bacterial reactive species probably play a role in overall virulence. The majority of pneumococcal H_2O_2 is a product of pyruvate oxidase. A study has shown that when this enzyme was deleted by mutagenesis, the resulting strain of pneumococci had massively reduced virulence in vivo (24). One of the explanations for this observed lack of virulence was the reduced production of H_2O_2, and that would fit with the observations here. However, disruption of pyruvate oxidase has multiple effects on the bacterial phenotype, making interpretation of data difficult (24).

Our studies have supported the view that virulence of *S. pneumoniae* is multifactorial and in order to develop therapeutic interventions we must take into account all potential bacterial virulence factors. The findings suggest that the role(s) of both pneumolysin and H_2O_2 in the pathophysiology of pneumococcal meningitis require further investigation.

REFERENCES

1. MR Del Bigio. The ependyma: a protective barrier between brain and cerebrospinal fluid. Glia 14:1–13, 1995.
2. BJ Mohamed, TJ Mitchell, PW Andrew, RA Hirst, C O'Callaghan. The effect of the pneumococcal toxin, pneumolysin on brain ependymal cilia. Microb Pathogen 27(5):303–309, 1999.
3. V Quagliarello, MW Scheld. Bacterial meningitis; pathogenesis, pathophysiology and progress. N Engl J Med 327:864–872, 1992.
4. V Bohr, O Paulson, N Rasmussen. Pneumococcal meningitis: late neurological sequelae and features of prognostic impact. Arch Neurol 41:1045–1049, 1984.
5. CB Johansson, S Momma, DL Clarke, M Reisling, U Lendahl, J Frisen. Identification of a neural stem cell in the adult mammalian central nervous system. Cell 96: 25–34, 1999.
6. AJ Winter, SD Comis, MP Osborne, MJ Tarlow, J Stephen, PW Andrew, J Hill, TJ Mitchell. A role for pneumolysin but not neuramidase in the hearing loss and cochlear damage induced by experimental pneumococcal meningitis in guinea pigs. Infect Immun 65(11):4411–4418, 1997.
7. AR Friedland, MM Paris, S Hickey, S Shelton, K Olsen, JC Paton, HC McCracken. The limited role of pneumolysin in the pathogenesis of pneumococcal meningitis. J Infect Dis 172:805–809, 1995.
8. M Weibel, B Pettmann, JC Artault, M Sensenbrenner, G Labourdette. Primary culture of rat ependymal cells in serum-free defined medium. Dev Brain Res 25:199–209, 1986.
9. RA Hirst, A Rutman, KS Sikand, PW Andrew, TJ Mitchell, C O'Callaghan. Effect of pneumolysin on rat brain ciliary function: comparison of brain slices with cultured ependymal cells. Paediatr Res 47(3):381–384, 2000.
10. RA Hirst, KS Sikand, A Rutman, TJ Mitchell, PW Andrew, C O'Callaghan. The relative roles of pneumolysin and hydrogen peroxide (H_2O_2) from *Streptococcus pneumoniae* in the inhibition of ependymal ciliary beat frequency. Infect Immun 68(3):1557–1562, 2000.
11. AM Berry, J Yother, DE Briles, JC Paton. Reduced virulence of a defined pneumolysin-negative mutant of *Streptococcus pneumoniae*. Infect Immun 57:2037–2042, 1989.
12. TJ Mitchell, JE Walker, FK Saunders, PW Andrew, GJ Boulnois. Expression of the pneumolysin gene in *Escherichia coli*: rapid purification and biological properties. Biochem Biophy Acta 1007:67–72, 1989.
13. PS Jackett, PW Andrew, VR Aber, DB Lowrie. Hydrogen peroxide and superoxide release by alveolar macrophages from normal and BCG-vaccinated guinea pigs after intravenous challenge with mycobacterium tuberculosis. Br J Exp Path 62:419–428, 1981.
14. WJ Burman, WJ Martin. Oxidant-mediated ciliary dysfunction. Possible role in airway disease. Chest 89:410–413, 1986.
15. JT Jackowski, ZS Szepfalusi, DA Wanner, ZS Seybold, MW Sielczak, IT Lauredo, T Adams, WM Abraham, A Wanner. Effects of *P. aeruginosa*-derived bacterial

products on tracheal ciliary function: the role of O_2 radicals. Am J Physiol 260:61–L67, 1991.

16. K Kobayashi, M Salathe, MM Pratt, NJ Cartagena, F Soloni, ZU Seybold, A Wanner. Mechanism of hydrogen peroxide-induced inhibition of sheep airway cilia. Am J Respir Cell Mol Biol 6:667–673, 1992.
17. JL Carson, AM Collier. Ciliary defects: Cell biology and clinical perspectives. Adv Paediat 35:139–165, 1988.
18. PA Ward. Mechanisms of endothelial cell killing by H_2O_2 or products of activated neutrophils. Am J Med 91:89–94S, 1991.
19. E Bingen, N Lambert-Zechovsky, P Mariani-Kurkdjian, C Doit, Y Aujard, F Fournerie, H Mathieu. Bacterial counts in cerebrospinal fluid of children with meningitis. Eur J Clin Micro Infect Dis 9:278–281, 1990.
20. A Kantar, N Oggiano, PL Giorgi, PC Braga, C Fiorini. Polymorphoneuclear leukocyte-generated oxygen metabolites decrease beat frequency of human respiratory cilia. Lung 172:215–222, 1994.
21. EW Stommel, RE Stephens. Cyclic AMP and calcium in the differential control of mytillus gill cilia. J Comparat Physiol 157:451–459, 1985.
22. SL Leib, YS Kim, LL Chow, AR Sheldon, MG Tauber. Reactive oxygen intermediates contribute to necrotic and apoptotic neuronal injury in an infant rat model of bacterial meningitis due to group B streptococci. J Clin Invest 98:2632–2639, 1996.
23. HW Pfister, U Koedel, S Lorenzl, A Tomaz. Antioxidants attenuate microvascular changes in the early phase of experimental meningitis in rats. Stroke 23:1798–1804, 1992.
24. B Spellerberg, DR Cundell, J Sandros, BJ Pearce, L Idanpaan-Heikkila, C Rosenow, R Masure. Pyruvate oxidase, as a determinant of virulence in *Streptococcus pneumoniae*. Mol Microbiol 19:803–813, 1996.

14

The Cytoprotective Effects of Macrolides, Azalides, and Ketolides on Human Ciliated Epithelium In Vitro

Charles Feldman
Johannesburg Hospital and University of the Witwatersrand
Johannesburg, South Africa

Ronald Anderson and Annette J. Theron
MRC Unit for Inflammation and Immunity
Pretoria, South Africa

Peter Cole and Robert Wilson
National Heart & Lung Institute
London, United Kingdom

INTRODUCTION

It has been known for many years that in addition to their antimicrobial actions, macrolide antibiotics have a number of secondary effects on various host cells and their functions (1). For example, more than 40 years ago it was suggested that troleandomycin (TAO) may be of benefit in the treatment of inflammatory conditions of the airways (1), and in 1970 erythromycin and TAO were shown to improve the clinical status of steroid-dependent asthma (2). There followed a longstanding debate as to whether these beneficial effects in asthma were due to effects on corticosteroid metabolism/clearance (1,3,4); whether they were due to treatment of putative chronic infective etiologies of asthma, such as *Chlamydia pneumoniae* and *Mycoplasma pneumoniae* (5–7); or whether they were due to direct anti-inflammatory properties of the macrolides (1,8–10).

More recent studies have confirmed that macrolides do have anti-inflammatory activity, as well as anti-infective properties that are unrelated to their conventional antimicrobial activity and which may underlie their reported usefulness in a number of chronic inflammatory disorders of the airways, including conditions of both infective and noninfective origin. The clinical conditions for which a benefit has been shown from macrolide therapy include bronchial asthma (1,11–14), diffuse panbronchiolitis (DPB) (14–16), and disorders associated with chronic bronchial sepsis, such as cystic fibrosis and bronchiectasis (14,17).

Among their anti-inflammatory activities, macrolides have been shown to inhibit neutrophil chemotaxis and reactive oxidant production and to suppress the production of nitric oxide and various pro-inflammatory cytokines such as tumor necrosis factor alpha (TNF-α), interleukin (IL)-1, IL-6, IL-8, and granulocyte colony-stimulating factor, as well as other mediators such as neutrophil elastase and leukotriene B_4 (14,18–26). The macrolides also inhibit the expression of the leukocyte adhesion molecule, ICAM-1, the release of the potent vaso- and broncho-constrictor agent endothelin-1, and the production of defensins (14).

Conditions associated with chronic bronchial sepsis and DPB are airway disorders associated with chronic *Pseudomonas aeruginosa* colonization. Inflammation in the airways may be generated either directly through the virulence factors of the organism or indirectly through the host inflammatory response to chronic infection. The efficacy of the macrolides in the therapy of these disorders, as shown in various clinical studies, may be partly due to many of their anti-inflammatory activities detailed above, but may also be due to their direct inhibitory effects on the production and/or activity of many known or potential virulence factors of *P. aeruginosa* (14,18,27–34). These would include effects on exotoxin A, protease, elastase, phospholipase C, DNase, lecithinase, gelatinase, lipase, and pyocyanin. The macrolides have also been shown to interfere with biofilm formation, motility, and expression of pili by *P. aeruginosa*, all important factors associated with persistence of this organism (14,18,29–34).

MODEL OF AIRWAY INFLAMMATION/CYTOPROTECTION

We have been investigating mechanisms of injury to human respiratory epithelium and the ability of the macrolide/ketolide group of antimicrobial agents to attenuate effects on ciliary function and on structural integrity of the ciliated epithelium. We have been particularly interested in determining possible mechanisms of cytoprotection. In these studies the source of human ciliated epithelium has been nasal brushings obtained from the inferior nasal turbinate of healthy volunteers, which were sampled using a bronchoscopy cytology brush as described previously (35). Ciliary beat frequency was measured using a photo-transistor technique (36), and effects on structural integrity of the ciliated epithelium were assessed by means of a visual scoring system (27). Superoxide production by human polymorphonuclear leukocytes (PMNL) was measured using a lucigenin (bis-*N*-methylacridinium nitrate)-enhanced chemiluminescence (LECL) method (37), and membrane stabilization was determined using a hemolytic assay (38).

In this model of airway disease we used the bioactive phospholipids, platelet-activating factor (PAF), lyso-PAF, and lysophosphatidylcholine (LPC) as mediators of inflammation, in order to mimic pathophysiological conditions in vivo in various clinical conditions. For example, the bioactive phospholipids are important putative mediators of asthma, and PAF is the only potential mediator that can mimic all the important pathophysiological manifestations of this disease (39). The injurious effects of the bioactive phospholipids were studied in the absence and presence of PMNL.

RESULTS AND DISCUSSION

All three bioactive phospholipids—PAF, lyso-PAF, and LPC—caused significant, irreversible, dose-dependent slowing of ciliary beating and damage to the structural integrity of ciliated epithelium at concentrations ≥ 1 μg/mL, and these effects were progressive over 4 hours (40). There were no striking differences in the injurious effects of the individual PL. The effects were most likely mediated via nonspecific cytotoxicity as a result of the detergent-like, membrane disruptive activity of the PL. The injurious effects in this experimental system were not due to reactive oxidant production by contaminating leukocytes in the epithelial strips since neither catalase nor superoxide dismutase, either individually or in combination, was able to reduce the injury induced by PAF (40). The macrolide agents clarithromycin and roxithromycin, as well as the azalide azithromycin (final concentration 20 μg/mL) were unable to attenuate the toxic effects of the PL (40). However the ketolide agent, HMR 3004 (final concentration 20 μg/mL) was able to attenuate almost completely the ciliary slowing and damage to structural integrity induced by all three PL (final concentration 2.5 μg/mL) (Table 1) (41).

The amount of ciliary slowing and damage to structural integrity of ciliated epithelium induced by the PL was significantly increased in the presence of PMNL, such that the effects on ciliary function and structural integrity of the epithelium at concentrations of 1 μg/mL of the PL in the presence of PMNL (final concentration 1×10^6/mL) were equivalent to those of the PL alone at 5 μg/mL (40). In this experimental system direct epithelial injury may have been present, but injurious effects appeared to be largely due to reactive oxidant release

TABLE 1 Mean ± SEM Ciliary Beat Frequency and Percent Ciliary Slowing and Epithelial Damage in Epithelial Strips Exposed to Bioactive Phospholipids in the Absence and Presence of HMR 3004

	Control strips (in HBSS alone)		Epithelial strips exposed to PL (2.5 μg/mL)			Epithelial strips incubated with HMR 3004 (20 μg/mL) prior to exposure to PL		
	CBF (Hz)	%ED	CBF (Hz)	%ED	%CS	CBF (Hz)	%ED	%CS
PAF[a]	11.4 ± 0.17	0	8.8 ± 0.38*	35*	23	11.6 ± 0.18	0	0
Lyso-PAF[a]	11.5 ± 0.17	0	8.9 ± 0.38*	33*	23	11.8 ± 0.16	0	0
LPC[a]	11.1 ± 0.15	0	9.5 ± 0.40*	28*	14	11.1 ± 0.23	7	0

CBF: Ciliary beat frequency; %ED: percent epithelial damage; %CS: percent ciliary slowing; PL: phospholipids; PAF: platelet-activating factor; LPC: lysophosphatidylcholine.
[a] Added only to epithelial strips exposed to PL.
* $p < 0.005$ vs. control. All other values not significantly different from control.
Source: Ref. 41.

from PL-primed PMNL. Confirmatory evidence for this was that PAF and LPC were shown to cause dose-related activation of superoxide production of PMNL, while catalase alone or in combination with superoxide dismutase was able to attenuate both the slowing of CBF and epithelial damage induced by all three PL significantly (38,41). In addition, preincubation of the PMNL with any of the macrolide, azalide, or ketolide agents prior to their exposure to the PL, followed by co-incubation with the epithelial cells was associated with almost complete attenuation of these injurious effects (Table 2) (40,41). Further studies have documented that the ketolide agents, and in particular HMR 3004, cause dose-related inhibition of superoxide production by PMNL activated by four different stimuli of membrane-associated oxidative metabolism, namely, *N*-formyl-Met-Leu-Phe (FMLP), the calcium ionophore, A23187, phorbol 12-myristate-13 acetate (PMA), and opsonized zymosan, all of which use different transductional mechanisms to activate the superoxide-generating system of phagocytes, NAPDH-oxidase (42).

The cytoprotective effects of these agents appear to be related to their membrane-stabilizing ability. These effects were demonstrated in a hemolytic assay whereby pretreatment of erythrocytes with each of these agents in similar concentrations to those that inhibited superoxide production antagonized the lytic effects of all three PL (38,41,42). In terms of potency, HMR 3004 was much more effective than the other ketolide, HMR 3647, with both ketolides being considerably superior to roxithromycin (42). The latter has been shown to have membrane-stabilizing properties equal to those of clarithromycin but superior to

Table 2 Mean ± SEM Ciliary Beat Frequency and Percent Ciliary Slowing and Epithelial Damage in Human Ciliated Respiratory Epithelium Exposed to Platelet-Activating Factor–Sensitized Polymorphonuclear Leukocytes in the Presence and Absence of HMR 3004 (final concentration 10 μg/mL)

	CBF (Hz)	%ED	%CS
Control epithelial strips	11.5 ± 0.22	—	—
Epithelial strips exposed to PAF (1μg/mL)	9.5 ± 0.38*	33*	17
Preincubation of epithelial strips with HMR 3004	10.5 ± 0.27**	17**	9
Preincubation of PMNL with HMR 3004	11.4 ± 0.14***	3***	<1

CBF: Ciliary beat frequency; %ED: percent epithelial damage; %CS: percent ciliary slowing; PAF: platelet-activating factor; PMNL: polymorphonuclear leukocytes.
* $p < 0.005$ vs. control; ** $p < 0.05$ vs. control; *** not significantly different from control, but significantly different ($p < 0.05$) from all other values.
Source: Ref. 41.

those of azithromycin or erythromycin (42). These differences in potency would also account for differences in cytoprotection afforded by the various agents in the different experimental systems. The most potent membrane-stabilizing agents, the ketolides, are the only ones that appear to be able to stabilize the ciliated epithelial membrane against direct cytolytic effects of the PL, whereas all three classes of agents are able to decrease reactive oxidant-induced epithelial injury by stabilizing the PMNL membrane against sensitization by lower concentrations of PL.

We also attempted to establish the relative cytoprotective potencies of clarithromycin, roxithromycin, and azithromycin for ciliated epithelium, but no significant differences were noted in experiments with PMNL (40). However, the ketolides were more effective cytoprotective agents than the macrolides/azalides, with comparative results similar to those in the experiments described above. There was also a slight difference in the relative potencies of the two ketolide agents tested, in that the cytoprotection afforded by 3004 was almost complete at the concentrations used (20 μg/mL), whereas there were still mild residual injurious effects with HMR 3647 (41). The differences in membrane-stabilizing ability and cytoprotection of the different agents in experiments with and without PMNL are almost certainly related to differences in their levels of intracellular accumulation, with the accumulation of HMR 3004 being the most impressive (41–43).

Our most recent study appears to confirm the distinct relationship between the anti-inflammatory activities of the macrolides/azalides/ketolides and their ability to inhibit superoxide production via membrane-stabilization (44). In this study, it was confirmed that clarithromycin, a 14-membered macrolide, with anti-inflammatory activity, caused dose-related inhibition of superoxide production by activated PMNL and also protected erythrocytes against PL-induced hemolysis, whereas spiramycin, a 16-membered macrolide of the group, known not to have anti-inflammatory activity, did not inhibit superoxide production and possessed only weak membrane-stabilizing properties (44). These experiments have been complemented by additional data showing that spiramycin may in fact increase superoxide production by stimulated PMNL (45), as well as IL-6 production by human monocytes in vitro (46).

CONCLUSION

Our studies have contributed to the understanding of aspects of the nonantimicrobial, anti-inflammatory activities of the macrolide/azalide/ketolide group of antibiotics, and in particular their direct and indirect cytoprotective effects on human ciliated epithelium (39,41–43,45). The macrolides have been shown to have a potential role in a number of conditions affecting the airways, these effects being related mainly to anti-inflammatory rather than antimicrobial activity. In diffuse panbronchiolitis, macrolides have become firmly established as first-line treatment and have made a major impact on the outcome of this condition (14–16). A number of clinical studies using the macrolides have been completed or are ongoing in asthma and have shown improvement in the clinical condition of these patients (1,11–14). More recently data are emerging on the potential benefits of the macrolides in cases with chronic bronchial suppuration such as cystic fibrosis and bronchiectasis (14,17). Still to be studied are the ketolide agents, which based on in vitro data may have even better clinical activity than the macrolides in these various chronic inflammatory conditions of the airways.

REFERENCES

1. PN Black. Anti-inflammatory effects of macrolide antibiotics. Eur Respir J 10:971–972, 1997.
2. IW Itkin, ML Menzel. The use of macrolide antibiotic substances in the treatment of asthma. J Allergy 45:146–162, 1970.
3. SJ Szefler, JO Rose, EF Ellis, SL Spector, AW Green, WJ Jusko. The effect of troleandomycin on methylprednisolone elimination. J Allergy Clin Immunol 66: 447–451, 1980.
4. RS Zeiger, M Schatz, W Sperling, RA Simon, DD Stevenson. Efficacy of troleandomycin in outpatients with severe, corticosteroid-dependent asthma. J Allergy Clin Immunol 66:438–446, 1980.
5. JS Seggev, I Lis, R Siman-Tov, R Gutman, H Abu-Samara, G Schey, Y Naot. *Mycoplasma pneumoniae* is a frequent cause of exacerbation of bronchial asthma in adults. Ann Allergy 57:263–265, 1986.

6. C-C Kuo, LA Jackson, LA Campbell, T Grayston. *Chlamydia pneumoniae* (TWAR). Clin Microbiol Rev 8:451–461, 1995.
7. PJ Cook, P Davies, W Tunnicliffe, JG Ayres, D Honeybourne, R Wise. *Chlamydia pneumoniae* and asthma. Thorax 53:254–259, 1998.
8. LS Groes, SJ Szefler, CG Irvin, GL Larsen, MR Hill. Does erythromycin prevent airway inflammation? Ann Allergy 64:A81, 1990.
9. LS Groes, SJ Szefler, GL Larsen, CG Irvin, MR Hill. Troleandomycin reduces airways inflammation. Am Rev Respir Dis 141:A933, 1990.
10. SK Goswami, S Kivity, Z Marom. Erythromycin inhibits respiratory glycoconjugate secretion from human airways in vitro. Am Rev Respir Dis 141:72–78, 1990.
11. H Miyatake, F Taki, H Taniguchi, R Suzuki, K Takagi, T Satake. Erythromycin reduces the severity of bronchial hyperresponsiveness in asthma. Chest 99:670–673, 1991.
12. PN Black. The use of macrolides in the treatment of asthma. Eur Respir Rev 6:240–243, 1996.
13. T Shimizu, M Kato, H Mochizuki, K Tokuyama, K Morikawa, T Kuruome. Roxithromycin reduces the degree of bronchial hyperresponsiveness in children with asthma. Chest 166:458–461, 1994.
14. D Wales, M Woodhead. The anti-inflammatory effects of macrolides. Thorax 54:(suppl 2):S58–S62, 1999.
15. S Kudoh. Erythromycin treatment in diffuse panbronchiolitis. Curr Opin Pulm Med 4:116–121, 1998.
16. H Koyama, DM Geddes. Erythromycin and diffuse panbronchiolitis. Thorax 52: 915–918, 1997.
17. PM Beringer. New approaches to optimizing antimicrobial therapy in patients with cystic fibrosis. Curr Opin Pulm Med 5:371–377, 1999.
18. RA Howe, RC Spencer. Macrolides for the treatment of *Pseudomonas aeruginosa* infections? J Antimicrob Chemother 40:153–155, 1997.
19. R Anderson. Erythromycin and roxithromycin potentiate human neutrophil locomotion in vitro by inhibition of leukoattractant-activated superoxide generation and autooxidation. J Infect Dis 159:966–973, 1989.
20. S Umeki. Anti-inflammatory action of erythromycin. Its inhibitory effect on neutrophil NADPH oxidase activity. Chest 104:1191–1193, 1993.
21. MT Labro, J El Benna, C Babin-Chevaye. Comparison of the in-vitro effect of several macrolides on the oxidative burst of human neutrophils. J Antimicrob Chemother 24:561–572, 1989.
22. E Kita, M Sawaki, K Mikasa, K Hamada, S Takeuchi, K Maeda, N Narita. Alterations of host response by a long-term treatment of roxithromycin. J Antimicrob Chemother 32:285–294, 1993.
23. MJ Schultz, P Speelman, S Zaat, SJH van Deventer, T van der Poll. Erythromycin inhibits tumor necrosis factor alpha and interleukin 6 production induced by heat-killed *Streptococcus pneumoniae* in whole blood. Antimicrob Agents Chemother 42:1605–1609, 1998.
24. T Kohyama, H Takizawa, S Kawasaki, N Akiyama, M Sato, K Ito. Fourteen-member macrolides inhibit interleukin-8 release by human eosinophils from atopic donors. Antimicrob Agents Chemother 43:907–911, 1999.

25. S Kawasaki, H Takizawa, T Ohtoshi, N Takeuchi, T Kohyama, H Nakamura, T Kasama, K Kobayashi, K Nakahara, Y Morita, K Yamamoto. Roxithromycin inhibits cytokine productions by and neutrophil attachment to human bronchial epithelial cells in vitro. Antimicrob Agents Chemother 42:1499–1502, 1998.
26. H Nakamura, S Fujishima, T Inoue, Y Ohkubo, K Soejima, Y Waki, M Mori, T Urano, F Sakamaki, S Tasaka, A Ishizaka, M Kanazawa, K Yamaguchi. Clinical and immunoregulatory effects of roxithromycin therapy for chronic respiratory tract infection. Eur Respir J 13:1371–1379, 1999.
27. E Tanaka, K Kanthakumar, DR Cundell, KWT Tsang, GW Taylor, F Kuze, PJ Cole, R Wilson. The effect of erythromycin on *Pseudomonas aeruginosa* and neutrophil mediated epithelial damage. J Antimicrob Chemother 33:765–775, 1994.
28. R Mizukane, Y Hirakata, M Kaku, Y Ishii, N Furuya, K Ishida, H Koga, S Kohno, K Yamaguchi. Comparative in vitro exoenzyme-suppressing activities of azithromycin and other macrolide antibiotics against *Pseudomonas aeruginosa*. Antimicrob Agents Chemother 38:528–533, 1994.
29. H Yasuda, Y Ajiki, T Koga, H Kawada, T Yokota. Interaction between biofilms formed by *Pseudomonas aeruginosa* and clarithromycin. Antimicrob Agents Chemother 37:1749–1755, 1993.
30. T Ichimiya, T Yamasaki, M Nasu. In-vitro effects of antimicrobial agents on *Pseudomonas aeruginosa* biofilm formation. J Antimicrob Chemother 34:331–341, 1994.
31. M Ozeki, N Miyamoto, M Hashiba, S Baba. Inhibitory effect of roxithromycin on biofilm formation of *Pseudomonas aeruginosa*. Acta Otolaryngol (Stockh) 525:61–63, 1996.
32. K Kondoh, M Hashiba, S Baba. Inhibitory activity of clarithromycin on biofilm synthesis with *Pseudomonas aeruginosa*. Acta Otolaryngol (Stockh) 525:56–60, 1996.
33. H Kobayashi. Biofilm disease: its clinical manifestation and therapeutic possibilities of macrolides. Am J Med 99 (suppl 6A):26S–30S, 1995.
34. G Molinari, CA Guzman, A Pesce, GC Schito. Inhibition of *Pseudomonas aeruginosa* virulence factors by subinhibitory concentrations of azithromycin and other macrolide antibiotics. J Antimicrob Chemother 31:681–688, 1993.
35. J Rutland, PJ Cole. Non-invasive sampling of nasal cilia for the measurement of beat frequency and ultrastructure. Lancet ii:564–565, 1980.
36. C Feldman, A Voorvelt. A photo-transistor technique for the measurement of the ciliary beat frequency of human ciliated epithelium in vitro. S Afr J Sci 90:555–556, 1994.
37. I Minkenberg, E Ferber. Lucigenin-dependent chemiluminescence as a new assay for NADPH-oxidase activity in particulate fractions of human polymorphonuclear leukocytes. J Immunol Meth 71:61–67, 1984.
38. R Anderson, AJ Theron, C Feldman. Membrane stabilizing, anti-inflammatory interactions of macrolides with human neutrophils. Inflammation 20:693–705, 1996.
39. PJ Barnes, KF Chung, CP Page. Inflammatory mediators and asthma. Pharmacol Rev 40:49–84, 1988.
40. C Feldman, R Anderson, AJ Theron, G Ramafi, PJ Cole, R Wilson. Roxithromycin, clarithromycin, and azithromycin attenuate the injurious effects of bioactive phospholipids on human respiratory epithelium in vitro. Inflammation 21:655–665, 1997.

41. C Feldman, R Anderson, A Theron, I Mokgobu, PJ Cole, R Wilson. The effects of ketolides on bioactive phospholipid-induced injury to human respiratory epithelium in vitro. Eur Respir J 13:1022–1028, 1999.
42. I Mokgobu, AJ Theron, R Anderson, C Feldman. The ketolide antimicrobial agent HMR-3004 inhibits neutrophil superoxide production by a membrane-stabilizing mechanism. Int J Immunopharmacol 21:365–377, 1999.
43. D Vazifeh, H Abdelghaffar, MT Labro. Cellular accumulation of the new ketolide RU 64004 by human neutrophils: comparison with that of azithromycin and roxithromycin. Antimicrob Agents Chemother 41:2099–2107, 1997.
44. AJ Theron, C Feldman, R Anderson. Comparison of the anti-inflammatory and membrane-stabilising potential of clarithromycin and spiramycin in vitro. J Antimicrob Chemother 46:269–271, 2000.
45. I Moutard, B Gressier, C Brunet, T Dine, M Luyckx, F Templier, M Cazin, JC Cazin. In vitro interaction between spiramycin and polymorphonuclear neutrophils oxidative metabolism. Pharmacol Res 37:197–201, 1998.
46. S Bailly, J-J Pocidalo, M Fay, M-A Gougerot-Pocidalo. Differential modulation of cytokine production by macrolides: interleukin-6 production is increased by spiramycin and erythromycin. Antimicrob Agents Chemother 35:2016–2019, 1991.

15

Examining Mucociliary Differentiation of Human Nasal Epithelial Cells In Vitro

J. Laoukili, K. Million, O. Houcine, F. Marano, and Frédéric Tournier
Université Paris 7
Paris, France

E. Perret and D. Caput
Sanofi Recherche
Labège Innopole, France

INTRODUCTION

The airway epithelium is the first target for a wide range of inhaled molecules and particles. Mucociliary clearance is responsive for their partial, if not total, elimination as coordinated beating of cilia along the upper respiratory tract continuously drives mucus towards the nasopharynx. Ciliated cells are one of the three main cell types lining the surface of the upper airways, along with basal and secretory cells. Ciliated cell differentiation is a complex molecular process involving centriole/basal body assembly (centriologenesis) and cilia formation (ciliogenesis). Detailed ultrastructural features have been described in different species (1). However, most of the molecular events underlying the differentiation process remain to be understood. The formation of 100–200 cilia in each terminally differentiated ciliated cell requires the assembly of the same number of centriole/basal bodies. Procentrioles are assembled around electron-dense cytoplasmic granules near the Golgi area. The centrioles elongate, migrate through the cytoplasm, and anchor onto the apical plasma membrane, where they trigger axonemal microtubule polymerization and cilia formation (2,3). Tubulin and centrin are the two most widely studied gene families involved in centriole and cilia structure and functions. The heterogeneity of α- and β-tubulins is generated at a genetic level by several genes (4) and at a biochemical level by posttranslational modifications. Among them, polyglutamylation and polyglycylation are common, consisting of the addition of glutamyl and glycyl residues, respectively, to form polyglutamyl and polyglycyl chains at the C-terminal region of both α- and β-tubulins (5,6). These additional peptide chains could play an important role in cytoplasmic microtubule networks as well as in centriolar and axonemal structures. Centrin proteins are associated with the distal lumen of the centrioles (7) and belong to the family of Ca^{2+}-binding protein (8). In humans, three centrin genes have been described: HsCEN1, HsCEN2 and HsCEN3. But functionally, only two distinct families of centrin proteins seem to exist, one of which has been implicated in centrosome duplication (9,10).

We have developed a primary culture system of human nasal epithelial cells where the mucociliary differentiation process can be observed and quantified. We characterized markers of ciliated cell differentiation at the cellular and biochemical level by studying polyglutamylation and polyglycylation of α- and β-tubulins and expression and cellular localization of centrin isoforms. Immunoelectron microscopy revealed that polyglutamylation is an early event in the centriole/basal body assembly process, while polyglycylation is restricted to axonemal tubulin (11). In parallel, we have shown a correlation between ultrastructural localization and function of two human centrin isoforms (12).

A MODEL SYSTEM TO STUDY MUCOCILIARY DIFFERENTIATION IN VITRO

Several different methods have been developed to culture primary respiratory epithelial cells. The most frequently used technique is the creation of an air/liquid interface, which results in mucociliary differentiation in vitro (13–15). For our studies, we used the spheroid culture model of human nasal epithelial (HNE) cells initially described by Jorissen et al. (16). In these cultures, we have quantified the mucociliary differentiation process using specific monoclonal antibodies against secretory and ciliated cell components. HNE cells were dissociated from nasal polyps or turbinates and were seeded onto a type I collagen gel where undifferentiated epithelial cells proliferated to confluence. The collagen gel was digested and cell sheets were shaken in a rotary motion for 5–8 days to induce formation of still undifferentiated epithelial spheroids (Fig. 1a). In these spheroids, epithelial cells polarized and differentiated. During the differentiation pro-

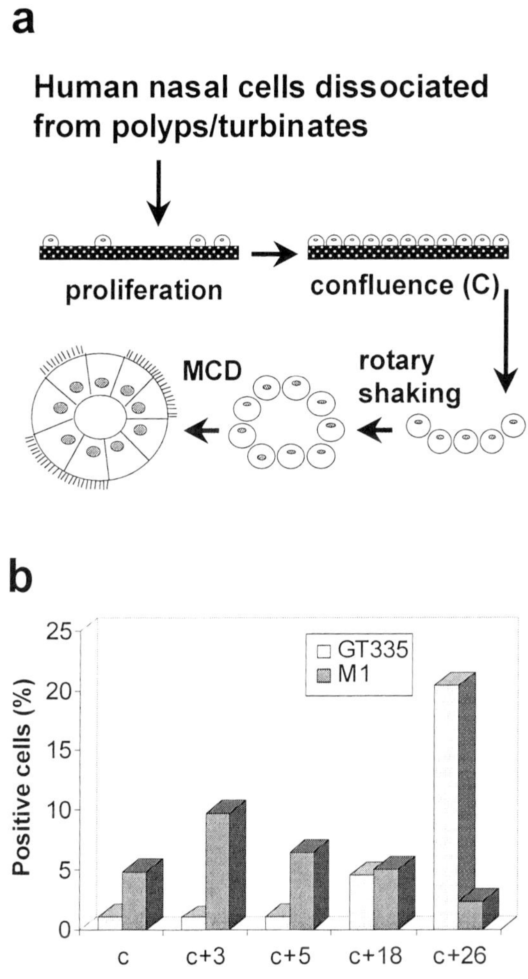

FIGURE 1 (a) Cell culture method: cells are dissociated from human nasal polyps or turbinates and platted on collagen gel matrix. Proliferation of epithelial cells is favored during a 2- to 3-week period. At cell confluence (C), the collagen gel is digested and epithelial sheets are further cultured for 5–8 days in a rotary shaker to obtain epithelial spheroids. The latters differentiate into secretory and ciliated cells (MCD = mucociliary differentiation). (b) Quantification of ciliated and secretory cells using FACS analysis: epithelial cells are dissociated from spheroids during the time-course of mucociliary differentiation and are immunostained using specific Abs. GT335, a monoclonal antibody raised against glutamylated peptides, specifically recognizes centriole/basal bodies and axonemal tubulin of ciliated epithelial cells. M1 mAb recognizes a subset of human mucin gene products including MUC5AC (c + n = n days after confluence).

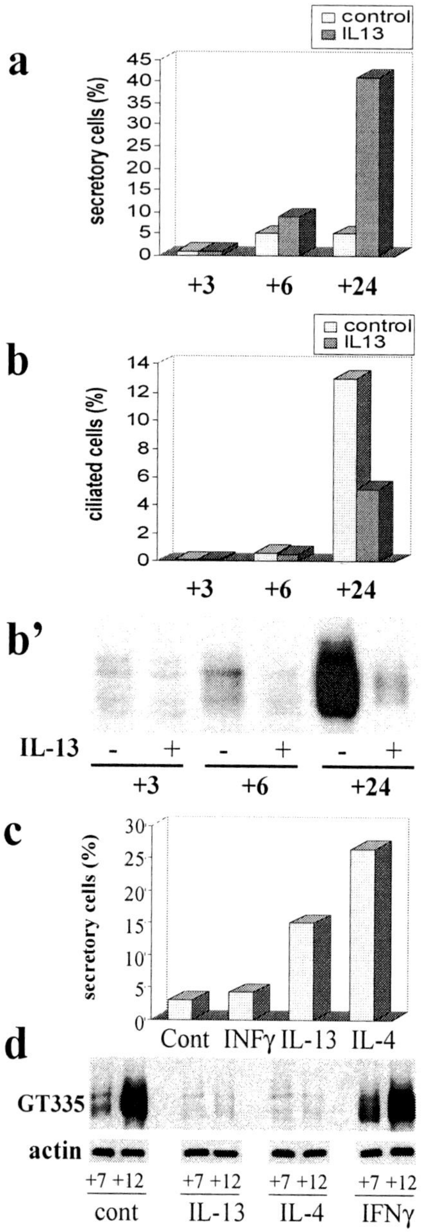
a
control
IL13
secretory cells (%)
45
40
35
30
25
20
15
10
5
0
+3
+6
+24
b
control
IL13
ciliated cells (%)
14
12
10
8
6
4
2
0
+3
+6
+24
b'
IL-13
-
+
-
+
-
+
+3
+6
+24
c
secretory cells (%)
30
25
20
15
10
5
0
Cont
INFγ
IL-13
IL-4
d
GT335
actin
+7 +12
+7 +12
+7 +12
+7 +12
cont
IL-13
IL-4
IFNγ

cess, spheroids were dissociated to analyze the phenotype distribution of the cells by flow cytometry using specific antibodies. We used M1 antibodies, which recognize MUC5AC gene product and potentially other human mucins (17), to estimate the relative proportion of secretory cells and GT335 antibodies, which recognize glutamylated tubulin (a marker of centrioles, basal bodies, and ciliary axonemes), to estimate the percentage of ciliated cells. Such experiments revealed that secretory cells are detected in significant numbers early during differentiation and subsequently decrease, while ciliated cells develop at a later stage and continue to increase in number throughout the differentiation process. A representative experiment is shown in Figure 1b.

We also studied the effects of two Th2-response interleukins, IL-4 and IL-13, on mucociliary differentiation. These interleukins play important roles in asthma (18,19), a disease known to change the epithelium to a more secretory phenotype. In particular, IL-13 has been shown to cause mucus cell metaplasia (20). When 10 ng/mL IL-13 was added to the cultures during differentiation, the percentage of secretory cells, compared to control, increased dramatically (Fig. 2a), while the percentage of ciliated cells remained low (Fig. 2b) (21). Western blot analysis confirmed the latter result (Fig. 2b′). Interferon gamma (IFNγ), a Th1 cytokine, had no effect on the percentages of secretory versus ciliated cells, while IL-4 had an effect similar to IL-13 (Fig. 2c and d).

CHARACTERIZATION OF CENTRIOLAR MARKERS DURING MUCOCILIARY DIFFERENTIATION

Both human centrin 1/2 protein (Hscen1/2p) and Hscen3p were associated with cytoplasmic granules (Fig. 3a,b) and enriched in the distal lumen of migrating centrioles (Fig. 3a,b) and mature centriole/basal bodies (Fig. 3a′,b′). Interestingly, the localization of Hscen3p was restricted to centriole/basal bodies (Fig.

Figure 2 IL-13 increases the percentage of secretory cells and inhibits ciliated cell differentiation. (a) The percentage of secretory cells (revealed with M1 mAb) remains low in control conditions while it largely increases for IL-13–treated spheroids (FACS analysis). (b) The percentage of ciliated cells (revealed with GT335 mAb) is significantly lower in IL-13–treated spheroids (FACS analysis). (b′) Western blot—the signal intensity using GT335 mAb reveals the proportion of ciliated cells from total protein extracts. (c) IL-4 (10 ng/mL), but not IFNγ (10 ng/mL), increases the percentage of M1-positive cells during MCD (FACS analysis). (d) Western blot—IL-4 and IL-13 largely impair the process of ciliated cell differentiation, as in both cases the signal detected by GT335 remains low (+7, +12 = 7 and 12 days after confluence).

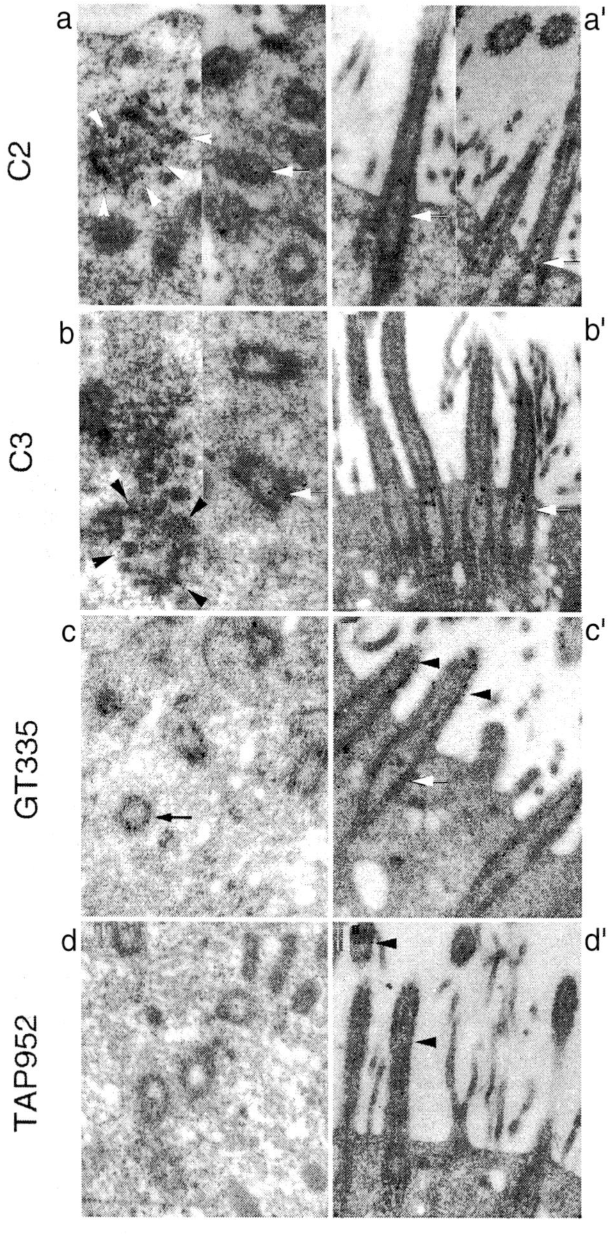
a
a'
C2
b
b'
C3
c
c'
GT335
d
d'
TAP952

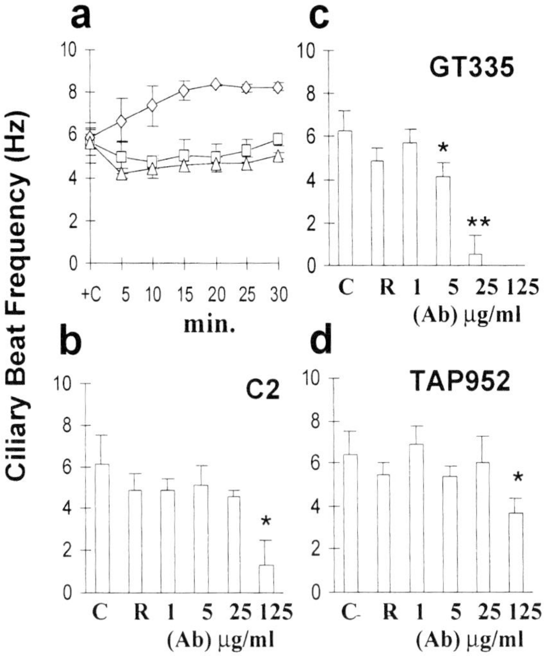

FIGURE 4 Effect of purified anti-centrin 2 and anti-tubulin (GT335 and TAP952) Igs on ciliary beat frequency (CBF). Epithelial spheroids differentiated in vitro were demembranated in a detergent-containing buffer, and cilia were rapidly immobile. A reactivation buffer (R) was then applied. Almost constant ciliary beating was maintained up to 30 minutes (a, three independent kinetics). Purified Igs diluted at 1, 5, 25, and 125 μg/mL in R buffer were applied on demembranated/reactivated spheroids. CBF was determined after a 5-minute application period. C2 strongly inhibits CBF at 125 μg/mL (b). GT335 has a dose-dependent inhibitory effect (c), while TAP952 has only a partial effect (d).

FIGURE 3 Immunolocalization of tubulin PTMs and centrin proteins in electronic microscopy during ciliated cell differentiation. Two centrin isoforms (C2 and C3) are detected with anti-centrin polyclonal antibodies in electron dense cytoplasmic granules involved in the early process of centriologenesis (arrowheads in a and b), and in migrating centrioles (arrows in a and b). C3 is strictly detected in the distal lumen of mature centriole/basal bodies (arrow in b′), while C2 localization extends to the proximal part of ciliary axonemes (arrows in a′). Glutamylated tubulin (GT335) is localized in migrating centrioles (arrow in c), in centriole/basal bodies (arrow in c′), and in ciliary axonemes (arrowheads in c′). By contrast, glycylated tubulin (revealed with TAP952 mAb) is not detected on centrioles (d), but strictly localized in mature axonemes (arrowheads in d′). Bar = 0.2 μm.

3b′), while Hscen1/2p was also detected in the proximal part of the axonemes (Fig. 3a′). Using a monoclonal antibody raised against glutamylated peptides (GT335, see above), we found glutamylated tubulin in assembling (11) and migrating centrioles (Fig. 3c). Glycylated tubulin was only detected in ciliary axonemes (Fig. 3d′) but not within migrating centrioles (Fig. 3d) or mature centriole/basal bodies (Fig. 3d′).

Finally, we measured ciliary beat frequency (CBF) using video-microscopy and ciliated epithelial spheroids to assess the role of centrins in ciliary beat regulation (Fig. 4). In ATP-reactivated demembranated epithelial spheroids, CBF was almost constant over a 30-minute incubation period (Fig. 4a). Purified anti-centrin 2 and GT335 antibodies reduced CBF in a dose-dependent manner (Fig. 4b,d). TAP952, a monoclonal antibody raised against glycylated tubulin, also reduced CBF at the highest concentration used, while C3 did not modify CBF (12).

CONCLUSION

The molecular events involved in the process of ciliated cell differentiation are poorly understood. The use of primary cell cultures of human respiratory epithelial cells allowed us to follow specific markers of ciliated cell differentiation and the percentage of secretory and ciliated cells during differentiation. Several data argue in favor of the existence of two functionally distinct centrin families: one implicated in centriole assembly, i.e., centrosome duplication or centriologenesis (22), and a second family that participates in other cellular events, such as cytokinesis in proliferating cells or ciliary beating in ciliated epithelial cells. In the latter case, the discrimination between centrin 1 and centrin 2 is interesting for future prospects. Among the different posttranslational modifications of tubulins, polyglutamylation and polyglycylation appear to be notably involved in the relationship between tubulin and microtubule-associated proteins or molecular motors. In nonciliated cells, polyglutamylation is involved in the maintenance of centriole integrity (23). Thus, the identification of specific enzymes (24) and the regulation of their activity are essential to understand their role in microtubule stabilization.

IL-13 largely enhances the expression of a secretory phenotype but does not totally inhibit the ciliated differentiation process. Nevertheless, alteration of epithelial cell shape and cell polarity probably interferes with the process of ciliogenesis. As IL-13 was recently shown to be specifically involved in the asthma phenotype, this culture model could provide new insights in the molecular mechanisms involved. We recently performed differential screenings using a cDNA subtraction method during mucociliary differentiation in the presence or absence of IL-13. We now analyze cDNA clones as potential molecular targets of the IL-13 downstream regulation. Because this Th2 cytokine, which shares a large homology with IL-4 but has distinct biological roles (18,19,25), is clearly associated with asthma and other pathologies, IL-13 is a potential key target for therapeutic intervention.

ACKNOWLEDGMENTS

We would like to thank the following collaborators for helpful discussions and technical advice: K. Adler, M. Bornens, D. Bourguignon, C. Guennou, M. Jorissen, J. C. Larcher, L. Martin, S. Middendorp, and A. Minty.

REFERENCES

1. ER Dirksen. Centriole and basal body formation during ciliogenesis revisited. Biol Cell 72:31–38, 1991.
2. SP Sorokin. Reconstruction of centriole formation and ciliogenesis in mammalian lungs. J Cell Sci 3:207–230, 1998.
3. M Lemullois, E Boisvieux-Ulrich, MC Laine, B Chailly, D Sandoz. Development and functions of the cytoskeleton during ciliogenesis in metazoa. Biol Cell 63:195–208, 1998.
4. KF Sullivan. Structure and utilization of tubulin isotypes. Ann Rev Cell Biol 4:687–716, 1988.
5. B Eddé, J Rossier, JP Le Caer, E Desbruyères, F Gros, P Denoulet. Posttranslational glutamylation of α-tubulin. Science 247:83–85, 1990.
6. V Redeker, N Levilliers, JM Schmitter, JP Le Caer, J Rossier, A Adoutte, MH Bré. Polyglycylation of tubulin: a posttranslational modification in axonemal microtubules. Science 266:1688–1690, 1994.
7. A Paoletti, M Moudjou, M Paintrand, J Salisbury, M Bornens. Most of centrin in animal cells is not centrosome-associated and centrosomal centrin is confined to the distal lumen of centrioles. J Cell Sci 109:3089–3102, 1996.
8. E Shiebel, M Bornens. In search of a function for centrins. Trends Cell Biol 5:197–201, 1995.
9. S Middendorp, A Paoletti, E Schiebel, M Bornens. Identification of a new mammalian centrin gene, more closely related to Saccharomyces cerevisiae CDC31 gene. Proc Natl Acad Sci USA 94:9141–9146, 1997.
10. S Middendorp, T Kuntziger, Y Abraham, S Holmes, N Bordes, M Paintrand, A Paoletti, M Bornens. A role for centrin 3 in centrosome reproduction. J Cell Biol 148:405–416, 2000.
11. K Million, JC Larcher, J Laoukili, D Bourguignon, F Marano, F Tournier. Polyglutamylation and polyglycylation of alpha- and beta-tubulins during in vitro ciliated cell differentiation of human respiratory epithelial cells. J Cell Sci 112:4357–4366, 1999.
12. J Laoukili, E Perret, S Middendorp, O Houcine, C Guennou, F Marano, M Bornens, F Tournier. Differential expression and cellular distribution of centrin isoforms during human ciliated cell differentiation in vitro. J Cell Sci 113:1355–1364, 2000.
13. F Tournier, J Laoukili, I Giuliani, MC Gendron, C Guennou, F Marano. Ciliated differentiation of rabbit tracheal epithelial cells in vitro. Eur J Cell Biol 77:205–213, 1998.

14. AB Clark, SH Randell, P Nettesheim, TE Gray, B Bagnell, LE Ostrowski. Regulation of ciliated cell differentiation in cultures of rat tracheal epithelial cells. Am J Respir Cell Mol Biol 12:329–338, 1995.
15. L Kaartinen, P Nettesheim, KB Adler, SH Randell. Rat tracheal epithelial cell differentiation in vitro. In Vitro Cell Dev Biol 29A:481–492, 1993.
16. M Jorissen, B Van Der Schueren, H Van Der Berghe, JJ Cassiman. Contribution of in vitro culture methods for respiratory epithelial cells to the study of the physiology of the respiratory tract. Eur Respir J 4:210–217, 1991.
17. J Bara, E Chastre, J Mahiou, RL Singh, ME Forgue-Lafitte, E Hollande, F Godeau. Gastric M1 mucin, an early oncofetalmarker of colon carcinogenesis, is encoded by the MUC5AC gene. Int J Cancer 75:767–773, 1998.
18. G Grünig, et al. Requirement for IL-13 independently of IL-4 in experimental asthma. Science 18(282):2261–2263.
19. M Wills-Karp, J Luyimbazi, X Xu, B Schofield, TY Neben, CL Karp, DD Donaldson. Interleukin-13: central mediator of allergic asthma. Science. 18(282): 2258–2261.
20. Z Zhu, RJ Homer, Z Wang, Q Chen, GP Geba, J Wang, Y Zhang, JA Elias. Pulmonary expression of interleukin-13 causes inflammation, mucus hypersecretion, subepithelial fibrosis, physiologic abnormalities, and eotaxin production. J Clin Invest 103:779–788, 1999.
21. F Tournier, J Laoukili, D Caput, F Marano. Enhanced secreted phenotype durind human respiratory epithelial cell differentiation in vitro. Am J Respir Crit Care Med 159(3):A177, 1999.
22. F Tournier, M Bornens. Cell cycle regulation of centrosome function. In: JS Hyams, CW Lloyd, eds. *Microtubules*. New York: John Wiley & Sons, 1994, pp 303–324.
23. Y Bobinnec, A Khodjakov, LM Mir, CL Rieder, B Edde, M Bornens. Centriole disassembly in vivo and its effect on centrosome structure and function in vertebrate cells. J Cell Biol 143:1575–1589, 1998.
24. C Regnard, E Desbruyeres, P Denoulet, B Edde. Tubulin polyglutamylase: isozymic variants and regulation during the cell cycle in HeLa cells. J Cell Sci 112:4281–4289, 1999.
25. T Murata, J Taguchi, RK Puri, H Mohri. Sharing of receptor subunits and signal transduction pathway between the IL-4 and IL-13 receptor system. Int J Hematol 69:13–20, 1999.

Part II

Mucus

Mucus in the airways plays an important role in clearance as part of the mucociliary system. In a number of disease states, including chronic bronchitis, asthma, and cystic fibrosis, overproduction of mucus can lead to airway obstruction, microbial colonization and/or infection, and compromised defense. Excess mucus in the airways can result from any of three different lesions, and in most cases various combinations of these three: (1) excess production through overexpression of mucin (MUC) genes; (2) excess production secondary to mucus cell hyperplasia, hypertrophy, or even metaplasia; (3) hypersecretion of formed and stored mucin by goblet cells or glands in the airways.

The chapters in this part all deal with mucus, and most are concerned with mechanisms related to these three lesions. What is striking when reading these chapters as a group is the tremendous increase in the use of modern-day technologies and expansion of mucus studies to encompass novel, state-of-the-art mechanistic pathways. This is apparent if one refers to the previous edition, which resulted from the meeting in Jerusalem only a few years ago.

Thus, after a thorough but concise review of respiratory tract mucins provided by Davies and Carlstedt (Chapter 16), giving an overall background of where we stand in relation to mucin structure, biochemistry, gene expression, oligomerization, etc., the remaining chapters branch out into areas that must be looked at as new and exciting pathways to investigate with regard to mucin and mucus biology and role in disease. Davis and Randell (Chapter 18) review potential pathways of differentiation of goblet and submucosal gland mucous cells in the airways, investigating possible markers of differentiation of one cell type versus the other. Rush and Rogers (Chapter 22) and Martin et al. (Chapter 23) look at mechanisms of goblet cell hyperplasia in vivo and in vitro, respectively. As pointed out by Rush and Rogers, models of goblet cell hyperplasia exist for allergic diseases, including the ovalbumin-sensitized mouse and the IL-13–exposed mouse, but animal models showing mucin hypersecretion in COPD are lacking. Martin et al., utilizing a well-differentiated human airway epithelial cell culture system in which proliferation and mucus differentiation are recreated in vitro, look at signaling pathways leading to IL-13–induced mucus cell hyperplasia. This focus on signaling mechanisms is also a theme of other contributions. Kim et al. (Chapter 20) take a novel approach, looking at MUC1, a surface mucin on airway epithelium, as a potential adhesion site for *Pseudomonas aeruginosa*, suggesting that the cytoplasmic tail of the MUC1 acts as a receptor and activates a MAP kinase pathway in the epithelium, leading to enhanced inflammation and expression of inflammation-associated genes. Voynow and Fischer (Chapter 19) suggest that elastase induces expression of MUC5AC in human airway epithelium and does so via a pathway involving reactive oxygen species that leads ultimately to increased mRNA stability of the MUC5AC gene. Finally, Li et al. (Chapter 17) provide a background for the potential role of MARCKS protein (myristoylated alanine-rich C-kinase substrate) as a key regulatory molecule controlling secretion of mucin granules in the airway as part of a novel paradigm for exocytotic secretion in general.

Wu et al. (Chapter 21) present some new and highly exciting work related to use of microarray membranes to look at patterns of expression of a plethora of novel and potentially important genes related to human airway epithelial differentiation. Finally, Yu and Chang (Chapter 24) and King et al. (Chapter 25) look at some potential therapeutic and/or diagnostic possibilities related to mucus. Yu and Chang describe the potential use of mucins expressed by specific lung tumors as potential diagnostic/prognostic tools and describe quantitative competitive (QC) PCR as a way to evaluate expression of these genes in clinical specimens. King et al. present new data in which charged oligosaccharides, such as low molecular weight dextran sulfate or heparin, show potential as effective mucolytics when administered by aerosol.

Thus, if one were to try to bunch these papers into categories, as I have done somewhat above, or reads them as single free-standing contributions, one can only be impressed by the novelty and sophistication of the techniques and the mechanistic pathways that are being addressed. I personally started to study mucus over 30 years ago, and I continue to be impressed by the new levels of sophistication and scientific advances in this field, advances that seem to come logarithmically. Although the field of mucus and mucociliary interactions involves a relatively small number of investigators, it is apparent from these contributed papers that the field does not suffer from a lack of quality science.

Kenneth B. Adler

16

Respiratory Tract Mucins

Julia R. Davies and Ingemar Carlstedt
Lund University
Lund, Sweden

INTRODUCTION

The respiratory tract mucosa is constantly bombarded with microorganisms as well as other extraneous material and is thus a major potential site for infection and the entry of foreign antigens into the body. Normally, mucosal protection is provided by a surface ''ectomatrix,'' the airway epithelial cell layer and the mucosal immune system in the submucosal connective tissue. The ectomatrix is composed of cell-associated glycoproteins and secreted mucins forming a ''glycocalyx'' and a mucus gel, respectively. The latter interacts with the cilia to form the mucocilliary transport system. In addition, many proteins involved in host defense such as secretory IgA, lysozyme, lactoferrin, and β-defensins are enriched at the respiratory tract surface. The wide variety of protective agents at the mucosal surface would appear to allow for a degree of adaptability to changing requirements, but how the different components interact to provide optimal protection under different circumstances is currently poorly understood. The focus of this review is on the large mucus-forming mucins, although the cell-associated ones will also be considered.

THE MUCIN SUPERFAMILY

Mucins are characterized by the presence of highly glycosylated ''mucin-like domains,'' and under the broadest definition the family contains both secreted and transmembrane glycoproteins including selectin ligands and glycophorin (1). However, 13 genes encoding molecules with mucin domains are so far considered to belong to the mucin family. These, designated *MUC1–MUC4*, *MUC5AC*, *MUC5B*, *MUC6–MUC12*, are numbered in order of their description. Although the apoproteins encoded by the genes differ significantly in structure, each contains one or several regions rich in serine and/or threonine residues to which large numbers of oligosaccharide side chains are attached via O-glycosidic bonds. The mucin domains often contain tandemly repeated amino acid sequences that appear to be unique for each mucin. The number of repeats within the tandem repeat regions, and thus the total lengths of the glycosylated regions, vary between individuals, and this is referred to as VNTR polymorphism. The complete sequences for *MUC1*, *MUC2*, *MUC4*, *MUC5B*, and *MUC7* (2–8) and large stretches of *MUC5AC* (9–12), as well as the C-terminal sequences of *MUC3* and *MUC6* (13–15), are now known. On the basis of their structures, MUC2, MUC5AC, MUC5B, and MUC6 are predicted to be ''oligomeric'' mucins, while MUC1, MUC3, MUC4, and MUC7 are considered to be monomeric species. Most of the latter appear to be membrane-bound. The MUC2, MUC5AC, and MUC5B mucins contain cysteine-rich domains in the C- and N-terminal regions with sequence homologies to each other as well as to the D-domains of the von Willebrand factor (vWF). The genes encoding these mucins, as well as that coding for the MUC6 mucin, are found as a gene cluster on chromosome 11p15.5 and are thought to have arisen from a common ancestral gene (12,16).

The deduced sequences for the MUC1, MUC3, and MUC4 mucin apoproteins indicate the presence of transmembrane domains but lack cysteine-rich domains with homology to those in the mucus-forming mucins, and they are thus predicted to be cell-associated, monomeric structures. Recently, two ''novel'' mucin cDNAs encoding molecules with transmembrane domains (*MUC11* and *MUC12*) have been identified, and both co-localize with *MUC3* on chromosome 7q22 suggesting the presence of a cluster of cell-associated mucin genes in this region (17,18). MUC7 is a monomeric, secreted mucin.

In situ hybridization and/or immunohistochemistry have revealed the expression of *MUC1*, *MUC2*, *MUC4*, *MUC5AC*, *MUC5B*, *MUC7*, and *MUC8* genes in the airways (19–23). *MUC2* and *MUC5AC* are expressed in the epithelial goblet cells, *MUC1* and *MUC4* in the epithelial ciliated cells, and *MUC5B*, *MUC7*, and *MUC8* are associated with the submucosal glands. Northern blotting has also indicated *MUC11* expression in the lungs (17).

OLIGOMERIC MUCUS-FORMING MUCINS

Previously, it has been demonstrated that gel-forming mucins from the respiratory tract are large (M_r 10–30 million daltons), oligomeric structures formed by subunits (monomers) joined end-to-end via disulfide bonds (24,25). Using antibodies raised against a series of peptides found within the unique, nonglycosylated regions of the MUC2, MUC5AC, and MUC5B apoproteins, it has been possible to assign a genetic and biochemical identity to the respiratory mucins as well as to undertake a more detailed characterization of their nature.

MUC5AC

Partial cDNA clones encoding the MUC5AC mucin (Fig. 1) have been obtained from several sources including respiratory tract, nasal polyps, and gastric epithelium (9,11,26), and sequence information is available for both the C- and N-terminal regions as well as part of the tandem repeat region. The sequences suggest that, like MUC2 (see below), domains with homology to the D1, D2, and D3 domains of the vWF are present within the N-terminal region with a D4-like domain in the C-terminal domain. The D3 and D4 domains are separated by a central tandem repeat region containing repeats of the serine/threonine rich sequence, TTSTTSAP, interrupted by cysteine-rich sequences (9,27). Recently, studies on the promoter region of the *MUC5AC* gene have demonstrated the presence of putative binding sites for NFκB and Spl (28).

Using an antiserum (LUM5-1) raised against a synthetic peptide with the sequence RNQDQQGPFKMC, present at several sites within the MUC5AC apoprotein, MUC5AC has been identified as a large, oligomeric, gel-forming mucin in secretions from healthy and chronic bronchitic airways (29). Immunohistochemical studies using the same antiserum have shown that the MUC5AC mucin is produced primarily by the goblet cells in the airway surface epithelium (30).

MUC5B

The entire sequence of the *MUC5B* gene is now known and has been shown to contain 48 exons, which encode a mucin apoprotein of 5662 amino acids (7).

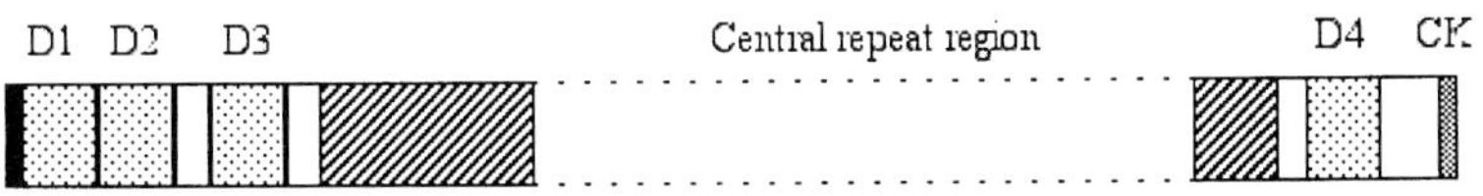

FIGURE 1 Schematic diagram of the domain structure of the MUC5AC apoprotein. D denotes regions with homology to the D domains of the von Willebrand factor. CK denotes the "cysteine knot" region. (Adapted from Ref. 11.)

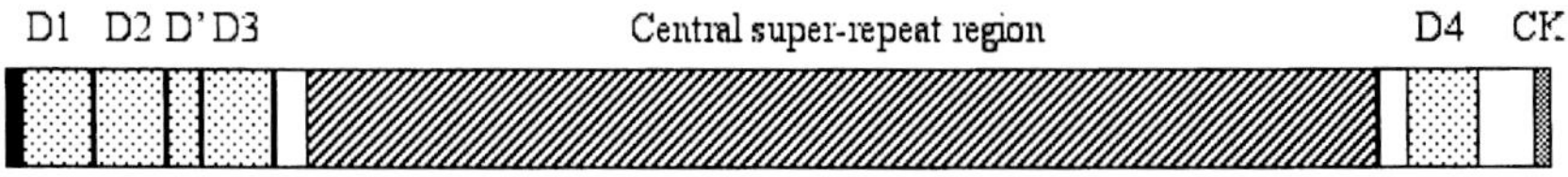

FIGURE 2 Schematic diagram of the domain structure of the MUC5B apoprotein. D denotes regions with homology to the D domains of the von Willebrand factor. CK denotes the "cysteine knot" region. (Adapted from Ref. 7.)

The N-terminal region of the molecule contains D-domains similar to those in the vWF and shows a high degree of homology with the N-terminal regions of both MUC5AC and MUC2. The whole central domain of the mucin, (Fig. 2), which is encoded within a single exon, contains four ''super-repeats,'' each consisting of cysteine-rich subdomains separated by serine/threonine-rich regions containing imperfect repeats of a 29-amino-acid sequence. The presence of five serine/threonine-rich regions, each giving rise to a highly glycosylated domain, is thus in good agreement with the structure previously predicted for the MUC5B molecule isolated from cervical secretions using biochemical and physical techniques (31).

Using the LUM5B-2 antiserum raised against a peptide sequence, RNREQVGKFKMC, present within the nonglycosylated regions in the central exon of MUC5B, it has been possible to identify this mucin as one of the major gel-forming species in secretions from both chronic bronchitic and cystic fibrotic individuals and to show that the glycoproteins originate mainly from the submucosal glands (32,33). In addition, MUC5B is a major component of cervical and gall bladder secretions as well as the MG1 mucin population from submandibular/sublingual saliva (32,34–37).

MUC2

MUC2 was the first mucin for which the cDNA was sequenced in its entirety, revealing a structure for the apoprotein containing domains with homology to the D1, D2, and D3 domains of vWF in the N-terminus and a D4-like domain in the C-terminal region (3). Between the D3 and the D4 domains, there is a serine/threonine/proline-rich repetitive region as well as a long tandem repeat region containing repeats of 23 amino acids: PTTTPITTTTTVTPTPTPTGTQT (Fig. 3).

Using the LUM2-3 antiserum raised against a peptide with the sequence NGLQPVRVEDPDGC found on the C-terminal side of the large tandem repeat domain of MUC2, the mucin has been identified as the dominant gel-forming mucin in secretions from human colon. Intestinal MUC2 is found as a glycoprotein complex, which cannot be solubilized by breaking noncovalent bonds and

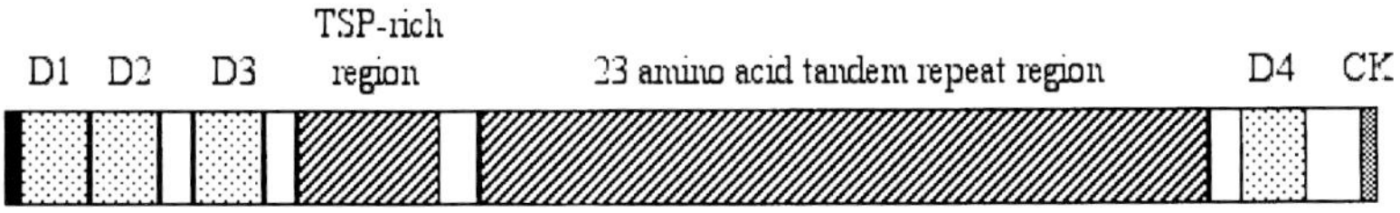

FIGURE 3 Schematic diagram of the domain structure of the MUC2 apoprotein. D denotes regions with homology to the D domains of the von Willebrand factor. CK denotes the "cysteine knot" region. (Adapted from Ref. 3.)

is referred to as the ''insoluble glycoprotein complex'' (38). Although evidence has been presented to demonstrate expression of MUC2 mRNA in the airways, only small amounts of MUC2 have been isolated biochemically from airway secretions (29,33).

MUCIN OLIGOMERIZATION

Treatment of MUC5AC, MUC5B, and MUC2 mucins with reducing agents leads to a reduction in molecular size showing that the molecules are oligomers composed of disulfide-bond–linked subunits. In vWF, disulfide-bond–mediated dimerization and oligomerization occur via the cysteine knot (CK) and D-domains, respectively (39), and the presence of sequences with homology to these motifs in the mucus-forming mucins has led to the suggestion that these proteins oligomerize in a similar fashion.

The MUC2 mucin has been identified as a mucin complex which cannot be solubilized by breaking noncovalent bonds with, for example, guanidinium chloride (38). Reduction of disulfide bonds in MUC2 generates discrete populations of ''subunits'' corresponding to monomers in addition to a series of oligomers of the primary gene product joined by a linkage that is insensitive to reduction (38). Although the presence of such a ''novel'' linkage has, as yet, only been demonstrated for the MUC2 molecule, a large proportion of the MUC5B mucins from submandibular/sublingual saliva (40), as well as MUC5AC secreted from the HT29-MTX cell-line, have also been shown to appear as ''insoluble complexes.''

CELL-ASSOCIATED MUCINS

In contrast to the MUC5AC, MUC5B, and MUC2 mucins, the products of the *MUC4*, *MUC8*, and *MUC11* genes have not yet been identified biochemically in airway secretions. From the sequences *MUC4*, like *MUC3*, is predicted to be a very large, monomeric mucin, which has a putative transmembrane domain and thus appears to be cell associated (4,13,14). The presence of EGF-like sequences within the C-terminal domains of both mucins, as well as in *MUC11* and *MUC12*, suggests a possible function for such molecules in cell signaling and growth (4,13,17,41), and the mucin-like subunit of the rat MUC4 homolog (SMC) has been identified as a ligand for the ErbB2 receptor tyrosine kinase (42). MUC3 occurs as different splice variants, possibly corresponding to membrane-associated and secreted forms of the mucin (14). The significance of cell-associated mucins for mucosal protection in the airways is not yet known, but it is possible that shedding of such molecules from the epithelium could play a role in preventing bacterial colonization at the mucosal surface. In addition, membrane-associated mucins may be important in regulating the properties of the mucus layer since CF mice that lack functional CFTR suffer from obstruction due to mucus in the intestinal lumen, an effect that can be abolished in mice that also lack the MUC1 mucin (43). Many studies have shown the release from airway tissues and cells of mucin-like macromolecules, which are distinctly different from the large, oligomeric species (see, e.g., Refs. 44,45), but the genetic nature of such material is currently unknown.

MUCIN GLYCOSYLATION

The large numbers of serine and threonine residues in mucin apoproteins represent a multitude of potential *O*-glycosylation sites, and mucins are characterized by stretches of dense glycosylation. *O*-glycans on mucins have been proposed to play an important role in mucosal defence by allowing bacterial binding via lectin-like structures—adhesins—on the bacterial surface. Bacterial binding to mucins may either lead to removal of potential pathogens or favour colonization if the secretions are not removed. In addition, a role in the recruitment and retention of inflammatory cells, e.g., neutrophils at the mucosal surface, has also been postulated (46). Several bacterial strains known to be present as colonizers and/or pathogens in the upper airways have been shown to bind to structures present in mucin oligosaccharides (for a review, see Ref. (47)). The discovery of distinct glycosylation variants or ''glycoforms'' of defined mucin apoproteins, even within a single tissue (32,48,49), suggests that mucin-producing cells express defined ''programs of glycosylation,'' which differ to the extent that such glycoforms can be recognized as discrete entities. Mucin glycosylation has been proposed to show changes associated with disease; for example, a relative increase in the Tn and sialyl-Tn antigen on mucins has been described in cancer (50,51), whereas a relative increase in mucin sulfation has been noted in cystic fibrosis (52). However, to date, no studies have conclusively shown changes in glycosylation connected to a specific secreted, oligomeric mucin apoprotein, and some degree of doubt still exists in this area. Possibly, changes in glycosylation associated with disease may influence the functioning of the mucosa in protection against microorganisms.

REGULATION OF MUCIN GENE EXPRESSION

Airway diseases such as cystic fibrosis, chronic bronchitis, and asthma are all characterized by hypersecretion of the large gel-forming mucins, and for this reason, many recent studies have focused upon the regulation of expression of the genes found within the 11p15.5 cluster. Many inflammatory mediators appear to cause mucin secretion, suggesting a link between mucin production and the immune system, but so far only a few substances, e.g., platelet-activating factor, IL-4, IL-9, and neutrophil elastase, have actually been shown to affect mucin gene transcription (53–56). The mechanisms controlling mucin gene transcription generally remain obscure, although an increase in MUC2 transcription in response to lipopolysaccharide from *Pseudomonas aeruginosa* has been demonstrated to occur through the activation of NFkB via the Src-Ras-MEK1/2-ERK1/2-pp90rsk pathway (28,56,57). Other airway irritants such as residual oil fly ash may also upregulate MUC5AC in a broadly similar manner (58,59). Studies on the promoter region of the MUC5AC gene demonstrate that, in addition to NFκB sites, Sp-1, GRE, and AP-2 sites are also present (28), suggesting a more complex regulation of gene activity. In addition to upregulation of specific genes, data from several laboratories have indicated that switching of transcription between different genes within the cluster may occur in disease. In colon cancer, for instance, normal expression of *MUC2* may be accompanied by *MUC5AC* expression, while in gastric cancer and/or intestinal metaplasia, *MUC5AC* and *MUC6* expression is replaced by that of *MUC2* and *MUC3* (60–62). Furthermore, we have recently observed that in airways showing an increased number of mucin-producing cells—possibly because of tobacco smoke exposure—goblet cells may express MUC5B and/or MUC2 in addition to MUC5AC (Fig. 4). Since both MUC2 and MUC5B can form insoluble complexes, such changes may drastically alter the physical properties—including "transportability"—of the mucus gel and thus enhance microbial colonization through binding to the mucins and inability to remove secretions.

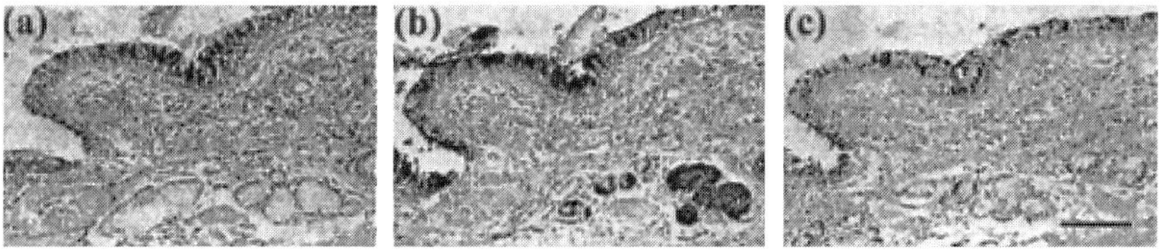

FIGURE 4 Serial sections (4 μm) of human trachea stained for (a) MUC5AC (LUM5-1 antiserum), (b) MUC5B (LUM5B-2 antiserum), or (c) MUC2 (LUM2-3 antiserum) and counterstained with Mayer's hematoxylin. The bar represents 150 μm.

PERSPECTIVES

The identification of the genes encoding at least 13 distinct mucin apoproteins has opened up possibilities for new approaches to the investigation of the roles played by these molecules in airway mucosal protection. Studies of individual mucin gene products at the transcriptional level will provide insight into how the production of each mucin is regulated and thus provide a basis for future studies of the molecular events leading up to mucus hypersecretion in airways disease. The expression of different mucin genes in the goblet cells and submucosal glands—cell types from which secretion appears to be differentially regulated—suggests that in vivo there may exist possibilities for regulating the properties of secretions to suit the prevailing conditions in the airways. A clearer understanding of how mucin synthesis and secretion are regulated, as well as which cells are active under different conditions, will provide a foundation upon which the development of new generations of drugs to manipulate these processes can be based.

ACKNOWLEDGMENTS

We thank the Swedish MRC (9711), the Medical Faculty of Lund, CFN, Council for Medical Tobacco Research, Swedish Match, Swedish Fund for Research Without Animal Experiments, Riksförbundet Cystisk Fibros, Alfred Österlunds Stiftelse, Crafoordska Stiftlesen, and Greta and Johan Kocks Stiftelse, the Swedish Foundation for Health Care Sciences and Allergy Research, and the Smokeless Tobacco Research Council, Inc., (USA) for financial support.

REFERENCES

1. KL Carraway, SR Hull. Cell surface mucin-type glycoproteins and mucin-like domains. Glycobiology 1:131–138, 1991.
2. SJ Gendler, CA Lancaster, J Taylor-Papadimitriou, T Duhig, N Peat, J Burchell, L Pemberton, E Lalani, D Wilson. Molecular cloning and expression of human tumor-associated polymorphic epithelial mucin. J Biol Chem 265:15286–15293, 1990.
3. JR Gum, JW Hicks, NW Toribara, B Siddiki, YS Kim. Molecular cloning of human intestinal mucin (MUC2) cDNA. Identification of the amino terminus and overall sequence similarity to prepro-von Willebrand factor. J Biol Chem 269:2440–2446, 1994.
4. N Moniaux, S Nollet, N Porchet, P Degand, A Laine, J-P Aubert. Complete sequence of the human mucin MUC4: a putative cell membrane-associated mucin. Biochem J 338:325–333, 1999.
5. J-L Desseyn, V Guyonnet-Dupérat, N Porchet, JP Aubert, A Laine. Human mucin gene *MUC5B*, the 10.7-kb large central exon encodes various alternate subdomains resulting in a super-repeat. J Biol Chem 272:3168–3178, 1997.
6. J-L Desseyn, J-P Aubert, I Van Seuningen, N Porchet, A Laine. Genomic organiza-

tion of the 3′ region of the human mucin gene *MUC5B*. J Biol Chem 272:16873–16883, 1997.
7. J-L Desseyn, M-P Buisine, N Porchet, J-P Aubert, A Laine. Genomic organization of the human mucin gene MUC5B. cDNA and genomic sequences upstream of the large central exon. J Biol Chem 273:30157–30164, 1998.
8. LA Bobek, H Tsai, AR Biesbrock, MJ Levine. Molecular cloning, sequence and specificity of expression of the gene encoding the low molecular weight salivary mucin (MUC7). J Biol Chem 268:20563–20569, 1993.
9. D Meerzaman, P Charles, E Daskal, MH Polymeropoulos, BM Martin, MC Rose. Cloning and analysis of cDNA encoding a major airway glycoprotein, human tracheobronchial mucin (MUC5). J Biol Chem 269:12932–12939, 1994.
10. T Lesuffleur, F Roche, AS Hill, M Lacasa, M Fox, DM Swallow, A Zweibaum, FX Real. Characterization of a mucin cDNA clone isolated from HT-29 mucus-secreting cells. J Biol Chem 270:13665–13673, 1995.
11. JHB Bovenkamp, CM van de Hau, GJAM Strous, HA Büller, J Dekker, AWC Einerhand. Molecular cloning of human gastric mucin MUC5AC reveals conserved cysteine-rich D-domains and a putative leucine zipper motif. Biochem Biophys Res Commun 245:853–859, 1998.
12. M-P Buisine, J-L Desseyn, N Porchet, P Degand, A Laine, J-P Aubert. Genomic organization of the 3′-region of the human *MUC5AC* mucin gene: additional evidence for a common ancestral gene for the 11p15.5 mucin gene family. Biochem J 332:729–738, 1998.
13. SJ Williams, DJ Munster, RJ Quin, DC Gotley, MA McGuckin. The MUC3 gene encodes a transmembrane mucin and is alternatively spliced. Biochem Biophys Res Commun 26:83–89, 1999.
14. SC Crawley, JR Gum, JW Hicks, WS Pratt, J-P Aubert, DM Swallow, YS Kim. Genomic organization and structure of the 3′ region of human *MUC3*: alternative splicing predicts membrane-bound and soluble forms of the mucin. Biochem Biophys Res Commun 263:728–736, 1999.
15. NW Toribara, AM Roberton, SB Ho, W-L Kuo, E Gum, JW Hicks, JR Gum, JC Byrd, B Siddiki, YS Kim. Human gastric mucin. Identification of a unique species by expression cloning. J. Biol Chem 266:5879–5885, 1993.
16. P Pigny, V Guyonnet Dupérat, AS Hill, WS Pratt, S Galiegue-Zouitina, MC d'Hooge, A Laine, I Van Seuningen, P Degand, JR Gum, YS Kim, DM Swallow, JP Aubert. Human mucin genes assigned to 11p15.5: identification and organization of a cluster of genes. Genomics 38:340–352, 1996.
17. SJ Williams, MA McGuckin, DC Gotley, HJ Eyre, GR Sutherland, TM Antalis. Two novel mucin genes down-regulated in colorectal cancer identified by differential display. Cancer Res 59:4083–4089, 1999.
18. MF Fox, F Lahbib, W Pratt, J Attwood, J Gum, Y Kim, DM Swallow. Regional localization of the intestinal mucin gene MUC3 to chromosome 7q22. Ann Hum Genet 56:281–287, 1992.
19. L Pemberton, J Taylor-Papidimitriou, S Gendler. Antibodies to the cytoplasmic domain of the MUC1 mucin show conservation throughout mammals. Biochem Biophys Res Commun 185:167–175, 1992.
20. A Dohrman, T Tsuda, E Escudier, M Cardone, B Jany, J Gum, Y Kim, C Basbaum.

Distribution of lysozyme and mucin (MUC2 and MUC3) mRNA in human bronchus. Exp Lung Res 20:367–380, 1994.
21. N Porchet, P Pigny, M-P Buisine, V Debailleul, P Degand, A Laine, J-P Aubert. Human mucin genes: genomic organization and expression of MUC4, MUC5AC and MUC5B. Biochem Soc Trans 23:800–805, 1995.
22. J-P Audie, A Janin, N Porchet, MC Copin, B Gosselin, J-P Aubert. Expression of human mucin genes in respiratory, digestive and reproductive tracts ascertained by in situ hybridization. J Histochem Cytochem 41:1479–1485, 1993.
23. V Shankar, P Pichan, RL Eddy, V Tonk, N Nowak, SN Sait, TB Shows, RE Schultz, G Gotway, RC Elkins, MS Gilmore, GP Sachdev. Chromosomal localization of a human mucin gene (MUC8) and cloning of the cDNA corresponding to the carboxy-terminus. Am J Respir Cell Mol Biol 16:232–241, 1997.
24. DJ Thornton, JR Davies, M Kraayenbrink, PS Richardson, JK Sheehan, I Carlstedt. Mucus glycoproteins from normal human tracheobronchial secretions. Biochem J 265:179–186, 1990.
25. JR Davies, HW Hovenberg, C-J Lindén, R Howard, PS Richardson, JK Sheehan, I Carlstedt. Mucins in airway secretions from healthy and chronic bronchitic subjects. Biochem J 313:431–439, 1995.
26. LWJ Klomp, L Van Rens, GJ Strous. Cloning and analysis of human gastric mucin cDNA reveals two types of conserved cysteine-rich domains. Biochem J 308:831–838, 1995.
27. V Guyonnet-Dupérat, J-P Audie, V Debailleul, A Laine, M-P Buisine, S Galiegue-Zouitina, P Pigny, P Degand, J-P Aubert, N Porchet. Characterization of the human mucin gene MUC5AC: a consensus cysteine-rich domain for the 11p15 mucin genes? Biochem J 305:211–219, 1995.
28. D Li, M Gallup, N Fan, DE Szymkowski, CB Basbaum. Cloning of the amino-terminal and 5′-flanking region of the human MUC5AC mucin gene and transcriptional up-regulation by bacterial exoproducts. J Biol Chem 273:6812–6820, 1998.
29. HW Hovenberg, JR Davies, I Carlstedt, MUC5AC, but not MUC2, is a prominent mucin in respiratory secretions. Glycoconjugate J 13:839–847, 1996.
30. HW Hovenberg, JR Davies, I Carlstedt. Different mucins are produced by the surface epithelium and the submucosa in human trachea: identification of MUC5AC as a major mucin from the goblet cells. Biochem J 318:319–324, 1996.
31. I Carlstedt, H Lindgren, JK Sheehan. The macromolecular structure of human cervical mucus glycoprotein. Biochem J 213:427–435, 1983.
32. C Wickström, JR Davies, G Eriksen, ECI Veerman, I Carlstedt. MUC5B is a major gel-forming, oligomeric mucin from human salivary gland, respiratory tract and endocervix: identification of glycoforms and C-terminal cleavage. Biochem J 334:685–693, 1998.
33. JR Davies, N Svitacheva, L Lannefors, R Kornfält, I Carlstedt. Identification of MUC5B, MUC5AC and small amounts of MUC2 in cystic fibrosis airway secretions. Biochem J 344:321–330, 1999.
34. AC Keates, DP Nunes, NH Afdhal, RF Troxler, GD Offner. Molecular cloning of a major human gall bladder mucin: complete C-terminal sequence and genomic organization of MUC5B. Biochem J 324:295–303, 1997.
35. PA Nielsen, EP Bennett, HH Wandall, MH Therkildsen, J Hannibal, H Clausen.

Identification of a major human high molecular weight salivary mucin (MG1) as tracheobronchial mucin MUC5B. Glycobiology 7:413–419, 1997.
36. RF Troxler, I Iontcheva, FG Oppenheim, DP Nunes, GD Offner. Molecular characterization of a high molecular weight mucin from human sublingual gland. Glycobiology 7:965–973, 1997.
37. DJ Thornton, N Khan, R Mehrotra, M Howard, E Veerman, NH Packer, JK Sheehan. Salivary mucin MG1 is comprised almost entirely of different glycosylated forms of the MUC5B gene product. Glycobiology 9:293–302, 1999.
38. A Herrmann, JR Davies, G Lindell, S Mårtensson, N Packer, DM Swallow, I Carlstedt. Studies on the 'insoluble' glycoprotein complex from human colon. J Biol Chem 274:15828–15836, 1999.
39. J Voorberg, R Fontijn, J Calafat, H Janssen, JA von Mourik, H Pannekoek. Assembly and routing of von Willeband factor variants: the requirements for disulphide-linked dimerization reside within the carboxy-terminal 151 amino acids. J Cell Biol 113: 195–205, 1991.
40. I Carlstedt, A Herrmann, HW Hovenberg, G Lindell, H Nordman, C Wickström, JR Davies. 'Soluble' and 'insoluble' mucins—identification of distinct populations. Biochem Soc Trans 23:845–851, 1995.
41. JR Gum, JJL Ho, WS Pratt, JW Hicks, AS Hill, LE Vinall, AM Roberton, DM Swallow, YS Kim. MUC3 human intestinal mucin. Analysis of gene structure, the carboxyl terminus, and a novel upstream repetitive region. J Biol Chem 272:26678–26686, 1997.
42. KL Carraway III, EA Rossi, M Komatsu, SA Price-Schiavi, D Huang, PM Guy, ME Carvajal, N Fregien, CA Carraway, KL Carraway. An intramembrane modulator of the ErbB2 receptor tyrosine kinase that potentiates neuregulin signaling. J Biol Chem 274:5263–5266, 1999.
43. RR Parmley, SJ Gendler. Cystic fibrosis mice lacking Muc1 have reduced amounts of intestinal mucus. J Clin Invest 102:1798–1806, 1998.
44. N Svitacheva, HW Hovenberg, JR Davies. Biosynthesis of mucins in bovine trachea: identification of the major radiolabelled species. Biochem J 333:449–456, 1998.
45. E Paul, DI Lee, SW Hyun, S Gendler, KC Kim. Identification and characterization of high molecular-mass mucin-like glycoproteins in the plasma membrane of airway epithelial cells. Am J Respir Cell Mol Biol 19:681–690, 1998.
46. A Prakobphol, K Tangemann, SD Rosen, CL Hoover, H Leffler, SJ Fisher. Separate oligosaccharide determinants mediate interactions of the low-molecular-weight salivary mucin with neutrophils and bacteria. Biochemistry 38:6817–6825, 1999.
47. A Scharfman, G Lamblin, P Roussel. Interactions between human respiratory mucins and pathogens. Biochem Soc Trans 23:836–839, 1995.
48. MR Price, A Ladanyi, K Uray, Y Ma, M Sekowski, LG Durrant. Separation of distinct MUC2 mucin glycoforms using two anti-peptide monoclonal antibodies. Int J Oncol 15:803–809, 1999.
49. DJ Thornton, M Howard, N Khan, JK Sheehan. Identification of two glycoforms of the MUC5B mucin in human respiratory mucus. J Biol Chem 272:9561–9566, 1997.
50. C Hanski, FG Hanisch, EO Riecken. Alteration of mucin-bound carbohydrate moieties in malignant transformation of colonic mucosa. Cancer J 5:332–342, 1992.

51. S Itzkowitz. Carbohydrate changes in colon carcinoma. APMIS Suppl 27:173–180, 1992.
52. Y Zhang, B Doranz, JR Yankaskas, JF Engelhardt. Genotypic analysis of respiratory mucous sulfation defects in cystic fibrosis. J Clin Invest 96:2997–3004, 1995.
53. YP Lou, K Takeyama, KM Grattan, JA Lausier, IF Ueki, C Agusti, JA Nadel. Platelet-activating factor induces goblet cell hyperplasia and mucin gene expression in the airways. Am J Respir Crit Care Med 157:1927–1934, 1998.
54. M Longphre, D Li, M Gallup, E Drori, CL Ordonez, T Redman, S Wenzel, DE Bice, JV Fahy, C Basbaum. Allergen-induced IL-9 directly stimulates mucin transcription in respiratory epithelial cells. J Clin Invest 104:1375–1382, 1999.
55. K Dabbagh, K Takeyama, HM Lee, IF Ueki, JA Lausier, JA Nadel. IL-4 induces mucin gene expression and goblet cell metaplasia in vitro and in vivo. J Immunol 162:6233–6237, 1999.
56. A Dohrman, S Miyata, M Gallup, JD Li, C Chapelin, A Coste, E Escudier, J Nadel, C Basbaum. Mucin gene (MUC2 and MUC5AC) upregulation by gram-positive and gram-negative bacteria. Biochim Biophys Acta 1406:251–259, 1998.
57. JD Li, W Feng, M Gallup, JH Kim, J Gum, Y Kim, C Basbaum. Activation of NF-1κB via a Src-dependent Ras-MAPK-pp90rsk pathway is required for *Pseudomonas aeruginosa*–induced mucin overproduction in epithelial cells. Proc Natl Acad Sci 95:5718–5723, 1998.
58. M Longphre, D Li, J Li, E Matovinovic, M Gallup, JM Samet, CB Basbaum. Lung mucin production is stimulated by the air pollutant residual oil fly ash. Toxicol Appl Pharmacol 162:86–92, 2000.
59. C Basbaum, H Lemjabbar, M Longphre, D Li, E Gensch, N McNamara. Control of mucin transcription by diverse injury-induced signalling pathways. Am J Crit Care Med 160:S44–S48, 1999.
60. SB Ho, LL Shekels, NW Toribara, YS Kim, C Lyftogt, DL Cherwitz, GA Niehans. Mucin gene expression in normal, preneoplastic, and neoplastic human gastric epithelium. Cancer Res 55:2681–2690, 1995.
61. J Bara, E Chastre, J Mahiou, RL Singh, ME Forgue-Lafitte, E Hollande, F Godeau. Gastric M1 mucin, an early oncofetal marker of colon carcinogenesis, is encoded by the MUC5AC gene. Int J Cancer 75:767–773, 1998.
62. CA Reis, L David, P Correa, F Carneiro, C de Bolos, E Garcia, U Mandel, H Clausen, M Sobrinho-Simoes. Intestinal metaplasia of human stomach displays distinct patterns of mucin (MUC1, MUC2, MUC5AC, and MUC6) expression. Cancer Res 59:1003–1007, 1999.

17

MARCKS Protein: A Potential Modulator of Airway Mucin Secretion

Yuehua Li, Linda D. Martin, and Kenneth B. Adler
North Carolina State University
Raleigh, North Carolina

INTRODUCTION

Mucus is a viscoelastic, gel-like substance produced and secreted by epithelium and submucosal glands in the mammalian respiratory, gastrointestinal, and reproductive tracts. It coats the epithelial surfaces to serve as a selective physical barrier between the extracellular milieu and the epithelial cell layer. Mucus consists of water, ions, mucin glycoproteins (mucins), and other secreted proteins including immunoglobulins; among them mucin is the major structural component and the basis for the viscous and gel-like property of mucus. In mammalian airways, the mucus layer serves as the first line of defense against inhaled pollutants and pathogens by trapping these foreign substances and clearing them out of airways via the mucociliary apparatus. Excess secretion of mucus can have deleterious effects on airways. It can lead to clogging of airways, compromising exchange of gases, increasing microbial infection, and even inducing severe inflammatory responses to cause destruction of airway walls and contiguous tissues. It has long been known that, in airway diseases such as asthma, chronic bronchitis, and cystic fibrosis, hypersecretion of mucus is a common lesion. Despite the obvious pathophysiological importance, mechanisms controlling secretion of airway mucin have not been elucidated.

MUCIN SECRETION

In order to understand the mechanisms controlling mucin secretion, a great deal of related research in the last two decades has focused on identifying mucin secretagogues and their corresponding signaling pathways in different species and model systems. As a result, a wide variety of agents and inflammatory/humoral mediators have been found capable of provoking mucin hypersecretion, including cholinergic agonists (1), lipid mediators (2,3), oxidants (3,4), cytokines (5), neuropeptides (6), purinergic agonists such as ATP and UTP (7–10), bacterial products (11), elastase (12), and inhaled pollutants (13,14). Interestingly, most of these secretagogues also are known as activators of protein kinases, especially protein kinase C (PKC) and cGMP-dependent protein kinase (protein kinase G, or PKG). Indeed, it has been found recurrently by many research groups that a variety of secretagogues induce mucin secretion via PKC or the nitric oxide (NO)-cGMP-PKG pathway (4,7–10,15–19). These findings have been remarkably consistent regardless of different experimental techniques and different species or model systems used. Given the apparent connection between mucin secretion and protein kinase activation elicited by various secretagogues, it appears that there could exist a universal signaling mechanism controlling mucin secretion, in which cellular substrates of either PKC and/or PKG could be pivotal, convergent molecules linking protein kinase activation to movement of mucin granules out of cells.

Recently, we have amassed evidence that MARCKS protein (myristoylated alanine-rich C kinase substrate) may be such a molecule. MARCKS is a substrate for PKC and is phosphorylated by activated PKC. Here we provide a review of MARCKS protein and a rationale for looking at MARCKS protein as a potentially important regulator of airway mucin secretion.

MARCKS PROTEIN

Myristoylated alanine-rich C kinase substrate, or MARCKS protein, is a widely distributed, specific cellular substrate for PKC. It is a myristoylated, membrane-associated protein, and also contains sites for calmodulin and actin filament binding. While the definitive functional role(s) of MARCKS is/are not yet clear, it has been implicated in a variety of biological processes that require actin cytoskeleton involvement, including cell motility, secretion, membrane trafficking, proliferation, and transformation (20–23).

Discovery

MARCKS protein was first described by Wu et al. in 1982 as the "87k protein" in rat brain synaptosomes (24). Shortly after this finding, Rozengurt and colleagues reported that this 80–87 kDa protein was a major substrate for PKC in Swiss 3T3 fibroblasts (25). In a later study, this group showed also that only agents stimulating phosphotidylinositol (PI) turnover would increase phosphorylation of this 80 kDa protein in intact cells, while cAMP-elevating agents, cGMP, or exogenous Ca^{2+} were unable to alter the phosphorylation state of the protein (26,27). In 1988, Aderem et al. reported that this PKC substrate was myristoylated and membrane-associated in mouse macrophages (28), and this observation has since been confirmed in a variety of cell types (29–31). In 1989, cDNA encoding the bovine "80–87 kDa" protein was cloned (32). The deduced primary sequence indicated that the protein was highly enriched in alanine and contained consensus sequences for amino-terminal myristoylation and PKC-dependent phosphorylation. Thus, this 80–87 kDa protein was named *M*yristoylated *A*lanine-*R*ich *C* *K*inase *S*ubstrate (MARCKS) (32).

Structure

Soon after bovine MARCKS cDNA was first isolated, cDNAs from other species were also cloned, including human (33,34), mouse (35,36), rat (37), and chicken (38), largely extending understanding of the biochemical features and structure of this protein. The deduced primary sequences reveal that MARCKS has an unusual amino acid composition highly rich in alanine, glycine, glutamic acid, and proline. There are no tyrosine or tryptophan residues in the protein. Alanine accounts for about 30% of the total residues. The only methionine is the initiator methionine that is cleaved in the process of amino-terminal myristoylation of the protein (39). Sedimentation and electron microscopy studies show that the protein has a flexible, rod-like shape (40), and this is thought to be responsible for its extremely anomalous migration in SDS-PAGE. MARCKS proteins from human, bovine, murine, and chicken have the sizes of 31.5, 31.9, 29.6, and 28.7 kDa, respectively, but migrate in SDS-PAGE with apparent molecular masses of 80, 87, 68, and 67 kDa (32–38).

Comparison of the primary sequences of MARCKS from different species reveals that MARCKS contains at least three distinct domains: an amino-terminal myristoylated domain that participates in anchoring the protein to the membrane (41), a conserved MH2 domain with unknown function, and a phosphorylation site domain (PSD) of 25 amino acid residues that contains the PKC phosphorylation sites, the calmodulin-binding site, and the actin-binding site (20). Remarkably, the conserved 25-amino-acid phosphorylation site domain is extremely basic, with a pI of 12.2, despite the fact that MARCKS is overall a very acidic protein with a determined pI of between 3.5 and 4.4 (23). Due to these various features, this 25-amino-acid sequence is also known as the basic effector domain.

PKC-Dependent Phosphorylation

Both in vivo and in vitro studies have shown MARCKS is a specific substrate for PKC, and all phosphorylation sites for PKC are clustered in the basic effector domain (26,27,42). Neither a synthetic peptide with the sequence identical to the PSD site (PSD peptide) nor the intact MARCKS is a good substrate for cAMP-dependent protein kinase (PKA), cGMP-dependent protein kinase (PKG), or calmodulin-dependent protein kinase I, II, and III (26,27). In addition, MARCKS protein contains no tyrosine residues, precluding it as a substrate for protein tyrosine kinases.

MARCKS protein from humans is phosphorylated at Ser-159, Ser-163, and Ser-170 when PKC is activated in intact cells. An additional serine residue (Ser-167), which is not in a good consensus sequence for PKC, also has been reported to be phosphorylated by PKC in vitro (43). Although MARCKS long has been known as a substrate for PKCα, its phosphorylation behavior as a substrate for other PKC isoenzymes has not been demonstrated until recent years. Studies show that diacylglycerol-activated PKCs, including conventional PKCs (α, βI, βII, γ) and novel PKCs (δ, ε, η, θ), can readily phosphorylate MARCKS in intact cells and in vitro; but the atypical PKCs (ζ, λ) do not appear to phosphorylate MARCKS in vivo or have significant affinity with MARCKS in vitro (44).

Membrane Association

Like other myristoylated, membrane-associated proteins, the myristate chain of MARCKS is absolutely required for its membrane binding. Mutational analysis in transfected cells has demonstrated that binding of MARCKS to the plasma membrane occurs via its N-terminal myristoylated domain (29). However, experiments with myristoylated peptides and purified MARCKS protein also show that the myristate chain provides, at maximum, barely enough hydrophobic energy to anchor a protein to a phospholipid bilayer (45). Thus, other factors must also contribute to the membrane-binding of MARCKS. A number of reports have demonstrated that, in addition to the N-terminal myristoylated domain, the basic effector domain is also involved in MARCKS-membrane attachment (46,47). It appears that association of MARCKS protein with the cytoplasmic surface of the plasma membrane requires two cooperative energy forces: the myristoyl moiety insertion into the hydrophobic interior of the lipid bilayer and electrostatic interaction between the basic effector domain and the polar head groups of acidic phospholipids of the plasma membrane. In response to this dual requirement for membrane binding, MARCKS cycles between the membrane and the cytoplasm dependent upon its phosphorylation state (31). In quiescent cells, MARCKS is not phosphorylated and is associated with the inner face of plasma membrane. When PKC is activated, phosphorylation of the serine residues in the basic effector domain introduces three negatively charged phosphate groups into this region, neutralizing the electrostatic interaction and consequently releasing MARCKS to the cytoplasm because the myristoyl moiety alone cannot confer sufficient membrane-binding energy to anchor the protein. Reversibly, MARCKS reassociates with the plasma membrane upon dephosphorylation in the cytoplasm by an okadaic acid–sensitive protein phosphatase(s), i.e., protein phosphatase 1 and/or 2A (48). This phosphorylation-dependent cycling of MARCKS between membrane and cytoplasm has been referred to as the ''myristoyl-electrostatic switch'' (49). It is thought to be one of the most striking characteristics of MARCKS protein and appears to be critical for its biological function(s) in cells.

Interactions with Calmodulin and Actin

The calmodulin-binding domain in MARCKS is identical to the phosphorylation site domain (50). Both the PSD peptide and intact MARCKS bind calmodulin with high affinity (51). The binding appears to be Ca^{2+}-dependent, as a decrease in intracellular Ca^{2+} concentration leads to loss of calmodulin binding (50). In addition, the MARCKS-calmodulin complex is rapidly disrupted by PKC-dependent phosphorylation (50). This phosphorylation-regulated calmodulin-binding ability of MARCKS has led to speculation that MARCKS may serve as a reservoir for calmodulin in intact cells. In resting cells, calmodulin is tethered to the plasma membrane through formation of a MARCKS-calmodulin complex. Upon PKC activation, MARCKS is phosphorylated, thereby disrupting the complex and releasing bound calmodulin to the cytoplasm. In this way, the concentration of cytoplasmic calmodulin is regulated by PKC through MARCKS phosphorylation. A similar role has been proposed for neuromodulin or GAP-43 (52,53). This idea is further supported by studies in cultured neuronal cells showing that activation of PKC results in an increase in cytosolic immunoreactive calmodulin, with a commensurate decrease in membrane-bound calmodulin (54,55). More intriguingly, since Ca^{2+}/calmodulin is the primary regulatory factor for activation of several isoforms of nitric oxide synthase (NOS) (56), MARCKS protein may also serve as a cross-bridge integrating the PKC and NO-cGMP-PKG pathways.

Electron microscopy with negative staining, and dynamic light scattering studies, have shown that nonphosphorylated MARCKS can bind to the sides of actin filaments and crosslink them (57). MARCKS phosphorylated by PKC in vitro can still bind actin filaments, but less tightly, and it cannot crosslink them. In addition, the actin crosslinking effect of nonphosphorylated MARCKS is inhibited by Ca^{2+}/calmodulin (57). The synthetic PSD peptide displays the same crosslinking capacity. Nonphosphorylated PSD peptide increases the light-scatter intensity of actin in solution and aggregates actin filaments into tight bundles observed by electron microscopy, while phosphorylation of serine residues in the peptide results in a loss of crosslinking ability of the peptide. Again, Ca^{2+}/calmodulin, at a 1:1 molar ratio to the peptide, completely inhibits the actin-crosslinking activity of the nonphosphorylated peptide (57). Taken together, the phosphorylation, calmodulin-binding, and actin-binding sites are all located within a single domain, and only one of the three events can occur at any one time. The structural basis of this reciprocal interaction is not clear. It has been suggested that the five lysine residues within the 25-amino-acid basic effector domain may serve as sites for calmodulin and actin filament binding, and phosphorylation introduces negatively charged phosphate groups into this region, thereby influencing the positively charged lysine residues and consequently disrupting the binding. Another intriguing speculation about MARCKS-actin interaction is that the crosslinking of actin filaments requires MARCKS protein either to have two actin-binding sites or to form a homodimer; phosphorylation or Ca^{2+}/calmodulin may either inactivate one actin-binding site or cause dissociation of the dimer (58). To date, the exact mechanism is not known.

Intracellular Targeting

Blackshear et al. have shown that neither boiling nor extensive trypsin digestion of a cellular membrane preparation affects associations between the membrane and MARCKS protein, suggesting the membrane-binding of MARCKS does not require a protein mediator or receptor (59). However, there is also strong evidence suggesting that protein-protein interaction is involved in specific membrane distributions of MARCKS in intact cells. For example, in mouse embryo fibroblasts, PKC-dependent phosphorylation regulates cycling of MARCKS between the plasma membrane and Lamp-1-positive lysosomes (60). In resting fibroblasts, MARCKS is predominantly associated with the plasma membrane. Activation of PKC results in MARCKS release from plasma membrane and redistribution to Lamp-1-positive lysosomes. This specific translocation appears to be regulated by an intracellular intermediate but not by the usual endocytic pathway, since lysosomotropic agents and microtubule destabilizing agents (such as NH_4Cl and nocodazole) promote accumulation of MARCKS on lysosomes despite their demonstrated effect on inhibition of endosome/lysosome fusion (61). Furthermore, in mouse macrophages, MacMARCKS, a MARCKS-like protein that has similar biochemical features and domain structure to those of MARCKS, also cycles between lysosomes and plasma membrane when the cells encounter aggregated bacterial lipopolysaccharides and following activation of PKC (22). In contrast to MARCKS in fibroblasts, PKC-dependent phosphorylation drives MacMARCKS from lysosomes to the plasma membrane in macrophages. Since MARCKS and MacMARCKS have almost identical myristoyl moiety and basic effector domains

(22), it appears that additional protein-protein interaction(s) is/are involved in the opposite subcellular redistribution of MARCKS and MacMARCKS, which would provide specificity for membrane-targeting.

More directly, $p60^{v\text{-}src}$, a myristoylated membrane-bound phosphoprotein that has similar membrane-binding domain to MARCKS, has been demonstrated to associate with the membrane via a 32 kDa receptor protein (62). In addition, Manenti et al. have reported that in the process of purification of MARCKS protein from the plasma membrane and the cytoplasmic fractions of calf brain, two preparations are eluted at markedly different positions from the gel filtration column (63). MARCKS from the cytoplasmic fraction is eluted much earlier than that from the membrane fraction. The determined Stokes radii are 85 and 45 Å for the cytoplasmic MARCKS and the membrane MARCKS, respectively. When this cytoplasmic MARCKS preparation is further purified using a calmodulin-affinity column and the purified MARCKS is then subjected to the same gel filtration column under the same conditions, the protein is eluted at a similar position to that of membrane MARCKS (63). The circular dichroism of the two MARCKS preparations shows that there is not a significant difference in the overall conformation, eliminating the possibility that the difference in Stokes radius is caused by conformational change during the purification. This also indicates that the cytoplasmic MARCKS is associated with a protein factor present in the cytoplasmic fraction that is removed by calmodulin-affinity chromatography. Furthermore, Aderem et al. (46) have shown, via a transfection study, that a mutant form of MARCKS, in which the intervening sequence between the myristoyl moiety and the basic effector domain is deleted, loses its specificity of membrane distribution. This mutated MARCKS can associate with a variety of intracellular membranes, and PKC activation does not influence its subcellular distribution (46), suggesting that the deleted intervening sequence may contain a motif involved in specific membrane targeting.

Expression

Although MARCKS is widely distributed in a variety of cell types and tissues, its expression is highly regulated. MARCKS is found in high levels in brain and inflammatory cells and is absent or barely detectable in skeletal muscle and liver of adult animals (64). MARCKS expression varies dramatically during the course of development (65). In neutrophils and macrophages, both mRNA and protein levels of MARCKS are increased acutely in response to bacterial lipopolysaccharide (LPS) and the inflammatory cytokine TNF-α, the two known major ''priming'' agents for phagocytes. Even more remarkably, over 90% of all newly synthesized proteins induced by either TNF-α or LPS in these cells is MARCKS (28,35,39). MARCKS genomic clones have been isolated and sequenced from human and mouse (66). The promoter regions of the two genes have a very high degree of sequence identity with at least 59 potential transcription factor binding sites in analogous locations (66). Both promoters lack a TATA box. In human MARCKS gene (MACS), a minimum of 213 base pairs upstream of the transcription initiation site is required for significant promoter activity (67). Further upstream sequences seem necessary for its tight regulation or enhanced transcription activity. For example, a potential binding site for NF-κB at site −1196 may be important for TNF-α–stimulated transcription. To date most studies indicate that cellular levels of MARCKS protein are tightly correlated to its mRNA levels (32). Additionally the mRNA is quite stable, at least in the cell lines tested, with a half-life of 4–6 hours (67), suggesting that regulation of expression occurs mainly at the level of gene transcription. However, the 5′-untranslated region (UTR) of MARCKS mRNA is predicted to form a stable secondary structure with ΔG −73.4. Similar UTR structures have been known to mediate regulation of translation or mRNA stability in proteins such as ferritin and the transferrin receptor (68,69). So far, no evidence supports the concept that MARCKS expression is regulated at the translational level.

POTENTIAL FUNCTIONAL ROLES OF MARCKS

The proposed functional roles of MARCKS protein are mainly based on its regulation of membrane-actin interactions and the tight correlation between its phosphorylation and/or synthesis and cellular responses observed. Direct evidence for its physiological function(s) in cells has not been established.

Motility

In macrophages, MARCKS has a punctate distribution and co-localizes with vinculin, talin, and PKC-α at the substrate-adherent surface of pseudopodia and filopodia. Activation of PKC by phorbol esters causes marked cell spreading and rounding that is accompanied by almost complete displacement of MARCKS from these focal contacts (70). Immunohistochemical studies at the EM level have further confirmed that MARCKS is clustered at points where actin filaments interact with the cytoplasmic surface of the plasma membrane (22). These observations have led Aderem et al. to propose that MARCKS might provide a PKC-sensitive regulation point for reversible interaction between the actin cytoskeleton and the plasma membrane, since a reversible association of actin with the membrane is a prerequisite for directional cellular locomotion during macrophage chemotaxis. In addition, as discussed previously, LPS and TNF-α greatly enhance synthesis of MARCKS in neutrophils and macrophages, and MARCKS constitutes over 90% of newly synthesized proteins induced by either LPS or TNF-α. Concomitantly, LPS and TNF-α prompt a marked increase in the number and prominence of lamellipodia, filopodia, and membrane veils in mouse macrophages (20). Furthermore, results of immunofluorescence microscopy and electron microscopy studies have shown that the total staining of membrane-bound MARCKS, as well as the staining density of MARCKS in the clusters, increases significantly after the macrophages are primed with LPS (20).

Secretion and Membrane Trafficking

MARCKS is found in high concentration in presynaptic junctions and is phosphorylated when synaptosomes are depolarized, coincident with neurotransmitter release provoked by membrane depolarization (71). In addition, arginine vasopressin causes a PKC-dependent increase in adrenocorticotropin secretion in ovine anterior pituitary cells, and this is correlated with an increase in MARCKS-phosphorylation (72). Again, both TNF-α and LPS induce the synthesis and phosphorylation of MARCKS (73), and concomitantly promote PKC-dependent secretion of inflammatory mediators and cytokines by macrophages and neutrophils. In guinea pig gastric chief cells, agents that stimulate pepsinogen secretion also lead to the phosphorylation of a MARCKS-like protein by activated PKC (74). To date, however, relationships between MARCKS phosphorylation and secretion have been correlative.

Proliferation and Transformation

It is reported that both the phosphorylation and protein levels of MARCKS are significantly reduced after NIH 3T3 fibroblasts are transformed to the tumorigenic state by the Ha-ras, Ki-v-ras, v-src, or v-fms oncogenes (75,76). Similar downregulation is also observed in other types of cell lines transformed with oncogenes or chemical agents (77,78). In addition, MARCKS protein levels are increased markedly in murine Swiss 3T3 fibroblasts when the cells move out of cycle and enter the G_0 state (79). More recently, Brooks et al. have shown that levels of MARCKS mRNA and protein are significantly lower in the spontaneously derived murine B16 melanoma cell line compared to syngeneic normal Mel-ab melanocytes (80). Transfection of B16 melanoma cells with MARCKS cDNA in the sense orientation leads to a growth suppression and a decreased anchorage-independent growth, whereas cells transfected with antisense MARCKS cDNA display enhanced growth and transforming potential compared to control cells transfected with vector alone (80). These results strongly suggest that MARCKS may function as a growth suppressor or have a role in tumor progression.

CONCLUSIONS

Since MARCKS protein was identified as a specific in vitro and in vivo substrate for PKC over a decade ago, it has been intensively studied. Some detailed understanding of this protein has been obtained, particularly its structure, PKC-dependent phosphorylation, and interactions with calmodulin and actin. However, many aspects of MARCKS function still remain obscure. We are investigating a potential role for MARCKS protein in the process of airway mucin secretion and have determined preliminarily that it is involved in the secretory pathway and could be a major convergent molecule in intracellular signaling leading to mucin granule transport and exocytosis in human goblet cells in vitro (81). Studies are in progress in our laboratory to elucidate the exact mechanism whereby MARCKS protein regulates secretion of airway mucin.

ACKNOWLEDGMENTS

This work was supported in part by grants #HL 36982 and 09689 from NIH.

REFERENCES

1. LW Chakrin, AP Baker, P Christian, JR Wardell Jr. Effect of cholinergic stimulation on the release of macromolecules by canine trachea in vitro. Am Rev Respir Dis 108:69–76, 1973.
2. KB Adler, NJ Akley, WC Glasgow. Platelet-activating factor provokes release of mucin-like glycoproteins from guinea pig respiratory epithelial cells via a lipoxygenase-dependent mechanism. Am J Respir Cell Mol Biol 6:550–556, 1992.
3. KB Adler, WJ Holden-Stauffer, JE Repine. Oxygen metabolites stimulate release of high-molecular-weight glycoconjugates by cell and organ cultures of rodent respiratory epithelium via an arachidonic acid-dependent mechanism. J Clin Invest 85: 75–85, 1990.
4. DT Wright, BM Fischer, C Li, LG Rochelle, NJ Akley, KB Adler. Oxidant stress stimulates mucin secretion and PLC in airway epithelium via a nitric oxide-dependent mechanism. Am J Physiol 271 (5 Pt 1):L854–L861, 1996.
5. SJ Levine, C Logun, P Larivee, JH Shelhamer. IL-1β induces secretion of respiratory mucous glycoprotein from human airways in vitro. Am Rev Respir Dis 147:A437, 1993.
6. SE Webber. The effects of peptide histidine isoleucine and neuropeptide Y on mucus volume output from the ferret trachea. Br J Pharmacol 95:49–54, 1988.
7. KC Kim, QX Zheng, I Van-Seuningen. Involvement of a signal transduction mechanism in ATP-induced mucin release from cultured airway goblet cells. Am J Respir Cell Mol Biol 8:121–125, 1993.
8. LH Abdullah, JD Conway, JA Cohn, CW Davis. Protein kinase C and Ca^{2+} activation of mucin secretion in airway goblet cells. Am J Physiol 273 (1 Pt 1):L201–L210, 1997.
9. LH Abdullah, SW Davis, L Burch, M Yamauchi, SH Randell, P Nettesheim, CW Davis. P2u purinoceptor regulation of mucin secretion in SPOC1 cells, a goblet cell line from the airways. Biochem J 316:943–951, 1996.
10. HA Brown, ER Lazarowski, RC Boucher, TK Harden. Evidence that UTP and ATP regulate phospholipase C through a common extracellular 5′-nucleotide receptor in human airway epithelial cells. Mol Pharmacol 40:648–655, 1991.

11. KB Adler, DD Hendley, GS Davis. Bacteria associated with obstructive pulmonary disease elaborate extracellular products that stimulate mucin secretion by explants of guinea pig airways. Am J Pathol 125:501–514, 1986.
12. KC Kim, K Wasano, RM Niles, JE Schuster, PJ Stone, JS Brody. Human neutrophil elastase releases cell surface mucins from primary cultures of hamster tracheal epithelial cells. Proc Natl Acad Sci USA 84:9304–9308, 1987.
13. JA Last, MD Jennings, LW Schwartz, CE Cross. Glycoprotein secretion by tracheal explants cultured from rats exposed to ozone. Am Rev Respir Dis 116:695–703, 1977.
14. B Mossman, KB Adler, L Jean, J Craighead. Mechanisms of hypersecretion in rodent tracheal explants after exposure to chrysotile asbestos: studies using lectins. Chest 81:23S–25S, 1982.
15. BM Fischer, LG Rochelle, JA Voynow, NJ Akley, KB Adler. Tumor necrosis factor-alpha stimulates mucin secretion and cyclic GMP production by guinea pig tracheal epithelial cells in vitro. Am J Respir Cell Mol Biol 20:413–422, 1999.
16. SJ Levine, P Larivee, C Logun, CW Angus, FP Ognibene, JH Shelhamer. Tumor necrosis factor-alpha induces mucin hypersecretion and MUC-2 gene expression by human airway epithelial cells. Am J Respir Cell Mol Biol 12:196–204, 1995.
17. KH Ko, M Jo, K McCracken, KC Kim. ATP-induced mucin release from cultured airway goblet cells involves, in part, activation of protein kinase C. Am J Respir Cell Mol Biol 16:194–198, 1997.
18. KB Adler, BM Fischer, H Li, NH Choe, DT Wright. Hypersecretion of mucin in response to inflammatory mediators by guinea pig tracheal epithelial cells in vitro is blocked by inhibition of nitric oxide synthase. Am J Respir Cell Mol Biol 13: 526–530, 1995.
19. P Larivee, SJ Levine, A Martinez, T Wu, C Logun, JH Shelhamer. Platelet-activating factor induces airway mucin release via activation of protein kinase C: evidence for translocation of protein kinase C to membranes. Am J Respir Cell Mol Biol 11: 199–205, 1994.
20. AC Nairn, A Aderem. Calmodulin and protein kinase C cross-talk: the MARCKS protein is an actin filament and plasma membrane cross-linking protein regulated by protein kinase C phosphorylation and by calmodulin. Ciba Found Symp 164: 145–161, 1992.
21. PJ Blackshear. The MARCKS family of cellular protein kinase C substrates. J Biol Chem 268:1501–1504, 1993.
22. A Aderem. The MARCKS family of protein kinase-C substrates. Biochem Soc Trans 23:587–591, 1995.
23. G Brooks. The role of 80K/MARCKS, a specific substrate of protein kinase C, in cell growth and tumor progression. Pigment Cell Res 7:451–457, 1994.
24. WC Wu, SI Walaas, AC Nairn, P Greengard. Calcium/phospholipid regulates phosphorylation of a Mr ''87k'' substrate protein in brain synaptosomes. Proc Natl Acad Sci USA 79:5249–5253, 1982.
25. E Rozengurt, M Rodriguez-Pena, KA Smith. Phorbol esters, phospholipase C, and growth factors rapidly stimulate the phosphorylation of a Mr 80,000 protein in intact quiescent 3T3 cells. Proc Natl Acad Sci USA 80:7244–7248, 1983.

26. A Rodriguez-Pena, E Rozengurt. Phosphorylation of an acidic mol. wt. 80,000 cellular protein in a cell-free system and intact Swiss 3T3 cells: a specific marker of protein kinase C activity. EMBO J 5:77–83, 1986.
27. JD Erusalimsky, I Friedberg, E Rozengurt. Bombesin, diacylglycerols, and phorbol esters rapidly stimulate the phosphorylation of an Mr = 80,000 protein kinase C substrate in permeabilized 3T3 cells. Effect of guanine nucleotides. J Biol Chem 263:19188–19194, 1988.
28. AA Aderem, KA Albert, MM Keum, JK Wang, P Greengard, ZA Cohn. Stimulus-dependent myristoylation of a major substrate for protein kinase C. Nature 332:362–364, 1988.
29. JM Graff, JI Gordon, PJ Blackshear. Myristoylated and nonmyristoylated forms of a protein are phosphorylated by protein kinase C. Science 246:503–506, 1989.
30. G James, EN Olson. Myristoylation, phosphorylation, and subcellular distribution of the 80-kDa protein kinase C substrate in BC3H1 myocytes. J Biol Chem 264: 20928–20933, 1989.
31. M Thelen, A Rosen, AC Nairn, A Aderem. Regulation by phosphorylation of reversible association of a myristoylated protein kinase C substrate with the plasma membrane. Nature 351:320–322, 1991.
32. DJ Stumpo, JM Graff, KA Albert, P Greengard, PJ Blackshear. Molecular cloning, characterization, and expression of a cDNA encoding the ''80- to 87-kDa'' myristoylated alanine-rich C kinase substrate: a major cellular substrate for protein kinase C. Proc Natl Acad Sci USA 86:4012–4016, 1989.
33. DM Harlan, JM Graff, DJ Stumpo, RL Eddy Jr, TB Shows, JM Boyle, PJ Blackshear. The human myristoylated alanine-rich C kinase substrate (MARCKS) gene (MACS). Analysis of its gene product, promoter, and chromosomal localization. J Biol Chem 266:14399–14405, 1991.
34. K Sakai, M Hirai, J Kudoh, S Minoshima, N Shimizu. Molecular cloning and chromosomal mapping of a cDNA encoding human 80K-L protein: major substrate for protein kinase C. Genomics 14:175–178, 1992.
35. JT Seykora, JV Ravetch, A Aderem. Cloning and molecular characterization of the murine macrophage ''68-kDa'' protein kinase C substrate and its regulation by bacterial lipopolysaccharide. Proc Natl Acad Sci USA 88:2505–2509, 1991.
36. SF Brooks, T Herget, JD Erusalimsky, E Rozengurt. Protein kinase C activation potently down-regulates the expression of its major substrate, 80K, in Swiss 3T3 cells. EMBO J 10:2497–2505, 1991.
37. JD Erusalimsky, SF Brooks, T Herget, C Morris, E Rozengurt. Molecular cloning and characterization of the acidic 80-kDa protein kinase C substrate from rat brain. Identification as a glycoprotein. J Biol Chem 266:7073–7080, 1991.
38. JM Graff, DJ Stumpo, PJ Blackshear. Molecular cloning, sequence, and expression of a cDNA encoding the chicken myristoylated alanine-rich C kinase substrate (MARCKS). Mol Endocrinol 3:1903–1906, 1989.
39. M Thelen, A Rosen, AC Nairn, A Aderem. Tumor necrosis factor alpha modifies agonist-dependent responses in human neutrophils by inducing the synthesis and myristoylation of a specific protein kinase C substrate. Proc Natl Acad Sci USA 87: 5603–5607, 1990.
40. KA Albert, AC Nairn, P Greengard. The 87-kDa protein, a major specific substrate

for protein kinase C: purification from bovine brain and characterization. Proc Natl Acad Sci USA 84:7046–7050, 1987.
41. DA Towler, JI Gordon, SP Adams, L Glaser. The biology and enzymology of eukaryotic protein acylation. Annu Rev Biochem 57:69–99, 1988.
42. PJ Blackshear, LA Witters, PR Girard, JF Kuo, SN Quamo. Growth factor-stimulated protein phosphorylation in 3T3-L1 cells. Evidence for protein kinase C-dependent and -independent pathways. J Biol Chem 260:13304–13315, 1985.
43. JM Graff, RR Rajan, RR Randall, AC Nairn, PJ Blackshear. Protein kinase C substrate and inhibitor characteristics of peptides derived from the myristoylated alanine-rich C kinase substrate (MARCKS) protein phosphorylation site domain. J Biol Chem 266:14390–14398, 1991.
44. F Uberall, S Giselbrecht, K Hellbert, F Fresser, B Bauer, M Gschwendt, HH Grunicke, G Baier. Conventional PKC-alpha, novel PKC-epsilon and PKC-theta, but not atypical PKC-lambda are MARCKS kinases in intact NIH 3T3 fibroblasts. J Biol Chem 272:4072–4078, 1997.
45. J Kim, T Shishido, X Jiang, A Aderem, S McLaughlin. Phosphorylation, high ionic strength, and calmodulin reverse the binding of MARCKS to phospholipid vesicles. J Biol Chem 269:28214–28219, 1994.
46. JT Seykora, MM Myat, LAH Allen, JV Ravetch, A Aderem. Molecular determinants of the myristoyl-electrostatic switch of MARCKS. J Biol Chem 271:18797–18802, 1996.
47. SL Swierczynski, PJ Blackshear. Membrane association of the myristoylated alanine-rich C kinase substrate (MARCKS) protein. Mutational analysis provides evidence for complex interactions. J Biol Chem 270:13436–13445, 1995.
48. PR Clarke, SR Siddhanti, P Cohen, PJ Blackshear. Okadaic acid-sensitive protein phosphatases dephosphorylate MARCKS, a major protein kinase C substrate. FEBS Lett 336:37–42, 1993.
49. S McLaughlin, A Aderem. The myristoyl-electrostatic switch: a modulator of reversible protein-membrane interactions. Trends Biochem Sci 20:272–276, 1995.
50. JM Graff, TN Young, JD Johnson, PJ Blackshear. Phosphorylation-regulated calmodulin binding to a prominent cellular substrate for protein kinase C. J Biol Chem 264:21818–21823, 1989.
51. BK McIlroy, JD Walters, PJ Blackshear, JD Johnson. Phosphorylation-dependent binding of a synthetic MARCKS peptide to calmodulin. J Biol Chem 266:4959–4964, 1991.
52. JH Skene. Axonal growth-associated proteins. Annu Rev Neurosci 12:127–156, 1989.
53. KA Alexander, BM Cimler, KE Meier, DR Storm. Regulation of calmodulin binding to P-57. A neurospecific calmodulin binding protein. J Biol Chem 262:6108–6113, 1987.
54. LA Mangels, ME Gnegy. Muscarinic receptor-mediated translocation of calmodulin in SK-N-SH human neuroblastoma cells. Mol Pharmacol 37:820–826, 1990.
55. M MacNicol, H Schulman. Cross-talk between protein kinase C and multifunctional Ca^{2+}/calmodulin-dependent protein kinase. J Biol Chem 267:12197–12201, 1992.
56. JC Salerno, DE Harris, K Irizarry, B Patel, AJ Morales, SM Smith, P Martasek, LJ Roman, BS Masters, CL Jones, BA Weissman, P Lane, Q Liu, SS Gross. An

autoinhibitory control element defines calcium-regulated isoforms of nitric oxide synthase. J Biol Chem 272:29769–29777, 1997.

57. JH Hartwig, M Thelen, A Rosen, PA Janmey, AC Nairn, A Aderem. MARCKS is an actin filament crosslinking protein regulated by protein kinase C and calcium-calmodulin. Nature 356:618–622, 1992.
58. A Aderem. The MARCKS brothers: a family of protein kinase C substrates. Cell 71:713–716, 1992.
59. DJ George, PJ Blackshear. Membrane association of the myristoylated alanine-rich C kinase substrate (MARCKS) protein appears to involve myristate-dependent binding in the absence of a myristoyl protein receptor. J Biol Chem 267:24879–24885, 1992.
60. LA Allen, A Aderem. Protein kinase C regulates MARCKS cycling between the plasma membrane and lysosomes in fibroblasts. EMBO J 14:1109–1120, 1995.
61. J Gruenberg, G Griffiths, KE Howell. Characterization of the early endosome and putative endocytic carrier vesicles in vivo and with an assay of vesicle fusion in vitro. J Cell Biol 108:1301–1316, 1989.
62. MD Resh, HP Ling. Identification of a 32K plasma membrane protein that binds to the myristoylated amino-terminal sequence of p60v-src. Nature 346:84–86, 1990.
63. S Manenti, O Sorokine, A Van Dorsselaer, H Taniguchi. Affinity purification and characterization of myristoylated alanine-rich protein kinase C substrate (MARCKS) from bovine brain. Comparison of the cytoplasmic and the membrane-bound forms. J Biol Chem 267:22310–22315, 1992.
64. KA Albert, SI Walaas, JK Wang, P Greengard. Widespread occurrence of ''87 kDa'', a major specific substrate for protein kinase C. Proc Natl Acad Sci USA 83: 2822–2826, 1986.
65. PJ Blackshear, WS Lai, JS Tuttle, DJ Stumpo, E Kennington, AC Nairn, KK Sulik. Developmental expression of MARCKS and protein kinase C in mice in relation to the exencephaly resulting from MARCKS deficiency. Brain Res Dev Brain Res 96: 62–75, 1996.
66. PJ Blackshear, JS Tuttle, RJ Oakey, MF Seldin, M Chery, C Philippe, DJ Stumpo. Chromosomal mapping of the human (MACS) and mouse (Macs) genes encoding the MARCKS protein. Genomics 14:168–174, 1992.
67. DM Harlan, JM Graff, DJ Stumpo, RL Eddy Jr, TB Shows, JM Boyle, PJ Blackshear. The human myristoylated alanine-rich C kinase substrate (MARCKS) gene (MACS). Analysis of its gene product, promoter, and chromosomal localization. J Biol Chem 266:14399–14405, 1991.
68. MW Hentze, TA Rouault, SW Caughman, A Dancis, JB Harford, RD Klausner. A cis-acting element is necessary and sufficient for translational regulation of human ferritin expression in response to iron. Proc Natl Acad Sci USA 84:6730–6734, 1987.
69. EW Mullner, B Neupert, LC Kuhn. A specific mRNA binding factor regulates the iron-dependent stability of cytoplasmic transferrin receptor mRNA. Cell 58:373–382, 1989.
70. A Rosen, KF Keenan, M Thelen, AC Nairn, A Aderem. Activation of protein kinase C results in the displacement of its myristoylated, alanine-rich substrate from punctate structures in macrophage filopodia. J Exp Med 172:1211–1215, 1990.

71. JK Wang, SI Walaas, TS Sihra, A Aderem, P Greengard. Phosphorylation and associated translocation of the 87-kDa protein, a major protein kinase C substrate, in isolated nerve terminals. Proc Natl Acad Sci USA 86:2253–2256, 1989.
72. JP Liu, D Engler, JW Funder, PJ Robinson. Arginine vasopressin (AVP) causes the reversible phosphorylation of the myristoylated alanine-rich C kinase substrate (MARCKS) protein in the ovine anterior pituitary: evidence that MARCKS phosphorylation is associated with adrenocorticotropin (ACTH) secretion. Mol Cell Endocrinol 101:247–256, 1994.
73. A Rosen, AC Nairn, P Greengard, ZA Cohn, A Aderem. Bacterial lipopolysaccharide regulates the phosphorylation of the 68K protein kinase C substrate in macrophages. J Biol Chem 264:9118–9121, 1989.
74. JP Raufman, R Malhotra, Q Xie, RD Raffaniello. Expression and phosphorylation of a MARCKS-like protein in gastric chief cells: further evidence for modulation of pepsinogen secretion by interaction of Ca^{2+}/calmodulin with protein kinase C. J Cell Biochem 64:514–523, 1997.
75. A Wolfman, TG Wingrove, PJ Blackshear, IG Macara. Down-regulation of protein kinase C and of an endogenous 80-kDa substrate in transformed fibroblasts. J Biol Chem 262:16546–16552, 1987.
76. M Oh-uchida, K Yano, S Kawamoto, K Shimizu. Transformation-specific decrease of phosphorylation of 80K protein, a substrate of protein kinase C, in NIH3T3 cells. Jpn J Cancer Res 81:799–806, 1990.
77. CK Joseph, SA Qureshi, DJ Wallace, DA Foster. MARCKS protein is transcriptionally down-regulated in v-Src-transformed BALB/c 3T3 cells. J Biol Chem 267: 1327–1330, 1992.
78. SL Simek, D Kligman, J Patel, NH Colburn. Differential expression of an 80-kDa protein kinase C substrate in preneoplastic and neoplastic mouse JB6 cells. Proc Natl Acad Sci USA 86:7410–7414, 1989.
79. T Herget, SF Brooks, S Broad, E Rozengurt. Expression of the major protein kinase C substrate, the acidic 80-kilodalton myristoylated alanine-rich C kinase substrate, increases sharply when Swiss 3T3 cells move out of cycle and enter G_0. Proc Natl Acad Sci USA 90:2945–2949, 1993.
80. G Brooks, SF Brooks, MW Goss. MARCKS functions as a novel growth suppressor in cells of melanocyte origin. Carcinogenesis 17:683–689, 1996.
81. Y Li, LD Martin, KB Adler. MARCKS protein: a key intracellular molecule controlling mucin secretion by human airway goblet cells (abstr). Am J Respir Crit Care Med 161:A259, 2000.

18

Airway Goblet and Mucous Cells: Identical, Similar, or Different?

C. William Davis and Scott H. Randell
University of North Carolina at Chapel Hill
Chapel Hill, North Carolina

The conducting airways have long been known to contain two sources of stored mucins: goblet and mucous cells (Figure 1).* Mucous cells reside in the secretory tubule epithelium of airway submucosal glands, which are present in the large, cartilaginous airways. Goblet cells, in contrast, reside in the superficial epithelium of the large, cartilaginous, *and* the small, noncartilaginous airways. Whether goblet and mucous cells represent a single cell type with a wide distribution, as well as whether they are distinct cell types with independent regulatory mechanisms and different distributions, are questions relevant to understanding the pathogenesis and the successful treatment of obstructive airway diseases. This paper examines the origins and properties of goblet and mucous cells in the airways and the roles in vivo and in vitro experimental models have played in efforts to elucidate their relationships.

* We define ''goblet cell'' and ''mucous cell'' as the mucin-secreting cells in superficial and submucosal gland epithelia, respectively, consistent with the trends expressed in recent literature. Readers should be aware that other authors, at different times, have used these terms with other specific meanings.

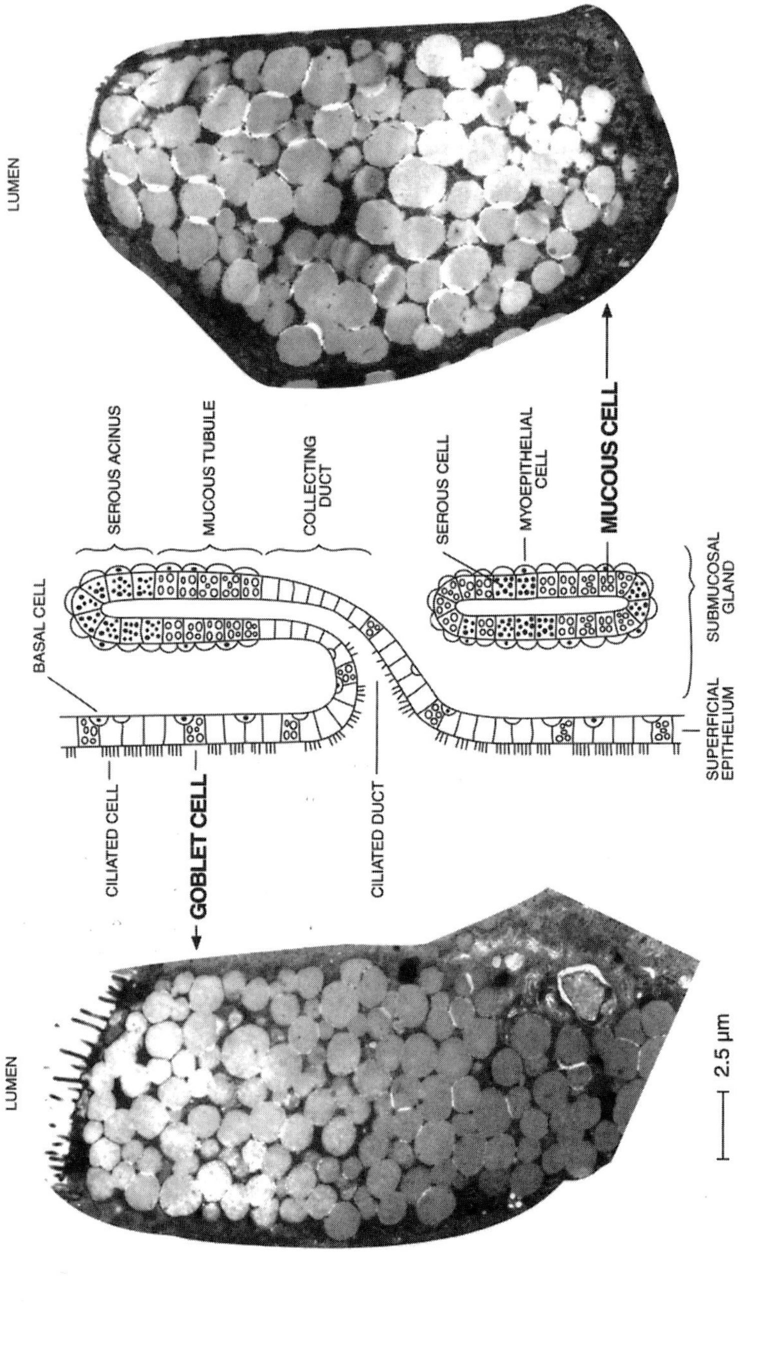

Figure 1 Organization and principal mucin-secreting cells of the airway epithelium. Goblet cells (left) and mucous cells (right) are very similar to one another in appearance. They differ most obviously by their relative location within the epithelium (center) and in the possession of apical membrane microvilli only by goblet cells.

HISTOLOGY OF THE AIRWAYS

The airways are commonly divided into the large, upper or proximal, and small, lower or distal, regions, which in humans correspond, roughly, with the presence or absence of cartilage. In normal humans, the tracheobronchial epithelium from the larynx down to 1 mm diameter bronchioles is pseudostratified (1), within which goblet cells comprise approximately 20% of columnar cells (2,3). Basal cells attach to the basement membrane and intercalate with the basal aspect of the overlying ciliated and goblet cells. Basal cell density, cross-sectional profile, and hemidesmosome keratin filament density correlate strongly with airway diameter and epithelial height, independent of species. Hence, basal cells apparently anchor ciliated and secretory cells to the basement membrane, and the need for anchorage increases with epithelial thickness (see Ref. 4). This notion is furthered by the empirical observation that ciliated and goblet cells isolated by protease digestion from the airways attach to collagen-coated substrata, in vitro, weakly, if at all, during an overnight incubation, whereas basal cells and clumps of surface epithelium with associated basal cells attach readily. Furthermore, basal cells strongly resist mechanical forces that easily damage the overlying columnar epithelial cells, and they quickly reconstitute the barrier function of the epithelium following injury (5).

In the cartilaginous airways, periodic openings in the superficial epithelium mark submucosal gland ducts. The ciliated ducts of mature submucosal glands connect to collecting ducts, which in turn are fed by a branched, tubular epithelium, comprised of mucous cells, the blind ends of which are comprised generally of serous cell* acini (see Refs. 6–8). During active secretion, fluid, driven by Cl^- and/or HCO_3^- transport across serous cells into the acinar lumen, washes mucins out of the gland and onto the surface of the airways (9). The movement of mucins through the gland lumen is also assisted by contraction of myoepithelial cells, which are basal cell–like, contractile cells that intercalate with the basal aspects of serous and mucous cells around gland periphery (see Refs. 8,10).

AIRWAY SECRETED-MUCIN GENE EXPRESSION

Three secreted, gel-forming mucins, MUC2, MUC5AC, and MUC5B, and a secreted, non–gel-forming mucin, MUC7, have been detected in adult human airway secretions by bio- and immunochemical methodologies (11–15).

MUC2 mRNA is expressed very early during human development (week 9) in all cells of the superficial epithelium, before the appearance of cilia and

* In some species serous cells are also found scattered in the superficial epithelium of the large and small airways, as well as in acini of submucosal glands. In adult humans, these cells appear to be restricted to glands (see Ref. 6) and to bronchioles (87).

differentiated goblet cells. Later (>week 18), as they differentiate in the superficial epithelium, basal cells express the highest levels of MUC2 mRNA; some expression also occurs at this time in submucosal glands (16). In the adult, the MUC2 gene is expressed in basal cells and some goblet cells and weakly in mucous cells and some serous cells (16–18).

MUC7 gene expression begins late in gestation (week 23) and is observed only in submucosal glands. In adults, MUC7 mRNA is expressed only in serous cells (14,16,18).

MUC5B mRNA appears early in gestation (week 13) in superficial epithelium and submucosal gland ducts. Later (week 18 and onward) stronger signals arise in the tubules of the glands. In adults, MUC5B gene expression is heavy in mucous cells and weak in scattered goblet cells both within the superficial epithelium and in submucosal gland ducts (14,16,18,19).

MUC5AC mRNA appears in goblet cells of the developing superficial epithelium and the ducts of submucosal gland at the same time as MUC5B. In adults, expression of the MUC5AC gene is restricted to goblet cells in these two locales (16,18,19).

AIRWAY EPITHELIAL CELL CULTURES

Primary cultures grown in defined media and derived from both superficial (20–27) and submucosal gland (28–30) epithelial cells have been used for a variety of experimental purposes over the past 2 decades; however, only primary cultures of superficial epithelium have been used to study regulated mucin secretion (see Ref. 31). Organ cultures of intact airways and isolated glands have played a large role in experiments that have contributed to our understanding of regulated mucin secretion from submucosal glands (see Refs. 7,8,10,32).

Submucosal gland epithelial cells are easily isolated by protease digestion from airways following removal of the superficial epithelium. When grown on human placental collagen the epithelial cell cultures differentiate into a mixed sero-mucous cell phenotype, but after a series of passages they appear to be purely serous in nature (30). When grown on a mixture of collagen I and III (Vitrogen), mature cultures appear to be comprised purely of mucous cells, suggesting that the cells may be exquisitely sensitive to the nature of the underlying substrata in terms of phenotype regulation (33). Significantly, there is no evidence of basal or ciliated cell differentiation in gland cell cultures as they are currently produced.

Fully differentiated, primary cultures of superficial airway epithelium have been developed from guinea pig, rat, and human airways (22,26,27,34; and see Ref. 31). In all these systems, mucociliary differentiation is highly dependent upon a permeable support to enabling epithelial polarization and the presence of retinoic acid in the culture medium. With rat and human airway epithelial cells, cultures are typically derived from the individual epithelial cells and small clumps of superficial epithelium provided by protease XIV digestion of airways tissue (24,27). Following attachment, the cells proliferate to form a confluent sheet, within which the cells polarize by forming tight junctions and distinct apical and basolateral membrane domains. Continued proliferation results in a pseudostratified to multilayered epithelium in which the luminal cells become columnar and differentiate into mature ciliated and goblet cells (24,35–37).

AIRWAY EPITHELIAL CELL PROGENITOR-PROGENY RELATIONSHIPS

Stem cell biology in the airways remains poorly understood. A conceptual framework for cell lineages and an appreciation of the vast body of knowledge derived from other tissues can be found in recent reviews (38,39). Classic cell lineage models predict a relatively rare population of stem cells in a specific location. However, a unique subpopulation of epithelial stem cells with very high growth capacities has not been identified in the airways. Cell lineages during development, steady-state renewal, and repair of injury still remain obscure. The report of a joint NCI-NHLBI workshop raised the concept that a stem cell, per se, may not exist in the airway epithelium (40). The epithelium is strikingly remodeled in disease, which is regarded as evidence for plasticity (41). Several cell types proliferate during remodeling, and daughter cells may follow different differentiation pathways than their parents. This great plasticity suggests an alternative model for cell lineages, in which many nonterminal airway epithelial cells are capable of nearly equivalent levels of growth and differentiation under stress. In this scenario, there is no specialized compartment containing true stem cells in the classical sense.

In the pseudostratified epithelium, basal and secretory cells are known to divide, whereas ciliated cells are considered terminally differentiated. During development, typical basal cells appear last and thus are not the ontogenic precursors of secretory and ciliated cells (42,43). In repair of injury, secretory cells are often highly proliferative and there is evidence that ciliated cells are derived from secretory cells (44). Progenitor-progeny relationships during steady-state renewal of airway epithelium in the adult are uncertain, but most investigations support a progenitorial role for basal cells, perhaps via an intermediate phenotype (45–47). One approach for studying cell lineages has been to dissociate the epithelium, separate the cells into subpopulations, and examine their growth and differentiation in tracheal grafts. When provided with an epithelial cell inoculum, the graft model reliably supports complete differentiation appropriate for the cells. Previous graft studies utilizing basal cell fractions obtained by centrifugal elutriation (48,49) suggested strongly that basal cells were multipotent progenitors. When flow cytometry was used to sort tracheal epithelial cells into subpopulations based on cell surface glycoconjugates, the results suggested that both basal and secretory cells regenerate a complete mucociliary epithelium (50,51). However, there are conflicting reports, with others concluding that only secretory cells (52) or only basal cells (53) are multipotent progenitors. The precise reason for the discrepancies in the literature is not clear, but the preponderance of the evidence is that cells within both the basal and secretory cell compartments of the pseudostratified portion of the airways divide and are multipotent progenitors.

Tracheal grafts have been adapted to the study of human bronchial epithelial cells (54–57). Human bronchial cells were retrovirally tagged while proliferating in vitro and then inoculated into tracheal grafts (58). A restricted pattern of colony types suggested that basal cells and intermediate cells were the predominant progenitors and that goblet cells did not undergo extensive self-renewal. However, a preliminary culture step in these experiments may have systematically caused the underestimation of the potential of certain cells. Additional evidence for enrichment of progenitors within a specific morphologic category is that the in vitro colony-forming efficiency of rat tracheal basal cell–enriched fractions was, on average, fivefold greater than that of nonbasal cells (59). However, cell culture may not have been truly indicative of the in vivo potential of the cells. Despite their limitations, these studies suggest that there is a spectrum of progenitorial capacities amongst proliferation-competent airway epithelial cells. While the results of these experiments do not conclusively distinguish between the theoretical cell lineage models presented above, they suggest that there is substantial proliferative reserve in many different airway epithelial cell types. However, this does not rule out the existence of true stem cells, which are only recruited under extreme conditions of epithelial damage.

Cell lineage models for large, cartilaginous airways must encompass the submucosal glands. It is known that during development the glands are derived from specialized outpockets of basally situated surface cells (60,61). Highly complex mesenchymal-epithelial interactions typical of the events governing morphogenesis (62) likely govern gland formation. Most gland development in normal humans occurs in utero, but gland size and complexity increases throughout childhood (3); new glands may form in the injured neonate. Specific gland progenitor cells have not been characterized, and there few data exist regarding ongoing patterns of cell renewal in the acinar and duct system. It is possible that an interaction of multiple epithelial progenitors (58), in combination with specific mesenchymal signals (63), may be necessary to form glands. Hence, it is important to identify the progenitor cells for submucosal glands in the surface epithelium and to determine cell lineages within the glands.

The pseudostratified columnar epithelium of the normal human airways becomes cuboidal in the distal portion of 1 mm diameter bronchioles (1). At this site, nonciliated bronchiolar cells (Clara cells) and ciliated cells predominate. Reepithelialization of oxidant gas injured bronchioles occurs by Clara cell division and subsequent differentiation into ciliated cells (64). Clara cells generate a bronchiolar epithelium of cuboidal ciliated and Clara cells when grown in tracheal grafts, while mixed tracheal cells generate a mucociliary epithelium (65,66). These observations strongly support the notion of distinct zones with different progenitor populations in the proximal and most distal airways. Recent studies following bronchiolar injury with the selective Clara cell toxin naphthalene suggest that the distal airway epithelial regenerative unit is determined by complex interactions of Clara cells with pulmonary neuroendocrine cells (67–69). These studies further suggest that there is a subset of Clara cells with extensive progenitorial capacity. To date, Clara cells have not been tested for clonal analysis of growth capacity. One problem is that Clara cells can only be enriched to 80–85% purity by density separation. In vitro culture methods supporting Clara cell proliferation and differentiation to a terminal bronchiolar phenotype are also not available. Thus, novel approaches are needed to help determine bronchiolar cell lineages, define whether there are unique stem cells among the Clara cell population, and determine if Clara cells can give rise to goblet cells in the most distal airways.

EPITHELIAL POLARITY IN THE REGULATION OF GOBLET AND MUCOUS CELLS

Epithelial cells are highly polarized, and concerted efforts by many laboratories over many years are now revealing the mechanisms by which the cells develop morphologically (e.g., see Ref. 70). Less well known but no less important are recent observations indicating an equally strong polarization in epithelial cell signaling. In cultured airway epithelial cells, for instance, it has been shown that the release of intracellular Ca^{2+} in response to the presentation of agonist is localized to the membrane stimulated: a luminal challenge with ATP results in the release of Ca^{2+} from endoplasmic reticulum associated with the apical, but not basolateral, membrane, and vice versa (71). For the secretory cells of the airways, however, there is very little information available on issues even as basic as the sidedness of agonist responses.* For the moment, we must therefore rely on analogous systems for guidance in polarized intracellular signaling mechanisms, such as the pancreatic acinar cell (see Ref. 70).

Mucous cells and serous cells in submucosal glands are situated with their basolateral membranes bathed on the outside by extracellular liquid, essentially an exudate of plasma, and exposed directly to the elaborations from innervating

* With isolated submucosal gland preparations, secretagogues are generally added to the bath or serosal side of the gland epithelium. With primary cultures of superficial epithelial cells, secretagogues are generally added bilaterally. In few cases have the two sides of the glandular or superficial epithelium been challenged selectively.

neurons. Their apical membranes, in contrast, are bathed by the contents of the gland lumen. Goblet cells are bathed luminally by airway surface liquid and serosally by extracellular liquid. The neuronal influence impinging upon the serosal aspects of submucosal glands and the superficial epithelium differs markedly. Whereas the glands are surrounded by a rich plexus of cholinergic, adrenergic, and noncholinergic, nonadrenergic (NANC) nerve fibers (see Refs. 72, 73), the innervation of the superficial epithelium is sparse: the human airway has but ~1 fiber/200 μm of basement membrane (74).

REGULATION OF AIRWAY MUCIN SECRETION

Macroscopically, many neural and humoral agents promote mucin secretion in the airways from submucosal gland mucous cells and from superficial epithelial goblet cells (see Refs. 7,8,10,75,76). Maneuvers such as stimulation of the vagus nerve, for instance, cause mucin secretion from both mucous and goblet cells (76); however, while cholinergic stimulation of mucin release from isolated submucosal glands has been demonstrated, there is no equivalent evidence for the direct effects of muscarinic agonists on goblet cells (see Refs. 10,31,77). Table 1 indicates the classes of secretagogue that are generally understood to promote mucin release selectively from submucosal glands and from goblet cells in the superficial epithelium. Notably, a diverse array of some two dozen neural and humoral agents promote submucosal gland mucin secretion, whereas very few stimulate goblet cells.

A major obstacle to our understanding of the regulation of mucin secretion in the airways is the internal complexity of its regulatory system. Interesting and instructive interactions that occur between signaling pathways within the epithelial tissue are presented by the effects of PAF in promoting mucin secretion from submucosal glands. In intact epithelium, PAF stimulates submucosal gland mucin secretion; however, when isolated glands are challenged PAF has no effect on mucin release. Only in the presence of platelets does PAF stimulate isolated gland mucin secretion as it does in the whole tissue. Hence, PAF appears to have no direct effects on submucosal mucin secretion; its effects in intact tissues are mediated by eicosanoids (see Refs. 10,77,78).

Similar interactions may also occur within the intact superficial epithelium. The stimulatory effects of PAF on goblet cell mucin release in primary epithelial cell cultures, for instance, are indirect and mediated by eicosanoids, as they are in submucosal glands (78). PAF has no effect on mucin release from SPOC1 mucin-secreting cells (77), a result that suggests that other cells of the epithelium are responsible for PAF-induced eicosanoid production. Similarly, ATP, ADP, and adenosine stimulate mucin release from goblet cells when presented to the serosal side of explanted superficial epithelium, but neither these, nor any other secretagogue tested have effects from the serosal side of SPOC1 cells (31,77,

TABLE 1 Regulation of Mucin Secretion from Submucosal Glands (isolated glands) and Superficial Epithelium (explants, primary cultures, or SPOC1 cells)

Secretagogue class	Submucosal glands (7,8,10)	Superficial epithelium (31,77,78,88)
Adrenergic		
α_1	Secretagogue and stimulates gland contraction	No effect
$\beta_?$	Secretagogue	No effect
Cholinergic		
Muscarinic (M_3, some M_1)	Potent secretagogue	No effect
Prostaglandins		
PGA_2, PGD_2, $PGF_{2\alpha}$	Secretagogues	No effect ($PGF_{2\alpha}$ stimulatory in guinea pig)
PGE_2	Inhibitory	No effect
Leukotrienes		
LTC_4, LTD_2, 15-HETE	Stimulatory in intact tissues	No effect
PAF	Stimulatory, but only in presence of platelets	No effect in SPOC1 cells; stimulatory in primary cultures via eicosanoids
Peptidergic		
NK_1: SP $>$ NKA	Potent secretagogue, stimulates gland contraction	No effect (SP), or not demonstrated
VIP	Secretagogue, augments muscarinic responses	Not demonstrated
Endothelin-1	Secretagogue	Not demonstrated
Purinergic		
P1 (adenosine)	No effect (89)	Secretagogue (basolateral aspect only, 90)
$P2Y_2$	Secretagogue	Potent secretagogue (apical aspect only)
Histaminergic		
H_2	Secretagogue	No effect
Bradykinin	Secretagogue	not demonstrated
Intracellular messengers		
Cyclic AMP	Secretagogue	No effect (SPOC1 cells)
Cyclic GMP	not demonstrated	No effect (SPOC1 cells)
Intracellular Ca^{2+}	Secretagogue	Secretagogue
Diacylglycerol (PMA)	Secretagogue	Secretagogue

and unpublished observations). In fact, mucin secretion from SPOC1 cells is presently known to be stimulated uniquely by luminal presentation of the $P2Y_2$ agonists ATP, UTP, and ATPγS (77,79,80). Furthermore, of the common methods for the receptor-independent activation cellular messenger pathways, only Ca^{2+} mobilization and PMA promote SPOC1 cell mucin release (Table 1), an observation that effectively excludes the possibility of direct effects by agonists whose actions generally are mediated by cyclic nucleotides.

CONCLUSIONS AND PERSPECTIVES

Determining whether goblet and mucous cells represent separate cell types from the data available is very difficult at present. The two cells are morphologically very similar, if not identical, they have common origins, yet they seem to be regulated by very different mechanisms (Table 1). Also contributing to the confusion is the nearly perfect separation in the distributions of MUC5AC and MUC5B gene expression: only the appearance of MUC5B expression in a subset of mucin-secreting cells in the superficial and submucosal gland ductal epithelia prevents a MUC5AC/goblet cell versus MUC5B/mucous cell distinction. Three scenarios that would explain the data are as follows.

1. *Goblet and mucous cells are different.* If MUC5AC and MUC5B prove to be expressed uniquely in mucin-secreting cells, this scenario would gain substantial weight. The only obvious alteration necessitated in our developing model would be their respective distributions. Goblet cells would be restricted to the superficial epithelium and submucosal gland ducts, but mucous cells would be present in submucosal glands and scattered in the superficial epithelium and gland ducts.
2. *Goblet and mucous cells are similar.* For this scenario to be true, we need only to imagine the mucin-secreting cell as possessing a phenotype spectrum for which goblet and mucous cells represent the extremes. Goblet cells in the superficial epithelium and gland ducts would express MUC5AC and be regulated uniquely by apical membrane $P2Y_2$ purinoceptors. Mucous cells in submucosal glands would express MUC5B and be regulated by a wide variety of neural and humoral mediators. In between would be mucin-secreting cells in the superficial epithelium and gland ducts, perhaps co-expressing MUC5AC and MUC5B and regulated by receptor-mediated pathways in addition to apical membrane $P2Y_2$ purinoceptors.
3. *Goblet and mucous cells are identical.* ''Identical'' in this scenario means in all respects *but* the MUC5AC/MUC5B expression pattern. Both goblet and mucous cells would be regulated by apical membrane $P2Y_2$ purinoceptors. Serous cells would be the most likely source of

ATP and/or UTP in the submucosal glands and would release the agonists when stimulated by neural and humoral mediators from the serosal compartment. Although the pathways and their regulation remain to be described, ATP secretion has been demonstrated recently from nonneuronal cells in culture (e.g., see Refs. 81–84), including secretion into the luminal compartment of cultured airway epithelial cells (85,86). In the gland ducts, the secreted nucleotides would be transported to mucous cells by the fluid secreted simultaneously by serous cells.

At present it is impossible to distinguish between these and other conceivable scenarios that explain the differences observed in mucin-secreting cells in the airway epithelium. Knowing whether MUC5AC and MUC5B are co-expressed in a subset of these cells in the superficial epithelium and submucosal gland ducts will add to our knowledge, but it will not allow a definitive discrimination between the different possibilities. Indeed, a final determination of the discriminating features of goblet and mucous cells promises to be a challenging opportunity.

ACKNOWLEDGMENTS

The authors acknowledge the generous funding of the work in their respective laboratories presented in this review. CWD received funding from the North American Cystic Fibrosis Foundation, The American Lung Association of North Carolina, and Glaxo-Wellcome Corporation. SHR received funding from the NHLBI (HL58345). The authors thank Ms. Kimberly Burns for electron micrographs and Mr. Edwin Staples for the illustration.

REFERENCES

1. RR Mercer, ML Russell, VL Roggli, JD Crapo. Cell number and distribution in human and rat airways. Am J Respir Cell Mol Biol 10:613–624, 1994.
2. JR Harkema, A Mariassy, JA St. George, DM Hyde, CG Plopper. Epithelial cells of the conducting airways. In: SG Farmer, DW Hay, eds. The Airway Epithelium: Physiology, Pathophysiology, and Pharmacology. New York: Marcel Dekker, 1991, pp 3–39.
3. PK Jeffery, D Gaillard, S Moret. Human airway secretory cells during development and in mature airway epithelium [review]. Eur Respir J 5:93–104, 1992.
4. MJ Evans, PC Moller. Biology of airway basal cells. Exp Lung Res 17:513–531, 1991.
5. JS Erjefalt, F Sundler, CG Persson. Epithelial barrier formation by airway basal cells. Thorax 52:213–217, 1997.
6. CB Basbaum, B Jany, WE Finkbeiner. The serous cell. Ann Rev Physiol 52:97–113, 1990.

7. WE Finkbeiner, Physiology and pathology of tracheobronchial glands. Respir Physiol 118:77–83, 1999.
8. DC Fung, DF Rogers. Airway submucosal glands: physiology and pharmacology. In: D Rogers, MI Lethem, eds. Airway Mucus: Basic Mechanisms and Clinical Perspectives. Basel: Birkhauser, 1997, pp 179–210.
9. SK Inglis, MR Corboz, ST Ballard. Effect of anion secretion inhibitors on mucin content of airway submucosal gland ducts. Am J Physiol 274:L762–L766, 1998.
10. S Shimura, T Takishima. Airway submucosal gland secretion. In: T Takishima, S Shimura, eds. Airway Secretion: Physiological Bases for the Control of Mucus Hypersecretion. New York: Marcel Dekker, 1994, pp 325–398.
11. DJ Thornton, I Carlstedt, M Howard, PL Devine, MR Price, JK Sheehan. Respiratory mucins: identification of core proteins and glycoforms. Biochem J 316:967–975, 1996.
12. DJ Thornton, M Howard, N Khan, JK Sheehan. Identification of two glycoforms of the MUC5B mucin in human respiratory mucus. Evidence for a cysteine-rich sequence repeated within the molecule. J Biol Chem 272:9561–9566, 1997.
13. C Wickstrom, JR Davies, GV Eriksen, EC Veerman, I Carlstedt. MUC5B is a major gel-forming, oligomeric mucin from human salivary gland, respiratory tract and endocervix: identification of glycoforms and C-terminal cleavage. Biochem J 334:685–693, 1998.
14. P Sharma, L Dudus, PA Nielsen, H Clausen, JR Yankaskas, MA Hollingsworth, JF Engelhardt. MUC5B and MUC7 are differentially expressed in mucous and serous cells of submucosal glands in human bronchial airways. Am J Respir Cell Mol Biol 19:30–37, 1998.
15. JR Davies, N Svitacheva, L Lannefors, R Kornfalt, I Carlstedt. Identification of MUC5B, MUC5AC and small amounts of MUC2 mucins in cystic fibrosis airway secretions. Biochem J 344 (Pt 2):321–330, 1999.
16. MP Buisine, L Devisme, MC Copin, M Durand-Reville, B Gosselin, JP Aubert, N Porchet. Development mucin gene expression in the human respiratory tract. Am J Respir Cell Mol Biol 20:209–218, 1999.
17. A Dohrman, T Tsuda, E Escudier, M Cardone, B Jany, J Gum, Y Kim, CB Basbaum. Distribution of lysozyme and mucin (MUC2 and MUC3) mRNA in human bronchus. Exp Lung Res 20:367–380, 1994.
18. MC Copin, L Devisme, MP Buisine, CH Marquette, A Wurtz, JP Aubert, B Gosselin, N Porchet. From normal respiratory mucosa to epidermoid carcinoma: expression of human mucin genes. Int J Cancer 86:162–168, 2000.
19. CJ Reid, S Gould, A Harris. Developmental expression of mucin genes in the human respiratory tract. Am J Respir Cell Mol Biol 17:592–598, 1997.
20. KC Kim. Possible requirement of collagen gel substratum for production of mucin-like glycoproteins by primary rabbit tracheal epithelial cells in culture. In Vitro Cell Dev Biol 21:617–621, 1985.
21. R Wu, E Nolan, C Turner. Expression of tracheal differentiated fucntions in serum-free hormone-supplemented medium. J Cell Physiol 125:167–181, 1985.
22. KB Adler, PW Cheng, KC Kim. Characterization of guinea pig tracheal epithelial cells maintained in biphasic organotypic culture: cellular composition and biochemi-

cal analysis of released glycoconjugates. Am J Respir Cell Mol Biol 2:145–154, 1990.
23. RD Rieves, J Goff, T Wu, P Larivee, C Logun, JH Shelhamer. Airway epithelial cell mucin release: immunologic quantitation and response to platelet-activating factor. Am J Respir Cell Mol Biol 6:158–167, 1992.
24. L Kaartinen, P Nettesheim, KB Adler, SH Randell. Rat tracheal epithelial cell differentiation in vitro. In Vitro Cell Dev Biol Animal 29A:481–492, 1993.
25. AB Clark, SH Randell, P Nettesheim, TE Gray, CR Bagnell, LE Ostrowski. Regulation of ciliated cell differentiation in cultures of rat tracheal epithelial cells. Am J Respir Cell Mol Biol 12:329–338, 1995.
26. N Emery, GA Place, S Dodd, M Lhermitte, G David, G Lamblin, JM Perini, AM Page, RL Hall, P Roussel. Mucous and serous secretions of human bronchial epithelial cells in secondary culture. Am J Respir Cell Mol Biol 12:130–141, 1995.
27. TE Gray, K Guzman, CW Davis, LH Abdullah, P Nettesheim. Mucociliary differentiation of serially passaged normal human tracheobronchial epithelial cells. Am J Respir Cell Mol Biol 14:104–112, 1996.
28. DJ Culp, DP Penney, MG Marin. A technique for the isolation of submucosal gland cells from cat trachea. J Appl Physiol Respir Environ Exercise Physiol 55:1035–1041, 1983.
29. A Paul, J Picard, M Mergey, D Veissiere, WE Finkbeiner, CB Basbaum. Glycoconjugates secreted by bovine tracheal serous cells in culture. Arch Biochem Biophys 260:75–84, 1988.
30. CP Sommerhoff, WE Finkbeiner. Human tracheobronchial submucosal gland cells in culture. Am J Respir Cell Mol Biol 2:41–50, 1990.
31. CW Davis, LH Abdullah. In vitro models for airways mucin secretion. Pulm Pharmacol Ther 10:145–155, 1997.
32. MG Marin. Update: pharmacology of airway secretion. Pharmacol Rev 46:35–65, 1994.
33. WE Finkbeiner, BQ Shen, JH Widdicombe. Chloride secretion and function of serous and mucous cells of human airway glands. Am J Physiol 267:L206–L210, 1994.
34. RR Scully, JR Dedman, SG Schultz. Localization of calmodulin in the enterocyte of Necturus small intestine. J Cell Sci 91:541–547, 1988.
35. H Matsui, SH Randell, SW Peretti, CW Davis, RC Boucher. Coordinated clearance of periciliary liquid and mucus from airway surfaces. J Clin Invest 102:1125–1131, 1998.
36. H Matsui, LG Johnson, SH Randell, RC Boucher. Loss of binding and entry of liposome-DNA complexes decreases transfection efficiency in differentiated airway epithelial cells. J Biol Chem 272:1117–1126, 1997.
37. H Matsui, CW Davis, R Tarran, RC Boucher. Osmotic water permeabilities of cultured, well-differentiated normal and cystic fibrosis airway epithelia [see comments]. J Clin Invest 105:1419–1427, 2000.
38. FM Watt, BL Hogan. Out of Eden: stem cells and their niches. Science 287:1427–1430, 2000.
39. JM Slack. Stem cells in epithelial tissues. Science 287:1431–1433, 2000.
40. RJ Mason, MC Williams, HL Moses, S Mohla, MA Berberich. Stem cells in lung development, disease, and therapy. Am J Respir Cell Mol Biol 16:355–363, 1997.

41. CB Basbaum, B Jany. Plasticity in the airway epithelium. Am J Physiol 259:L38–L46, 1990.
42. KP Keenan, TS Wilson, EM McDowell. Regeneration of hamster tracheal epithelium after mechanical injury. IV. Histochemical, immunocytochemical and ultrastructural studies. Virchows Arch B Cell Pathol Incl Mol Pathol 43:213–240, 1983.
43. CG Plopper, JL Alley, AJ Weir. Differentiation of tracheal epithelium during fetal lung maturation in the rhesus monkey *Macaca mulatta*. Am J Anat 175:59–71, 1986.
44. EM McDowell, C Newkirk, B Coleman. Development of hamster tracheal epithelium: II. Cell proliferation in the fetus. Anatom Rec 213:448–456, 1985.
45. WK Blenkinsopp. Proliferation of respiratory tract epithelium in the rat. Exp Cell Res 46:144–154, 1967.
46. R Breuer, G Zajicek, TG Christensen, EC Lucey, GL Snider. Cell kinetics of normal adult hamster bronchial epithelium in the steady state. Am J Respir Cell Mol Biol 2:51–58, 1990.
47. GM Donnelly, DG Haack, CS Heird. Tracheal epithelium: cell kinetics and differentiation in normal rat tissue. Cell Tissue Kinet 15:119–130, 1982.
48. Y Inayama, GE Hook, AR Brody, AM Jetten, T Gray, J Mahler, P Nettesheim. In vitro and in vivo growth and differentiation of clones of tracheal basal cells. Am J Pathol 134:539–549, 1989.
49. Y Inayama, GE Hook, AR Brody, GS Cameron, AM Jetten, LB Gilmore, T Gray, P Nettesheim. The differentiation potential of tracheal basal cells. Lab Invest 58: 706–717, 1988.
50. JK Sheehan, C Hanski, AP Corfield, C Paraskeva, DJ Thornton. Mucin biosynthesis and macromolecular assembly. Biochem Soc Trans 23:819–821, 1995.
51. JY Liu, P Nettesheim, SH Randell. Growth and differentiation of tracheal epithelial progenitor cells. Am J Physiol 266:L296–L307, 1994.
52. NF Johnson, AF Hubbs. Epithelial progenitor cells in the rat trachea. Am J Respir Cell Mol Biol 3:579–585, 1990.
53. JR Ford, M Terzaghi-Howe. Characteristics of magnetically separated rat tracheal epithelial cell populations. Am J Physiol 263:L568–L574, 1992.
54. JF Engelhardt, Y Yang, LD Stratford-Perricaudet, ED Allen, K Kozarsky, M Perricaudet, JR Yankaskas, JM Wilson. Direct gene transfer of human CFTR into human bronchial epithelia of xenografts with E1-deleted adenoviruses. Nat Genet 4:27–34, 1993.
55. JF Engelhardt, JR Yankaskas, JM Wilson. In vivo retroviral gene transfer into human bronchial epithelia of xenografts. J Clin Invest 90:2598–2607, 1992.
56. JM Pilewski, JF Engelhardt, JE Bavaria, LR Kaiser, JM Wilson, SM Albelda. Adenovirus-mediated gene transfer to human bronchial submucosal glands using xenografts. Am J Physiol 268:L657–L665, 1995.
57. Y Yang, JF Engelhardt, JM Wilson. Ultrastructural localization of variant forms of cystic fibrosis transmembrane conductance regulator in human bronchial epithelial of xenografts. Am J Respir Cell Mol Biol 11:7–15, 1994.
58. JF Engelhardt, H Schlossberg, JR Yankaskas, L Dudus. Progenitor cells of the adult human airway involved in submucosal gland development. Development 121:2031–2046, 1995.
59. SH Randell, CE Comment, FC Ramaekers, P Nettesheim. Properties of rat tracheal

epithelial cells separated based on expression of cell surface alpha-galactosyl end groups. Am J Respir Cell Mol Biol 4:544–554, 1991.
60. U Bucher, L Reid. Development of the mucus secreting elements in human lung. Thorax 16:219–225, 1961.
61. M Tos. Development of the mucous glands in the human main bronchus. Anat Anz 123:376–389, 1968.
62. BL Hogan. Morphogenesis. Cell 96:225–233, 1999.
63. D Duan, Y Yue, W Zhou, B Labed, TC Ritchie, R Grosschedl, JF Engelhardt. Submucosal gland development in the airway is controlled by lymphoid enhancer binding factor 1 (LEF1). Development 126:4441–4453, 1999.
64. MJ Evans, SG Shami, LJ Cabral-Anderson, NP Dekker. Role of nonciliated cells in renewal of the bronchial epithelium of rats exposed to NO2. Am J Pathol 123: 126–133, 1986.
65. AR Brody, GE Hook, GS Cameron, AM Jetten, CJ Butterick, P Nettesheim. The differentiation capacity of Clara cells isolated from the lungs of rabbits. Lab Invest 57:219–229, 1987.
66. GE Hook, AR Brody, GS Cameron, AM Jetten, LB Gilmore, P Nettesheim. Repopulation of denuded tracheas by Clara cells isolated from the lungs of rabbits. Exp Lung Res 12:311–329, 1987.
67. TP Stevens, JT McBride, JL Peake, KE Pinkerton, BR Stripp. Cell proliferation contributes to PNEC hyperplasia after acute airway injury. Am J Physiol 272:L486–L493, 1997.
68. JL Peake, SD Reynolds, BR Stripp, KE Stephens, KE Pinkerton. Alteration of pulmonary neuroendocrine cells during epithelial repair of naphthalene-induced airway injury. Am J Pathol 156:279–286, 2000.
69. SD Reynolds, A Giangreco, JH Power, BR Stripp. Neuroepithelial bodies of pulmonary airways serve as a reservoir of progenitor cells capable of epithelial regeneration. Am J Pathol 156:269–278, 2000.
70. C Yeaman, KK Grindstaff, MD Hansen, WJ Nelson. Cell polarity: versatile scaffolds keep things in place. Curr Biol 9:R515–R517, 1999.
71. AM Paradiso, SJ Mason, ER Lazarowski, RC Boucher. Membrane-restricted regulation of Ca^{2+} release and influx in polarized epithelia. Nature 377:643–646, 1995.
72. CF Culling, PE Reid. The apparent failure of sodium borohydride reduction to block further PAS reactivity in rat epithelial mucins. Histochem J 9:781–785, 1977.
73. PE Reid, CF Culling, WL Dunn. A histochemical method for the identification of 9-O-acyl sialic acid. An investigation of bovine submaxillary gland and intestinal mucins. J Histochem Cytochem 26:187–192, 1978.
74. A Laitinen. Ultrastructural organization of intraepithelial nerves in the human airway tract. Thorax 40:488–492, 1985.
75. DF Rogers. Airway goblet cells: responsive and adaptable front-line defenders. Eur Respir J 7:1690–1706, 1994.
76. DF Rogers, A Dewar. Neural control of airway mucus secretion [review]. Biomed Pharmacother 44:447–453, 1990.
77. CW Davis. Goblet cells: physiology and pharmacology. In: DF Rogers, MI Lethem, eds. Airway Mucus: Basic Mechanisms and Clinical Perspectives. Basel: Berkhauser, 1997, pp 150–177.

78. LA Cohn, KB Adler. Interactions between airway epithelium and mediators of inflammation. Exp Lung Res 18:299–322, 1992.
79. LH Abdullah, SW Davis, L Burch, M Yamauchi, SHXNP Randell, CW Davis. P2u purinoceptor regulation of mucin secretion in SPOC1 cells, a goblet cell line from the airways. Biochem J 316:943–951, 1996.
80. CW Davis, LH Abdullah, RC Boucher. Cellular basis for the purinergic regulation of mucin secretion in the airways. In: GL Baum, ed. Cilia, Mucus, and Mucociliary Interactions New York: Marcel Dekker, 1998, pp 153–166.
81. TK Harden, ER Lazarowski. Release of ATP and UTP from astrocytoma cells. Prog Brain Res 120:135–143, 1999.
82. H Sauer, J Hescheler, M Wartenberg. Mechanical strain-induced Ca(2+) waves are propagated via ATP release and purinergic receptor activation. Am J Physiol Cell Physiol 279:C295–C307, 2000.
83. K Dezaki, T Tsumura, E Maeno, Y Okada. Receptor-mediated facilitation of cell volume regulation by swelling-induced ATP release in human epithelial cells. Jpn J Physiol 50:235–241, 2000.
84. RS Ostrom, C Gregorian, PA Insel. Cellular release of and response to ATP as key determinants of the set-point of signal transduction pathways. J Biol Chem 275: 11735–11739, 2000.
85. WC Watt, ER Lazarowski, RC Boucher. Cystic fibrosis transmembrane regulator-independent release of ATP. Its implications for the regulation of P2Y2 receptors in airway epithelia. J Biol Chem 273:14053–14058, 1998.
86. ER Lazarowski, RC Boucher, TK Harden. Constitutive release of ATP and evidence for major contribution of ectonucleotide pyrophosphatase and nucleoside diphosphokinase to extracellular nucleotide concentrations. J Biol Chem: 2000.
87. AV Rogers, A Dewar, B Corrin, PK Jeffery. Identification of serous-like cells in the surface epithelium of human bronchioles. Eur Respir J 6:498–504, 1993.
88. KC Kim, K McCracken, BC Lee, CY Shin, MJ Jo, CJ Lee, KH Ko. Airway goblet cell mucin: its structure and regulation of secretion. Eur Respir J 10:2644–2649, 1997.
89. S Shimura, T Sasaki, M Nagaki, T Takishima, K Shirato. Extracellular ATP regulation of feline tracheal submucosal gland secretion. Am J Physiol 267:L159–L164, 1994.
90. CW Davis, ML Dowell, M Lethem, M Van Scott. Goblet cell degranulation in isolated canine tracheal epithelium: response to exogenous ATP, ADP, and adenosine. Am J Physiol 262:C1313–C1323, 1992.

19

Regulation of Respiratory Mucin Gene Expression by Neutrophil Elastase

Judith A. Voynow and Bernard M. Fischer
Duke University Medical Center
Durham, North Carolina

INTRODUCTION

Neutrophil-predominant airway inflammation and mucus obstruction of airways are major pathological features of cystic fibrosis, chronic bronchitis, and acute exacerbations of asthma. The severity of airflow obstruction and respiratory morbidity in patients with chronic inflammatory airway diseases have been directly related to the intensity of neutrophilic inflammation (1). Neutrophils release elastase (NE), a serine protease, that impairs mucociliary clearance by several mechanisms. NE activates mucin secretion (2), triggers goblet cell metaplasia (3), and increases mucin granule production (4). Thus, NE may be a key molecule linking neutrophilic inflammation and overproduction of mucus in chronic inflammatory airway diseases.

The molecular mechanisms by which NE increases mucus production are beginning to be elucidated. The major macromolecular constituents of mucus are the mucin glycoproteins. Of the mucin genes expressed in the respiratory tract, *MUC*4 and *MUC*5AC are two of the predominant mucin genes expressed in superficial tracheobronchial epithelium. *MUC*4 is expressed in both superficial secretory and ciliated tracheobronchial cells (5). The *MUC*4 cDNA encodes for both a transmembrane domain and a proteolytic cleavage site, suggesting that it may be present as a cell surface mucin and/or a component of airway secretions (6). *MUC*5AC is primarily expressed in goblet cells but also can be detected in submucosal gland cells (5). MUC5AC glycoprotein is present in airway secretions from normal subjects, patients with bronchial asthma (7) and with cystic fibrosis (8). Therefore, we have investigated the effect of NE on the regulation of *MUC*5AC and *MUC*4.

METHODS AND RESULTS

To study the effect of NE on airway epithelial cell mucin gene expression, we have used two cell culture models: A549 cells and normal human bronchial epithelial cells (NHBE). A549 (ATCC) is a lung carcinoma cell line that expresses both MUC5AC mRNA and glycoprotein (9). NHBE (Clonetics), grown in submerged conditions on plastic tissue culture dishes (undifferentiated NHBE), express *MUC*4 mRNA. When NHBE are cultured in an air/liquid interface culture system (10), the cells develop a mucociliary phenotype (differentiated NHBE) and express both *MUC*4 and *MUC*5AC.

Regulation of *MUC*5AC and *MUC*4 gene expression by NE was assessed by northern analysis. In A549 cells, NE increased *MUC*5AC mRNA expression in a concentration- and time-dependent manner. NE also increased *MUC*5AC expression in a time-dependent manner in differentiated NHBE (11). In undifferentiated NHBE, NE increased *MUC*4 mRNA levels. A comparison of NE-regulated mucin expression revealed that both *MUC*4 and *MUC*5AC mRNA levels increased in response to the same NE treatment conditions (50 nM, 24 hours).

To determine the mechanism of NE-regulated mucin expression, we tested whether NE increased *MUC*5AC gene expression by transcriptional or posttranscriptional regulation. NE did not increase transcription of *MUC*5AC as determined by nuclear run-on assay (11). In contrast, NE did regulate both *MUC*5AC and *MUC*4 by a posttranscriptional mechanism. A549 and undifferentiated NHBE cells were stimulated with neutrophil elastase or control vehicle and then treated with an RNA transcription inhibitor, actinomycin D. The decay rate of *MUC*5AC and *MUC*4 mRNA was followed over time. Treatment with NE tripled the half-lives of both *MUC*5AC and *MUC*4 mRNA. Thus, these studies suggest that NE increases expression of both *MUC*5AC and *MUC*4 by a novel regulatory mechanism for mucin genes: stabilization of mRNA transcripts.

The stability of mRNA transcripts is controlled by the interaction of RNA-binding proteins and mRNA *cis* stability sequences. This interaction affects the activity of RNases. Several intracellular signaling molecules, such as protein kinase C, calcium, iron, and reactive oxygen species, alter mammalian gene mRNA stability (12). We hypothesized that NE may regulate the stability of both

*MUC*5AC and *MUC*4 mRNA by activating similar intracellular signals, specifically reactive oxygen species. To test this hypothesis, A549 and undifferentiated NHBE were treated with antioxidants prior to and during NE stimulation. The effect of antioxidants on NE-induced mucin expression was evaluated by Northern analyses for *MUC*5AC and *MUC*4. Dimethylthiourea, a broad-spectrum antioxidant that scavenges hydroxyl radical, hydroxylated products, and peroxynitrite, attenuated the NE-induced increase in both *MUC*5AC and *MUC*4 mRNA. Furthermore, in A549, polyethylene glycol–conjugated catalase, a hydrogen peroxide scavenger, inhibited the NE-induced increase in *MUC*5AC mRNA levels. Together, these experiments suggest that reactive oxygen species are critical intracellular signals that mediate NE-induced mucin gene expression.

To confirm the presence of reactive oxygen species in respiratory epithelial cells following NE treatment, A549 and both differentiated and undifferentiated NHBE were evaluated by fluorescent microscopy for changes in fluorescence of an oxidation marker, dichlorodihydrofluorescein (DCF). Cells were loaded with DCF, a molecule that is retained intracellularly and fluoresces when exposed to hydrogen peroxide or hydroxylated products (13). Following DCF loading, cells were stimulated with NE or control vehicle and then examined by fluorescent microscopy. Cells fluoresced only after NE treatment and not after control vehicle treatment. In the absence of DCF, there was no increase in autofluorescence following NE or control vehicle treatment. Thus, NE treatment increases oxidant stress in epithelial cells.

DISCUSSION AND CONCLUSIONS

NE regulates *MUC*5AC and *MUC*4 by a novel mechanism, stabilization of mRNA transcripts. There is a growing body of evidence that mucin genes are regulated by posttranscriptional as well as transcription mechanisms. Rat Muc4 is regulated at the posttranscriptional level by transforming growth factor β in mammary epithelial cells (14). *MUC*2 is regulated posttranscriptionally by phorbol ester in a human colon carcinoma cell line (15). Tumor necrosis factor α increases the mRNA stability of *MUC*5AC in lung carcinoma cells (16). Our data support the concept that posttranscriptional regulation is an important mechanism to amplify mucin gene expression. Furthermore, this is the first report to demonstrate that two different mucin genes, localized to different chromosomes, are regulated by the same mediator with similar kinetics and mechanisms.

Although NE has been reported to regulate expression of several mammalian genes (reviewed in Ref. 11), little is known about the molecular mechanisms of NE action. We demonstrate that NE triggers oxidant stress in respiratory epithelial cells, and this oxidant stress is an important signal in mucin gene regulation. There are several potential sources of oxidant stress that are triggered by NE activity. NE has been reported to catalyze the conversion of xanthine dehydrogenase to xanthine oxidase, a superoxide generator (17). Alternatively, NE may promote oxidative stress in respiratory epithelial cells by proteolytically degrading antioxidant defenses such as extracellular superoxide dismutase (18), thus diminishing the antioxidant capacity of the cell.

In conclusion, by defining the molecular pathways utilized by NE to regulate respiratory mucin genes, we will identify specific targets for more effective therapies to treat mucus obstruction in chronic airway inflammatory diseases.

REFERENCES

1. A DiStefano, A Capeli, M Lusuardi, P Balbo, C Vecchio, P Vastrelli, CE Mapp, LM Fabbri, CF Donner, M Saetta. Severity of airflow limitation is associated with severity of airway inflammation in smokers. Am J Respir Crit Care Med 158:1277–1285, 1998.
2. KC Kim, K Wasano, RM Niles, JE Schuster, PJ Stone, JS Brody. Human neutrophil elastase releases cell surface mucins from primary cultures of hamster tracheal epithelial cells. Proc Natl Acad Sci USA 84:9304–9308, 1987.
3. TG Christensen, R Breuer, LJ Hornstra, EC Lucey, PJ Stone, GL Snider. An ultrastructural study of the response of hamster bronchial epithelium to human neutrophil elastase. Exp Lung Res 13:279–297, 1987.
4. R Breuer, TG Christensen, EC Lucey, PJ Stone, GL Snider. An ultrastructural morphometric analysis of elastase-treated hamster bronchi shows discharge followed by progressive accumulation of secretory granules. Am Rev Respir Dis 136:969–703, 1987.
5. JP Audie, A Janin, N Porchet, MC Copin, B Gosselin, JP Aubert. Expression of human mucin genes in respiratory, digestive and reproductive tracts ascertained by in situ hybridization. J Histochem Cytochem 41:1479–1485, 1993.
6. N Moniaux, S Nollet, N Porchet, P Degand, A Laine, JP Aubert. Complete sequence of the human mucin MUC4: a putative cell membrane-associated mucin. Biochem J 338:325–333, 1999.
7. MC Rose, B Kaufman, BM Martin. Proteolytic fragmentation and peptide mapping of human carboxyamidomethylated tracheobronchial mucin. J Biol Chem 264:8193–8199, 1989.
8. JR Davies, N Svitacheva, L Lannefors, R Kornfalt, I Carlstedt. Identification of MUC5B, MUC5AC, and small amounts of MUC2 mucins in cystic fibrosis airway secretions. Biochem J 344:321–330, 1999.
9. JT Berger, JA Voynow, KW Peters, MC Rose. Respiratory carcinoma cell lines: MUC genes and glycoconjugates. Am J Respir Cell Mol Biol 20:500–510, 1999.
10. TE Gray, K Guzman, CW Davis, LH Abdullah, P Nettesheim. Mucociliary differentiation of serially passaged normal human tracheobronchial epithelial cells. Am J Respir Cell Mol Biol 14:104–112, 1996.
11. JA Voynow, LR Young, Y Wang, T Horger, MC Rose, BM Fischer. Neutrophil elastase increases MUC5AC mRNA and protein expression in respiratory epithelial cells. Am J Physiol 176:L835–L843, 1999.
12. J Ross. mRNA stability in mammalian cells. Microbiol Rev 59:423–450, 1995.

13. JA Royall, H Ischiropoulos. Evaluation of 2′,7′-dichlorofluorescin and dihydrorhodamine 123 as fluorescent probes for intracellular H_2O_2 in cultured endothelial cells. Arch Biochem Biophys 302:348–355, 1993.
14. SA Price-Schiavi, CAC Carraway, N Fregien, KL Carraway. Post-transcriptional regulation of a milk membrane protein, the sialomucin complex (ascites sialoglycoprotein ASGP-1/ASGP-2 (rat Muc4)) by transforming growth factor beta. J Biol Chem 273:35228–35237, 1998.
15. A Velcich, LH Augenlicht. Regulated expression of an intestinal mucin gene in HT29 colonic carcinoma cells. J Biol Chem 268:13956–13961, 1993.
16. MT Borchers, MP Carty, GD Leikauf. Regulation of human airway mucins by acrolein and inflammatory mediators. Am J Physiol 276:L549–L555, 1999.
17. SH Phan, DE Gannon, J Varani, RS Ryan, PA Ward. Xanthine oxidase activity in rat pulmonary artery endothelial cells and its alteration by activated neutrophils. Am J Pathol 134:1201–1211, 1989.
18. JM McCord, B Gao, J Leff, SC Flores. Neutrophil-generated free radicals: possible mechanisms of injury in adult respiratory distress syndrome. Environ Health Perspect 102(suppl) 120:57–60, 1994.

20

Pseudomonas Adhesion to MUC1 Mucins: A Potential Role of MUC1 Mucins in Clearance of Inhaled Bacteria

Kwang Chul Kim, Sang Won Hyun, Beom Tae Kim, Daoud Meerzaman, Min Ki Lee, and Erik Lillehoj
University of Maryland
Baltimore, Maryland

Mucus lining the airway luminal surface serves as a primary physicochemical barrier for the lung against airborne particles and chemicals. The major constituent of mucus responsible for this property is a mixture of glycoproteins collectively referred to as mucins. Mucins are produced by two types of cells in the airway—goblet cells in the surface epithelium and mucous cells in the submucosal gland. At present, 13 genes have been identified as mucin-producing genes, five of which have been shown to be expressed in the airway (1)—MUC1, MUC2, MUC4, MUC5/5AC, and MUC5B. Mucins derived from MUC genes are secreted into the airway lumen, except for MUC1 mucin, which is present as a transmembrane glycoprotein on the surface of secretory cells (2). Although it seems likely that MUC1 mucin is also released into the lumen, as previously shown for other cell types including various tumor cells (3) and primary uterine epithelial cells (4), its contribution to the total amount of the secreted mucins in the airway appears to be negligible (5). Excellent reviews are available elsewhere

(2,6). In this chapter, we will focus on the potential role of MUC1 mucins in the mucociliary clearance of the airway with some background information on the structure and function of MUC1 mucins. By convention, human mucin genes will be referred to as MUC and nonhuman mucin genes as Muc throughout this chapter.

STRUCTURE AND FUNCTION OF MUC1 MUCIN

MUC1 mucins are highly glycosylated transmembrane proteins with high molecular mass and are widely expressed on the apical surface of most secretory epithelial cells. The human MUC1 gene is localized on chromosome 1q21-24 (7), and its deduced amino acid sequence indicates four characteristic domains—the N-terminal signal sequence, the extracellular (EC) domain containing tandem repeats, the transmembrane (TM) domain, and the C-terminal cytoplasmic (CT) domain (8–12). A schematic structure of MUC1 mucin is seen in Fig. 1. The EC domain contains 20-amino-acid tandem repeats that seem to occur 21–125 times as result of genetic polymorphism (13). In contrast, the EC domain of the hamster Muc1 gene contains only 12 tandem repeats of 20 amino acids (14). This repetitive region and the regions adjacent to it comprise most of the EC portion of the molecule, extending 200–500 nm above the plasma membrane (15). The C-terminal CT domain contains 69 (human) or 68 (hamster) amino acids with a high percentage of tyrosine, serine, and threonine residues, which are potential phosphorylation sites. While the EC domain shows low sequence identity among species, the amino acid sequence of the CT domain is very similar (88%) (14). Based on their anatomical location and numerous phosphorylation sites, MUC1 mucins have been suggested to function as a receptor. Cell surface

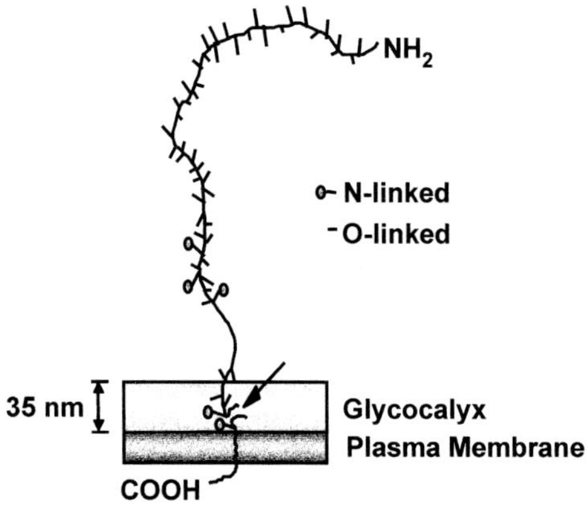

FIGURE 1 A proposed structure of MUC1 mucin.

mucins are abundantly expressed in most carcinoma cells and have been shown to reduce intercellular (15,16) as well as matrix adhesion (17) through their extended and rigid structure (15). They are also expressed by normal epithelial cells in various glandular epithelial tissues, including the respiratory, gastrointestinal, and female reproductive tracts (18–20), and have been shown to be associated with cell differentiation in mammary (21), uterine epithelial (20), and airway epithelial cells (14). Similar to cancer cells, Muc1 mucins in uterine epithelial cells have been shown to exhibit an antiadhesive property, suggesting that they play an important role in maintaining the prereceptive phase in the uterus (22). The role of MUC1 mucins in the airway, however, is unknown.

MUC1 MUCINS AND INTRACELLULAR SIGNALING

The deduced amino acid sequence of MUC1 mucin reveals that some of the tyrosine residues present in the CT domain of MUC1 mucins, once phosphorylated, constitute docking sites for proteins known to be involved in signal transduction, such as PI-3′-kinase, PLCγ1, Src and Grb2. Despite the presence of potential adaptor protein docking sites, MUC1 mucins are different from the typical receptor tyrosine kinase (RTK) in that they do not contain autophosphorylation sites. In this aspect, MUC1 mucins rather resemble the structure of a prototypical cytokine receptor known to have no autophosphorylation activity. Unlike the cytokine receptors, however, MUC1 mucins do not contain consensus sequence motifs known for JAKs (Janus-kinases). Furthermore, judging from their structure, it seems quite unlikely that MUC1 mucins can form a dimer as do RTK molecules following ligand-receptor binding. Collectively, it seems unclear at this point to which of the known receptor types MUC1 mucins belong.

Recently, Zrihan-Licht et al. (23) demonstrated, for the first time, the presence of phosphorylation on the CT domain of MUC1 mucins in a breast cancer cell line. Using the same cell line, Pandey et al. (24) showed that MUC1 mucins can directly interact with the SH2 domain of an adaptor protein, Grb2. They also demonstrated that the MUC1/Grb2 complex associates with a guanine nucleotide exchange protein, Sos, supporting a role for MUC1 mucin in intracellular signaling. A ligand responsible for tyrosine phosphorylation of the CT domain of MUC1 mucins has been identified in the breast cancer cell as the EC domain of MUC1 mucins containing a tandem repeat array (25). The signaling mechanism of MUC1 mucins, however, remains to be uncovered.

MUC1 MUCINS PRODUCED BY AIRWAY EPITHELIAL CELLS

Primary hamster tracheal surface epithelial (HTSE) cells grown on a thick collagen gel synthesize and secrete mucins at confluence (26,27). This HTSE cell culture system has been used extensively for studying the regulation of goblet cell mucin secretion (28). Many years ago we made an interesting observation that human neutrophil elastase can release mucins from HTSE cells (29). Some of these released mucins were derived from the apical surface of goblet cells, indicating for the first time the presence of mucins on the surface of airway goblet cells. Further characterization of the plasma membrane of HTSE cells revealed the presence of two different types of membrane mucins—one almost indistinguishable from secreted mucins and the other immunoprecipitable with anti-MUC1 antibody (30). We then cloned a full-length Muc1 cDNA from our HTSE cell cDNA library and found that expression of Muc1 mRNA correlated with goblet cell differentiation (14). The degree of its expression in HTSE cells was very high and almost comparable to that of MCF-7 cells (a breast cancer cell line). Based on both the localization and molecular structure of Muc1 mucins in the airway, it seemed highly likely that they serve a receptor function in airway epithelial cells.

PSEUDOMONAS ADHESION TO EPITHELIAL CELLS

Pseudomonas aeruginosa (PA) is an opportunistic pathogen responsible for a wide range of infections, one of the most debilitating being chronic pulmonary infection in cystic fibrosis (CF) patients. In CF, nonmucoid strains of PA initially colonize the upper respiratory tract of patients before converting into mucoid alginate-producing variants (31). The latter are almost exclusively associated with hyperviscous bronchial secretions of CF patients. A plethora of information has amassed with respect to the interactions between PA and airway mucus (32). Although the exact pathophysiology of PA infection in CF is still unclear, it is currently thought that the initial stage of infection involves adhesion to airway epithelial cells (33) through asialoglycolipids present on the cell surface (34). It was recently shown that adhesion of PA to these glycolipids on airway epithelial cells results in translocation of NF-κB and initiation of IL-8 expression (35), indicating an important role in initiating an epithelial proinflammatory response to PA adhesion.

MUC1 MUCINS ARE ADHESION SITES FOR *PSEUDOMONAS AERUGINOSA*

Based on the structure of MUC1 mucins and their location in the airway, we hypothesized that Muc1 mucins may serve as an adhesion site for PA. In testing the hypothesis, we first established a cell line that expresses both Muc1 mRNA and protein by stable transfection of a full-length cDNA of hamster Muc1 (14) into CHO cells. In contrast to CHO-Muc1 cells, the parent CHO cells do not express Muc1. Our binding experiments revealed that PA adhesion to CHO-Muc1

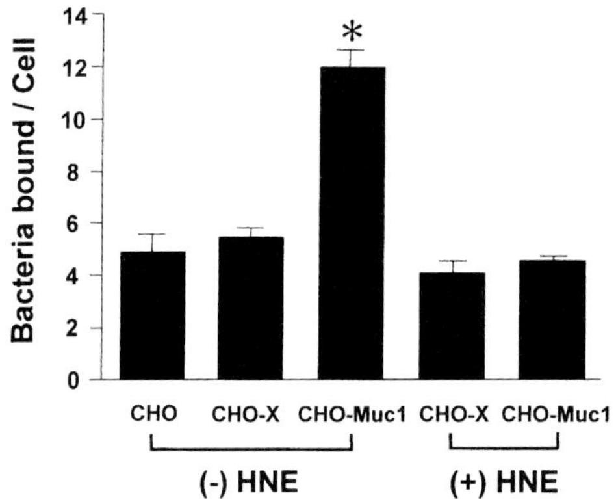

FIGURE 2 *Pseudomonas* adhesion to CHO, CHO-X, and CHO-Muc1 cells in the presence or absence of human neutrophil elastase (HNE). (Detailed methods for bacterial growth and adhesion are described in Ref. 36.) Briefly, a PA strain (CF-3, mucoid) isolated from a cystic fibrosis patient was grown in LB broth and then metabolically radiolabeled in sulfate-free M9 medium containing $Na_2{}^{35}SO_4$ as described. CHO, CHO-X (CHO cells transfected with the vector alone), and CHO-Muc1 cells (CHO cells transfected with the vector plasmid containing the hamster Muc1 cDNA) were incubated with ^{35}S-PA in the presence or absence of 5 μg/mL of HNE, and the numbers of PA adhered were measured. Each data point represents a mean $\pm$ SEM of 6 wells. * Significantly different ($p < 0.01$)

cells was significantly greater than to CHO cells and the increase in PA adhesion to CHO-Muc1 cells was completely abolished following treatment with neutrophil elastase (Fig. 2). Since neutrophil elastase cleaves the EC domain of Muc1 mucins (data not shown), as predicted from our previous publication (29), these results indicate that Muc1 mucins serve as an adhesion site for PA.

A POTENTIAL ROLE OF MUC1 MUCINS IN THE AIRWAY

The physiological significance of PA adhesion to MUC1 mucins remains unknown. Transiently inspired PA are normally trapped by airway secretions and removed by mucociliary clearance mechanisms. It might be possible that MUC1 mucins serve as a secondary defense barrier against airborne bacteria that manage to escape the gel phase of airway mucus, the primary physical barrier. Bacteria "trapped" by MUC1 mucins might be cleared from the airway by some as yet unidentified proteolytic cleavage. In other membrane receptor-ligand systems, it has been demonstrated that ligand binding induces intracellular signaling events, ultimately leading to release of the receptor from the cell surface. For example,

Fan and Derynck (37) recently showed that growth factor activation of cell surface RTKs induces protease-mediated release of the EC domain of TGF-α through activation of the ERK MAP kinase signaling pathway. As discussed above, MUC1 mucin interaction with Grb2 and Sos proteins similarly implicates its association with a MAP kinase pathway. It is tempting to speculate that PA adhesion to cell-associated MUC1 mucin activates one or more intracellular MAP kinase cascades and that one of the resulting consequences is release of the MUC1 mucin-PA complex from epithelial cells and clearance from the airway. Our preliminary data showed that PA adhesion to CHO-Muc1 cells results in an increase in tyrosine phosphorylation of Muc1 mucins (data not shown), strongly suggesting a role of Muc1 mucins in airway infection and inflammation.

REFERENCES

1. M Rose, S Gendler. Airway mucin genes and gene products. In: DF Rogers, MI Lethem, eds. Airway Mucus. Basic Mechanisms and Clinical Prospectives. Basel: Birkhauser Verlag, 1997, pp 41–66.
2. SJ Gendler, AP Spicer. Epithelial mucin genes. Annu Rev Physiol 57:607–634, 1995.
3. M Boshell, EN Lalani, L Pemberton, J Burchell, S Gendler, J Taylor-Papadimitriou. The product of the human MUC1 gene when secreted by mouse cells transfected with the full-length cDNA lacks the cytoplasmic tail. Biochem Biophys Res Commun 185:1–8, 1992.
4. RA Pimental, J Julian, SJ Gendler, DD Carson. Synthesis and intracellular trafficking of Muc-1 and mucins by polarized mouse uterine epithelial cells. J Biol Chem 271: 28128–28137, 1996.
5. D Meerzaman, X Zhang, MJ Jo, KC Kim. Muc1 mucins are not a major type of secreted mucins from airway goblet cells in primary culture (abstr). Am J Respir Crit Care Med 157:A728, 1998.
6. J Hilkens, MJL Ligtenberg, HL Vos, SV Litvinov. Cell membrane-associated mucins and their adhesion-modulating property. Trend Biol Soc 17:359–363, 1992.
7. DM Swallow, S Gendler, B Griffiths, A Kearney, S Povey, D Sheer, RW Palmer, J Taylor-Papadimitriou. The hypervariable gene locus PUM, which codes for the tumour-associated epithelial mucins, is located on chromosome 1, within the region 1q21-24. Ann Hum Genet 51:289–294, 1987.
8. MS Lan, SK Batra, WN Qi, RS Metzgar, MA Hollingsworth. Cloning and sequencing of a human pancreatic tumor mucin cDNA. J Biol Chem 265:15294–15299, 1990.
9. MJL Ligtenberg, HL Vos, AMC Gennissen, J Hilkens. Episialin, a carcinoma-associated mucin, is generated by a polymorphic gene encoding splice variants with alternative amino termini. J Biol Chem 265:5573–5578, 1990.
10. DH Wreschner, M Hareuveni, H Tsarfaty, N Smorodinsky, J Horev, J Zaretsky, P Kotkes, M Weiss, R Lathe, A Dion, I Keydar. Human epithelial tumor antigen cDNA sequences. Differential splicing may generate multiple protein forms. Eur J Biochem 189:463–473, 1990.

11. AP Spicer, G Parry, S Patton, SJ Gendler. Molecular cloning and analysis of the mouse homologue of the tumor-associated mucin, MUC1, reveals conservation of potential O-glycosylation sites, transmembrane, and cytoplasmic domains and a loss of minisatellite-like polymorphism. J Biol Chem 266:15099–15109, 1991.
12. HL Vos, Y Vries, J Hilkens. The mouse episialin (MUC1) gene and its promoter: rapid evolution of the repetitive domain in the protein. Biochem Biophys Res Commun 181:121–130, 1991.
13. SJ Gendler, CA Lancaster, J Taylor-Papadimitriou, T Duhig, N Peat, J Burchell, L Pemberton, EN Lalani, D Wilson. Molecular cloning and expression of human tumor-associated polymorphic epithelial mucin. J Biol Chem 265:15286–15293, 1990.
14. H Park, SW Hyun, KC Kim. Expression of MUC1 mucin gene by hamster tracheal surface epithelial cells in primary culture. Am J Respir Cell Mol Biol 15:237–244, 1996.
15. E Van Kemenade, MJL Ligtenberg, AJ de Boer, F Buijs, HL Vos, CJM Melief, J Hilkens, CG Figdor. Episialin (MUC1) inhibits cytotoxic lymphocyte-target cell interaction. J Immunol 151:767–776, 1993.
16. K Kondo, N Kohno, A Yokoyama, K Hiwada. Decreased MUC1 expression induces E-cadherin-mediated cell adhesion of breast cancer cell lines. Cancer Res 58:2014–2019, 1998.
17. J Wesseling, SW van der Valk, HL Vos, A Sonnenberg, J Hilkens. Episialin (MUC1) overexpression inhibits integrin-mediated cell adhesion to extracellular matrix components. J Cell Biol 129:255–265, 1995.
18. L Pemberton, J Taylor-Papadimitriou, SJ Gendler. Antibodies to the cytoplasmic domain of the MUC1 mucin show conservation throughout mammals. Biochem Biophys Res Comm 185:167–175, 1992.
19. MA Hollingsworth, SK Batra, WN Qi, JP Yankaskas. MUC1 mucin mRNA expression in cultured human nasal and bronchial epithelial cells. Am J Respir Cell Mol Biol 6:516–520, 1992.
20. NA Hey, RA Graham, MW Weif, JD Aplin. The polymorphic epithelial mucin MUC1 in human endometrium is regulated with maximal expression in the implantation phase. J Clin Endocrinol Metab 78:337–342, 1994.
21. G Parry, JL Stubbs, MJ Bissell, C Schmidhauser, AP Spicer, SJ Gendler. Studies of Muc-1 mucin expression and polarity in the mouse mammary gland demonstrate developmental regulation of Muc-1 glycosylation and establish the hormonal basis of mRNA expression. J Cell Sci 101:191–199, 1992.
22. S Hild-Petito, AT Fazleabas, J Julian, DD Carson. Mucin (Muc-1) expression is differentially regulated in uterine luminal and glandular epithelia of the baboon (Papio anubis). Biol Reproduction 54:939–947, 1996.
23. S Zrihan-Licht, A Baruch, O Elroy-Stein, I Keydar, DH Wreschner. Tyrosine phosphorylation of the MUC1 breast cancer membrane proteins. Cytokine receptor-like molecules. FEBS Lett 356:130–136, 1994.
24. P Pandey, S Kharbanda, D Kufe. Association of the DF3/MUC1 breast cancer antigen with Grb2 and the Sos/Ras exchange protein. Cancer Res 55:4000–4003, 1995.
25. A Baruch, M Hartmann, M Yoeli, AS Greenstein, Y Stadler, Y Skornik, J Zaretsky, NI Smorodinsky, I Keydar, DH Wreshner. The breast cancer-associated MUC1 gene

generates both a receptor and its cognate binding protein. Cancer Res 59:1552–1561, 1999.
26. KC Kim, JI Rearick, P Nettesheim, AM Jetten. Biochemical characterization of mucous glycoproteins synthesized and secreted by hamster tracheal epithelial cells in primary culture. J Biol Chem 260:4021–4027, 1985.
27. R Wu, E Nolan, C Turner. Expression of tracheal differentiation functions in serum-free hormone-supplemented medium. J Cell Physiol 125:167–181, 1985.
28. KC Kim, JS Brody. Use of primary cell culture to study regulation of airway surface epithelial mucus secretion. In: EN Chantler, NA Ratcliffe, eds. Mucus and Related Topics. Cambridge: The Company of Biologists Limited, 1989, pp 231–239.
29. KC Kim, K Wasano, RM Niles, JE Schuster, PJ Stone, JS Brody. Human neutrophil elastase releases cell surface mucins from primary cultures of hamster tracheal epithelial cells. Proc Natl Acad Sci USA 84:9304–9308, 1987.
30. E Paul, DI Lee, SW Hyun, S Gendler, KC Kim. Identification and characterization of high molecular-mass mucin-like glycoproteins in the plasma membrane of airway epithelial cells. Am J Respir Cell Mol Biol 19:681–690, 1998.
31. N Hoiby. *Pseudomonas aeruginosa* infection in cystic fibrosis: relationship between mucoid strains of *Pseudomonas aeruginosa* and the humoral immune response. Acta Pathol Microbiol Scand 82:551–558, 1974.
32. R Ramphal, C Guay, GB Pier. *Pseudomonas aeruginosa* adhesins for tracheobronchial mucin. Infect Immun 55:600–603, 1987.
33. L Imundo, J Barash, A Prince, Q Al-Awqati. Cystic fibrosis epithelial cells have a receptor for pathogenic bacteria on their apical surface. Proc Natl Acad Sci USA 92:3019–3023, 1995.
34. HC Krivan, DD Roberts, V Ginsburg. Many pulmonary pathogenic bacteria bind specifically to the carbohydrate sequence GalNAc$_{\beta 1-4}$ Gal found in some glycolipids. Proc Natl Acad Sci USA 85:6157–6161, 1998.
35. E DiMango, AJ Ratner, R Bryan, S Tabibi, A Prince. Activation of NF-κB by adherent *Pseudomonas aeruginosa* in normal and cystic fibrosis respiratory epithelial cells. J Clin Invest 101:2598–2605, 1998.
36. KS Rostand, JD Esko. Cholesterol and cholesterol esters: host receptors for *Pseudomonas aeruginosa* adherence. J Biol Chem 268:24053–24059, 1993.
37. H Fan, R Derynck. Ectodomain shedding of TGF-α and other transmembrane proteins is induced by receptor tyrosine kinase activation and MAP kinase signaling cascades. EMBO J 18:6962–6972, 1999.

21

High-Density DNA Microarray Membranes to Study Gene Expression Patterns Associated with Human Airway Epithelial Cell Differentiation in Culture

Mary M. J. Chang, Yin Chen, Yu Hua Zhao, and Reen Wu
University of California, Davis
Davis, California

Ching Li and Konan Peck
Sinica Academy
Nankang, Taiwan

INTRODUCTION

The mucociliary epithelium plays an important role in regulating the homeostasis of conducting airways (1). An aberrant expression of mucociliary functions is frequently associated with various lung and airway diseases. The nature of the regulation of this aberrant expression is not known. In vitro culture of primary airway epithelial cells is an invaluable tool used to study, manipulate, and identify conditions that regulate cell differentiation. Our laboratory was first involved in the development of serum-free, hormone-supplemented medium to grow airway epithelial cells from various species in culture, including those from humans (2). Subsequently and together with several other laboratories, we were able to demonstrate the importance of several factors contributing to airway epithelial cell differentiation in culture. These factors included the thickness of the collagen gel substratum (3), the calcium level in the medium (4), the vitamin A level, and an air/liquid interface culture condition (5–7). How these factors contribute to airway mucociliary differentiation in vitro is still unknown.

We previously utilized differential and/or subtractive hybridization and differential display methods to identify those genes that are differentially expressed in the presence of vitamin A (8–10). These studies have established that several differentially expressed genes can be identified. For instance, the expression of a small proline-rich protein gene (SPRR1B), whose expression was shown to be elevated in cells and tissues with a squamous phenotype, was also enhanced in airway epithelial cell culture under vitamin A–depleted conditions (9). In addition, this elevation was consistently seen with other agents able to elevate various squamous cell functions (13), such as high calcium, UV light (11,12), and phorbol ester. Thus, these substances and vitamin A deprivation activate SPRR1B gene expression. We also demonstrated that, in addition to SPRR1B, several other genes were affected. One example was keratin gene expression, which was elevated in vitamin A–depleted culture condition (14). There should be several additional, not yet identified genes, whose expression is affected by these treatments. Since many genes are involved in each step of cell differentiation, an essential question for understanding cell differentiation is whether all these differentially expressed genes are coordinately regulated. The traditional approach using Northern blots to examine a small number of genes is insufficient to understand the differentiation process comprehensively. One of the research challenges in the postgenomic period is to have a high-throughput system to identify all genes involved and the pattern of the gene expression associated with the process. The recent development of high-density DNA microarray membranes (15) and glass slides (16,17) is quite appropriate for this challenge. With microarrays, the expression level of thousands of genes can be investigated simultaneously using a small area, which can be easily handled for hybridization, quantitation, and analysis. Thus, thousands of genes can be analyzed together, and genes that are differentially expressed can be identified.

The purpose of this chapter is to utilize the newly developed technology of microarray membranes to analyze genes whose expression is associated with mucociliary differentiation of human airway epithelial cells in vitro. Two types of nylon membranes were used. One contains 884 sequence-verified expression sequence tag (EST) clones, the other contains 576 uni-EST clones. Data obtained from these membranes were further characterized by Northern blot hybridization.

MATERIALS AND METHODS

Sources of Human Airway Tissues and Primary Epithelial Cell Culture System

Human tracheobronchial and lung tissues were obtained from the University of California at Davis Medical Center or the Anatomic Gift Foundation (Laurel, MD) with consent (4,5). The Human Subject Review Committee of the University of California at Davis approved all procedures involved in tissue procurement. Excised tissues were transported to the lab in ice-cold minimal essential medium (MEM, Sigma, St. Louis, MO). Primary airway epithelial cells were isolated by the protease method as described before (4,5). Epithelial cells were plated on tissue culture dishes under a serum-free hormone-supplemented medium, which was Ham's F12/DME (1 : 1), supplemented with insulin (5μg/mL), transferrin (5μg/mL), EGF (10 ng/mL), dexamethasone (0.1μM), cholera toxin (20 ng/mL), and bovine hypothalamus extract (15 μg/mL). After confluence, primary cultures were passaged and plated onto four different culture substrates: tissue culture dishes (TC), dishes coated with collagen gel substratum (CG), Transwell™ chambers (BI), and Transwell™ chambers coated with collagen gel substratum (BICG). Medium used for all cultures was the same as described above, except that all-*trans*-retinoic acid (0.3 μM) was added. For both TC and CG culture conditions, cultures were immersed under the culture medium, while for BI and BICG, cultures were lifted up to the air/liquid interface after one week of immersed conditions. All these passage-1 cultures were terminated on day 21 for RNA isolation and ultrastructural analysis.

RNA Isolation and Northern Blot Hybridization

At day 21 after plating, RNA was isolated from cultures by a single-step phenol-chloroform extraction method (18). For Northern blot hybridization, equal amounts of total RNA (20 μg/lane) were subjected to electrophoresis on a 1.2% agarose gel in the presence of 2.2 mM formaldehyde and transblotted onto Nytran membranes. The RNA was cross-linked to the membrane by a UV Stratalinker 2400 (Stratagene, La Jolla, CA). Membrane prehybridization and hybridization with ^{32}P-labeled cDNA probes was carried out (10). The relative abundance of the specific gene message in Northern blots was normalized using the 18S ribosomal RNA (rRNA) band (10).

High-Density DNA Microarray Membrane Preparation and Hybridization

EST clone DNA was PCR amplified and concentrated (15). An Arrayer-03™ was used to spot DNA in high density onto nylon membranes. The pin size for the

Arrayer was 75 μm, spotting 100 μm apart. Two types of microarray membranes were used in this study. One contained 576 EST clones using a 0.3 × 0.5 cm membrane, the other contained 884 EST clones using a 0.6 × 0.9 cm membrane (Details of the EST clones included in these two membranes can be found on the following web sites: *ftp://genestamp.sinica.edu.tw/marray/arrayinfo* and *http://140.116.58.123/microarray/*, respectively.)

Total RNA isolated from these cultures was used for poly(A) mRNA preparation using a kit from the Qiagen Inc., following the manufacturer's suggested protocol. cDNA probes labeled with biotin-dUTP or with Dig-dUTP were generated from mRNA with random primers by reverse transcription.

Membranes were prehybridized and hybridized with these cDNA probes at 60°C for 12–18 hours. Microarray membranes were washed twice with 2X SSC-0.1% SDS and three times in 0.1X SSC-0.1% SDS. After washing, membranes were treated with streptavidin-β-galactosidase or alkaline phosphatase–conjugated anti-DIG antibody and substrates, X-gal, and Fast Red TR salts, respectively. Color-stained membranes were scanned in a flatbed scanner with 3000 dpi. Images were stored as Photoshop files for quantitative analysis of the color intensity of each spot (15).

RESULTS AND DISCUSSION

Effects of Culture Conditions on Human TBE Cell Differentiation In Vitro

Previously, we have demonstrated that both high calcium (3) and thick collagen gel substratum (4) enhanced mucous cell differentiation in primary cultures of monkey airway epithelial cells. Similar results can be demonstrated in human TBE cells. We have analyzed mucin secretion (19) and MUC gene expression in these cultures. For the steady state rate of mucin secretion, we observed a production of mucin at 30–200 ng/million cells/day in primary BICG cultures. The level of secretion decreased in an order of BICG > BI > CG > TC (data not shown). A similar trend was also seen for MUC gene expression in these cultures. Figure 1 shows the expression of MUC4 message in these culture conditions. After normalizing the data with the 18S rRNA band, the level of expression was BICG > BI > CG > TC. Similar results were obtained for the MUC5AC and MUC5B gene expression (data not shown). These analyses therefore demonstrated that the expression of these mucus genes was concordant with the level of differentiation of the airway epithelial cells in cultures.

To further characterize epithelial cell differentiation in culture, both transmission and scanning electron microscopy (TEM and SEM) were used. At the TEM level, mucus-secreting granules were seen in CG, BI, and BICG cultures. Examples of this feature in CG and BICG cultures are shown in Figure 2. We

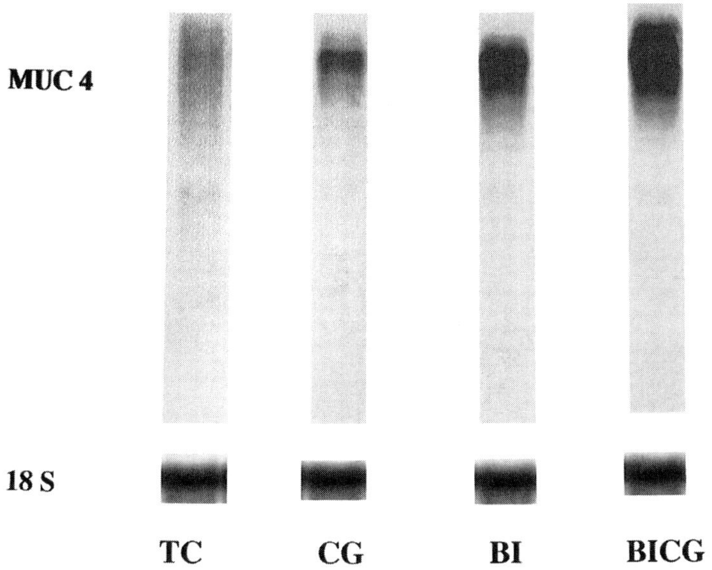

FIGURE 1 Northern blot analysis of MUC4 gene expression in primary human tracheobronchial epithelial (TBE) cells cultured under various culture conditions. TBE cells were cultured under TC, CG, BI, or BICG culture conditions for 21 days as described in the text. RNA was prepared for Northern blot hybridization with a ^{32}P-labeled oligomer probe specific to human MUC4 message. The 18S rRNA band was used to normalize for RNA loading on the gel.

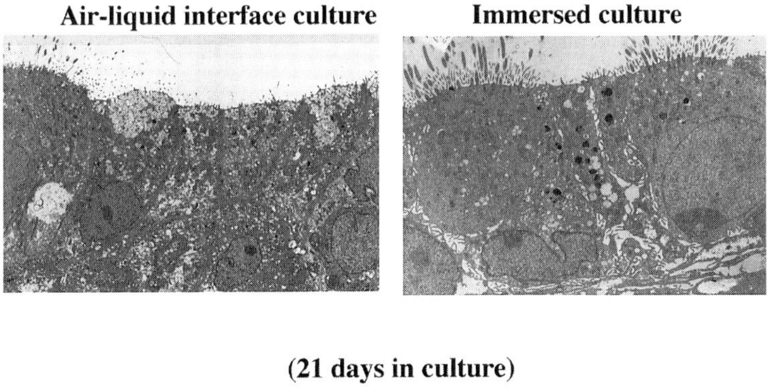

FIGURE 2 Transmission EM analysis of cultured primary human TBE cells culture. TBE cells were grown on collagen gel-coated Transwell™ chambers and maintained either immersed or at an air/liquid interface for 21 days as described in text. Note: ciliated feature and mucus-secreting granules can be seen in both cultures.

Air-liquid interface culture **Immersed culture**

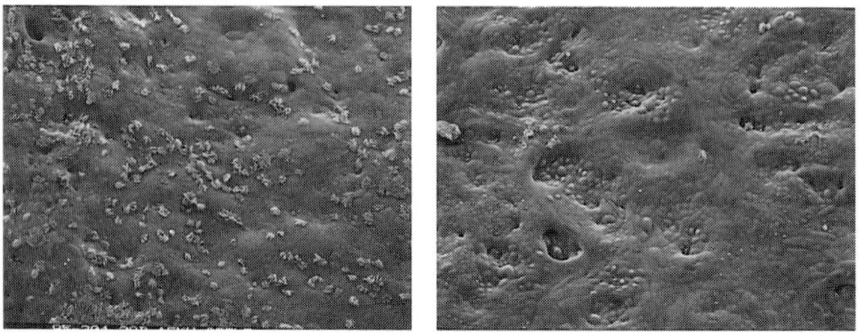

(21 days in culture)

Figure 3 Scanning EM analysis of cultured cells of human primary TBE cultures. TBE cells were cultured as described in Figure 2. *Note*: Ciliated cells are seen in cultures grown at the air/liquid interface, while few ciliated cells are seen in immersed cultures.

have also examined the presence or absence of ciliated features under these culture conditions. SEM analysis demonstrated the presence of numerous cilia on the surface of BI (data not shown) and BICG (Fig. 3) cultures. In contrast, immersed cultures, such as TC and CG conditions, had ciliated cells (Fig. 2) but in a low number (Fig. 3). These data are consistent with publications from other laboratories. Overall, we have demonstrated a tendency of mucociliary cell differentiation associated with cultures that are maintained biphasically and on collagen gel substratum. The extent of cell differentiation in these culture conditions is BICG > BI > CG > TC.

Analysis of Differential Gene Expression Associated with Mucociliary Differentiation

mRNA isolated from these four culture conditions at day 21 was used to generate cDNA probes for hybridization to DNA microarray membranes. Using the single color approach, Figures 4 and 5 show hybridization results with the 576 and 884 DNA microarray membranes, respectively. More than 50% of EST clones in these 576 and 884 EST DNA microarray membranes had positive hybridization. Quantitative analyses of the hybridization to these spots (using intensity measurements) revealed no significant expression change of a majority of these EST between the four different culture conditions. However, there were changes in the expression between different culture conditions. For instance, in one of the hybridization experiments using the 884 DNA microarray membrane, the scatter plot between BICG and TC showed five genes, which changed their hybridization

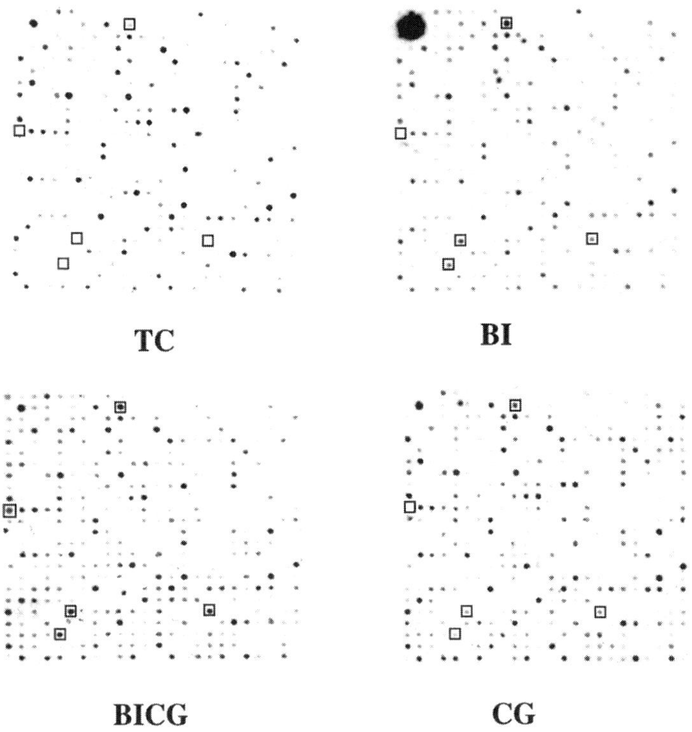

FIGURE 4 Microarray analysis of gene expression in primary cultures of human TBE cells grown under various culture conditions. Culture conditions are described in the text. At day 21, RNA was isolated and used to generate Dig-dUTP–labeled cDNA for hybridization. The microarray membrane contained 576 uni-EST clones. The hybridization and color development was carried out as described in the text. Squares cover genes whose expression is elevated in concordance with cell differentiation.

intensity significantly, while six genes changed between BICG and CG and five genes between BICG and BI cultures (Fig. 6). For the purpose of this chapter, we will focus on those changes that were consistently associated with differentiation to the mucociliary cell phenotype. Using these criteria, we observed 11 EST clones whose corresponding messages were changed in culture according to the level of mucociliary differentiation as described above. Among these genes, 8 were found to increase expression levels in parallel with differentiation by showing the following expression pattern: BICG > BI > CG > TC. Three genes on the other side decreased in expression in parallel with differentiation. Table 1

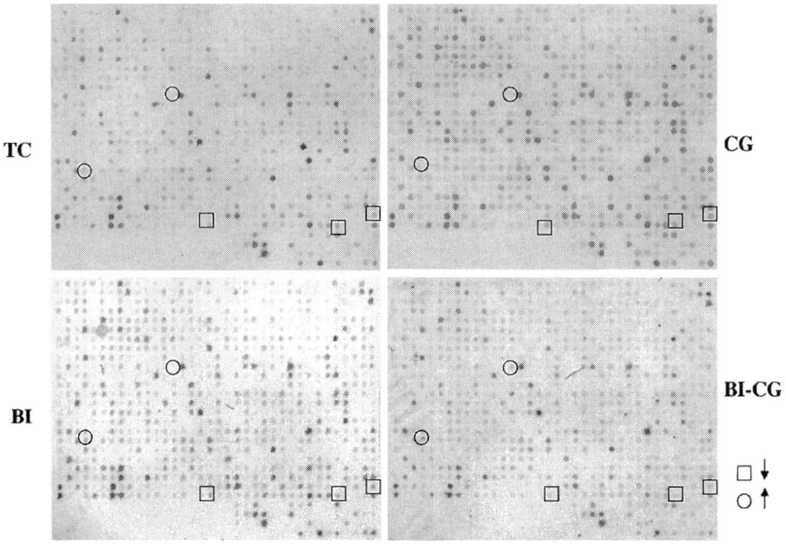

FIGURE 5 Microarray analysis of gene expression in primary cultures of human TBE cells grown under various culture conditions. The culture conditions and procedure of hybridization are the same as in Figure 4, except that cDNA probes were labeled with biotin-dUTP and the microarray membrane contained 884 sequence-verified EST clones. Squares cover genes whose expression is elevated in concordance with cell differentiation, while circles cover genes whose expression is suppressed relative to cell differentiation.

lists the information regarding these genes. Six of the eight upregulated genes have been identified to be involved in signal transduction, and the remaining two are related to the family of surface receptor proteins. Among the suppressed genes, one is involved in transcriptional regulation, and the other two are cytokine and protein receptors.

Verification of Differentially Expressed Genes by Northern Blot Hybridization

To verify the significance of the above finding, Northern blot hybridization was used to verify the changes in expression of some of the 11 identified genes in these cultures. As shown in Figure 7, the changes observed by the microarray analysis could be reproduced by Northern blot hybridization. In this figure, the expression of Smad6, a ribonuclease/angiogenin inhibitor, and Jun-B proto-oncogene was altered by the culture condition. The expression of these two genes was low in CG culture condition, while the level of expression was elevated

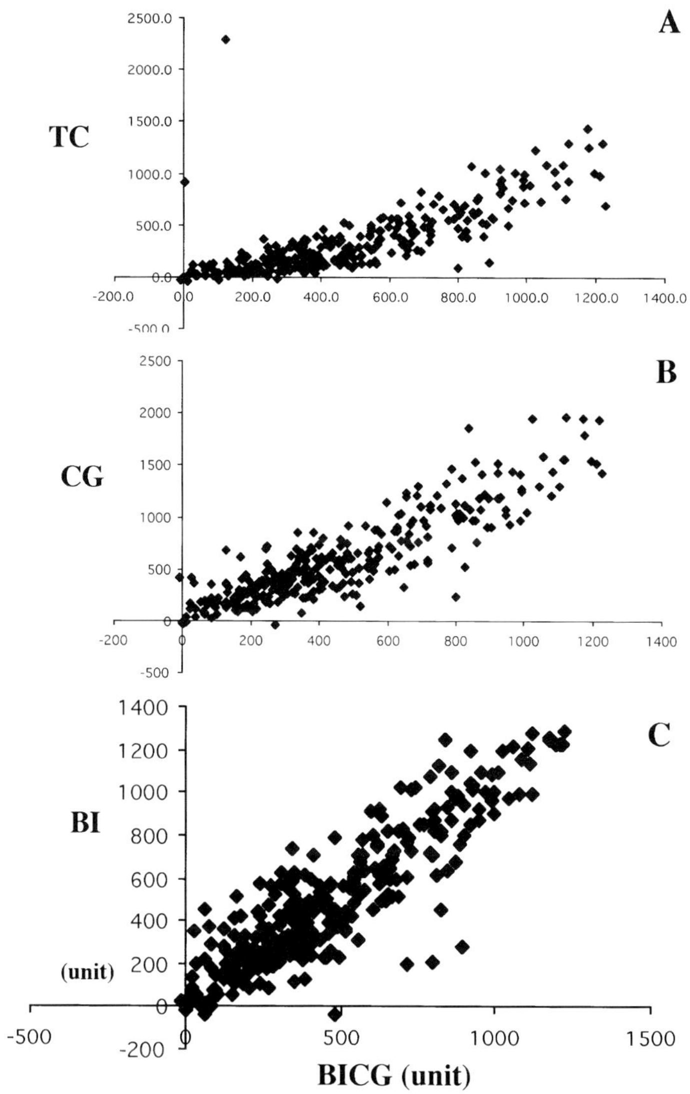

Figure 6 Scatter plot analysis of relative hybridization intensity on microarray membranes. Identical DNA microarray membranes were hybridized with cDNA probes of mRNA isolated from human airway epithelial cells cultured under TC, CG, BI, or BICG condition as described in text. Image analysis and quantitation of these microarray membranes were carried out as described in a previous publication (15). The image data (units) of the microarray membranes of TC, CG, and BI cultures were plotted against the image data of the corresponding DNA spots on BICG microarray membranes. This type of scatter plot provides the information about the degree of differential gene expression between the BICG culture condition (x-axis) and the rest of three culture conditions (y-axis). (A) TC vs. BICG; (B) CG vs. BICG; (C) BI vs. BICG.

TABLE 1 Differentially Expressed Genes Characterized by Microarray Analysis

Gene	Relative to mucociliary differentiation[a]
Jun-B proto-oncogene	Elevated
JNK2 protein kinase	Elevated
Mitogen-activated protein kinase 3	Elevated
Homo sapiens Smad6	Elevated
Sarcomeric mitochondrial creatine kinase	Elevated
Ribonuclease/angiogenin inhibitor	Elevated
IL-13 receptor	Elevated
CD5/T-cell surface glycoprotein CD5 precursor	Elevated
Gro-1	Suppressed
Transcriptional activator hSNF2b	Suppressed
Bone morphogenetic protein receptor	Suppressed

[a] Based on the relationship between relative gene expression and extent of mucociliary differentiation seen in various culture conditions (BICG > BI > CG > TC).

three- to fivefold in BI and BICG culture conditions. Thus, the level of expression of these genes was BICG > BI > CG. This is consistent with the microarray data described above.

The Smad6 result was surprising as vitamin A treatment decreases its expression in culture. As described above, however, many published results demonstrate that vitamin A and its derivatives (retinoids) are important mediators for

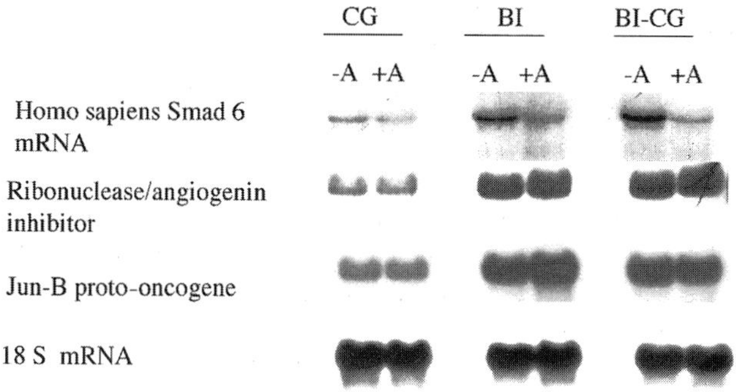

FIGURE 7 Northern blot analysis of differentially expressed genes identified using microarray analysis. Experiments were carried out as described in Figure 1. Cultures were treated with (+A) or without (−A) all-*trans*-retinoic acid (0.3 μM).

airway epithelial cell development and maturation (1,6,7,20,21). Based on published data, however, this apparent inconsistency maybe be explained in the following way. Smad6 is known for its antagonistic effects on TGF-β action (22,23). Human airway epithelial cells are capable of producing TGF-β, and the production is elevated in vitamin A–depleted culture conditions. Since TGF-β also has an inhibitory effect on airway mucous cell differentiation, an elevation of Smad6 may be necessary to counteract this activity in order to facilitate cell differentiation. However, TGF-β and bone morphogenic factor are known inducers of Smad6 gene expression (23,24). An elevation of TGF-β, especially in vitamin A–depleted conditions can stimulate Smad6 gene expression. These results suggest a complex auto-regulation of TGF-β associated with airway cell differentiation.

In contrast to Smad6, the expression of ribonuclease/angiogenin inhibitor and Jun-B proto-oncogene is elevated by vitamin A and by the culture conditions that favor cell differentiation. These results suggest positive roles for these genes in the regulation of airway cell differentiation. Ribonuclease/angiogenin inhibitor is known for its inhibition of angiogenesis (25). The nature of this activity is unclear in airway epithelial cells and further studies are needed. Jun-B proto-oncogene is an important transcriptional factor that regulates AP-1–dependent transcriptional activity. The elevation of Jun-B gene expression has been shown to be associated with skin cell differentiation (11,12,26), but its association with airway mucociliary differentiation is a new finding. Interestingly, we also observed an elevation of JNK2 protein kinase gene expression in association with culture conditions that favor mucociliary differentiation. JNK2 protein kinase plays an essential role in the activation of Jun-B. The coordinated activation of JNK2 kinase and Jun-B gene product suggests an important role for this signaling pathway in the regulation of airway mucociliary differentiation.

In summary, we have demonstrated that the differentiation of primary human TBE cells can be manipulated in culture. The best culture condition for human TBE cells to differentiate is the BICG condition, and the tendency of TBE cell differentiation is BICG > BI > CG > TC. Utilizing these culture conditions, we initiated an approach based on high-density DNA microarray technology to study the gene expression pattern and to identify differentially expressed genes that are associated with TBE cell differentiation. Although only 1400 genes were chosen for the initial experiments, the experience demonstrated the feasibility of this technology to study the expression of thousands of genes simultaneously and to identify those that show a differential expression. We have observed at least 11 genes whose expression patterns coincided with TBE cell differentiation. This result was further verified by Northern blot hybridization. Although the role of these genes in airway cell differentiation is not clear, we have found several interesting regulatory events involved, including TGF-β– and JNK 2 protein kinase–dependent pathways. Further studies are needed to understand the regulation of airway mucociliary differentiation.

ACKNOWLEDGMENTS

This paper is supported in part by grants from NIH (HL35635, ES06230, ES09701) and the California Tobacco-Related Disease Research Program (7RT-0145).

REFERENCES

1. R Wu, Growth and differentiation of tracheobronchial epithelial cells. In: JA McDonald, ed. Lung Growth and Development. New York: Marcel Dekker, 1997, pp 211–241.
2. R Wu. In vitro differentiation of airway epithelial cells. In: LJ Schiff, ed. In Vitro Models of Respiratory Epithelium. Boca Raton, FL: CRC Press, 1986, pp 1–26.
3. WR Martin, C Brown, YJ Zhang, R Wu. Growth and differentiation of primary tracheal epithelial cells in culture: regulation by extracellular calcium. J Cell Physiol 147:138–148, 1991.
4. CB Robinson, R Wu. Mucin synthesis and secretion by cultured tracheal cells: effects of collagen gel substratum thickness. In Vitro Cell Dev Biol 29A:469–477, 1993.
5. R Wu, YH Zhao, MM Chang. Growth and differentiation of conducting airway epithelial cells in culture. Eur Respir J 10(10):2398–2403, 1997.
6. JS Koo, JH Yoon, T Gray, D Norford, AM Jetten, P Nettesheim. Restoration of the mucous phenotype by retinoic acid in retinoid-deficient human bronchial cell cultures: changes in mucin gene expression. Am J Respir Cell Mol Biol 20:43–52, 1999.
7. SH Bernacki, AL Nelson, L Abdullah, JK Sheenhan, A Harris, CW Davis, SH Randell. Mucin gene expression during differentiation of human airway epithelia in vitro. Am J Respir Cell Mol Biol 20:595–604, 1999.
8. DK Ann, MMJ Wu, T Huang, DM Carlson, R Wu. Retinol-regulated gene expression in human tracheobronchial epithelial cells: enhanced expression of elongation factor EF-1a. J Biol Chem 263:3546–3549, 1988.
9. G An, TH Huang, J Tesfaigzi, J Garcia-Heras, DH Ledbetter, DM Carlson, R Wu. An unusual expression of a squamous cell marker, small proline-rich protein (spr1) gene in tracheobronchial epithelium: differential regulation and gene mapping. Am J Respir Cell Mol Biol 7:104–111, 1992.
10. SPM Reddy, G An, YH Zhao, R Wu. Isolation and characterization of nucleolin gene expression in airway epithelium by a palindromic primer-based mRNA differential display method. Am J Respir Cell Mol Biol 15:398–403, 1996.
11. T Kartasova, P van de Putte. Isolation and characterization and UV-stimulated expression of two families of genes encoding polypeptides of related structure in human epidermal keratinocytes. Mol Cell Biol 8:2195–2203, 1988.
12. T Kartasova, GNP van Muijen, H van Pelt-Heerschap, P van de Putte. Novel protein in human epidermal keratinocytes: regulation of expression during differentiation. Mol Cell Biol 8:2204–2210, 1988.
13. SPM Reddy, YJ Chuu, PN Lao, J Donn, DK Ann, R Wu. Expression of human

squamous cell differentiation marker, spr1, in tracheobronchial epithelium depends on Jun and TRE motifs. J Biol Chem 270:26451–26459, 1995.
14. TH Huang, DK Ann, YJ Zhang, AT Chang, JW Crabb, R Wu. Control of keratin gene expression by vitamin A in tracheobronchial epithelial cells. Am J Respir Cell Mol Biol 10:192–201, 1994.
15. JJW Chen, R Wu, PC Yang, JY Huang, YP Sher, MH Han, WC Kao, PJ Lee, TF Chiu, F Chang, YW Chu, CW Wu, K Peck. Profiling expression patterns and isolating differentially expressed genes by cDNA microarray system with colorimetry detection. Genomics 51:313–324, 1998.
16. M Schena, D Shalon, RW Davis, PO Brown. Quantitative monitoring of gene expression patterns with a complementary DNA microarray. Science 270:567–570, 1995.
17. DJ Lockhart, H Dong, MC Byrne, MT Follettie, MV Gallo, MS Chee, M Mittmann, C Wang, M Kobayashi, H Horton, EL Brown. Expression monitoring by hybridization to high-density oligonucleotide arrays. Nat Biotechnol 14:1675–1680, 1996.
18. P Chomczynski, N Sacchi. Single-step method of RNA isolation by acid guanidinium thiocyanate-phenol-chloroform extraction. Anal Biochem 162:156–159, 1987.
19. H Lin, DM Carlson, JA St. George, CG Plopper, R Wu. An ELISA method for the quantitation of tracheal mucins from human and non-human primates. Am J Respir Cell Mol Biol 1:41–48, 1989.
20. SB Wolbach, PR Howe. Tissue changes following deprivation of fat-soluble A vitamin. J Exp Med 42:753–781, 1926.
21. YC Wong, RC Buck. An electronic microscopic study of metaplasia of the rat tracheal epithelium in vitamin A deficiency. Lab Invest 24:55–66, 1971.
22. KM Galvin, MJ Donovan, CA Lynch, RI Meyer, RJ Paul, JN Lorenz, V Fairchild-Huntress, KL Dixon, JH Dunmore, MA Gimbrone Jr, D Falb, D Huszar. A role for smad6 in development and homeostasis of the cardiovascular system. Nature Genetics 24:171–174, 2000.
23. L Choy, J Skillington, R Derynck. Roles of autocrine TGF-beta receptor and Smad signaling in adipocyte differentiation. J Cell Biol 149:667–682, 2000.
24. K Matsuzaki, M Date, F Furukawa, Y Tahashi, M Matsushita, K Sakitani, N Yamashiki, T Seki, H Saito, M Nishizawa, J Fujisawa, K Inoue. Autocrine stimulatory mechanism by transforming growth factor beta in human hepatocellular carcinoma. Cancer Res 60:1394–1402, 2000.
25. DJ Strydom. The angiogenins. Cell Mol Life Sci 54:811–824, 1998.
26. K Ariizumi, PR Bergstresser, A Takashima. A wavelength-specific induction of immediate early genes by ultraviolet radiation. J Dermatol Sci 12:147–155, 1996.

22

In Vivo Models of Airway Goblet Cell Hyperplasia and Mucin Gene Expression

Alinka K. Smith and Duncan F. Rogers
National Heart & Lung Institute (Imperial College)
London, United Kingdom

INTRODUCTION

Markedly increased numbers of goblet cells in the conducting airways is a pathophysiological feature of asthma and chronic bronchitis, two common severe respiratory conditions (1,2). Chronic bronchitis is a clinical definition based upon long-standing sputum production with the implication that this is associated with mucus hypersecretion in the respiratory tract. Hypersecretion, chronic bronchiolitis (small airways disease), and emphysema (alveolar destruction) comprise chronic obstructive pulmonary disease (COPD). The relative contribution of each component to disease progression varies between patients. Airway submucosal gland hypertrophy is also a feature of asthma and COPD. However, hypersecretion associated with gland hypertrophy is likely to be in the more proximal airways, where cough aids mucociliary clearance to expel excess mucus. Consequently, gland hypertrophy has received only scant attention in experimental models of airway hypersecretion (3). The present chapter considers goblet cell hyperplasia in asthma and COPD and examines data from animal models of these two respiratory conditions in terms of induction mechanisms of hyperplasia and of changes in mucin (MUC) gene expression.

AIRWAY MUCUS

Under normal circumstances, a thin film of viscoelastic liquid protects the epithelial surface of the airways. The liquid is often referred to as ''mucus'' and is a dilute and complex aqueous solution of electrolytes, mucous glycoproteins (termed mucins), proteoglycans, enzymes, antienzymes, oxidants, antioxidants, bacterial products, antibacterial agents, lipids, cellular mediators, and plasma-derived proteins and mediators (4). The liquid forms an upper gel layer, for entrapment of inhaled particles, and a lower sol layer in which cilia beat. Intimate interaction between mucus and beating cilia facilitates removal of inhaled particles from the lung, a process termed mucociliary clearance. The efficiency of mucociliary clearance is directly related to the elasticity and inversely proportional to the viscosity of the gel layer and is dependent upon the depth of the sol layer. Chronic increases in volume and viscosity of the mucus layer impair clearance and precipitate hypersecretory conditions of the airways, for example, asthma and chronic bronchitis.

RESPIRATORY MUCINS IN HEALTH AND DISEASE

The viscoelastic properties of airway mucus are attributed largely to high molecular weight mucins that are secreted by epithelial goblet cells and submucosal gland mucous cells. Other cells in the airways that may be secretory, although not necessarily of mucin, are the epithelial serous cell and its equivalent in the glands, the ciliated cell and the Clara cell. In airway hypersecretory conditions, goblet cell hyperplasia is associated with reductions in serous, ciliated, and Clara cells. Mucins comprise a peptide backbone, termed apomucin, to which multiple oligosaccharide side chains are bound (5). Apomucins are encoded by several genes (see next section) and are expressed in goblet cells and submucosal glands.

Goblet cells and submucosal glands may produce different mucins. Postmortem analysis of the trachea from a single individual indicated that the product of mucin gene 5AC (MUC5AC) was a goblet cell mucin rather than a gland mucin (6). In contrast, MUC5B was found mainly in glands, but also in goblet cells, and there was a suggestion that there were populations of gland cells containing different glycoforms of MUC5B (7). It is vital to determine whether or not this differential expression of mucins between goblet cells and glands is retained after analysis of further samples because it has implications for categorizing different diseases, for developing appropriate disease models, and for rationale design of therapeutic drugs (see below).

MUC5AC is the major gel-forming species in pooled secretions from young healthy nonsmokers aspirated into tracheal cannulae at minor dental surgery under general anaesthesia (8–10). These aspirates are the most ''normal'' respiratory secretions described to date. Analysis of the viscid secretions collected postmortem from an asthmatic patient demonstrates both MUC5AC (9) and a low charge glycoform of MUC5B (11). MUC5B is also a major mucin species in sputum from a patient with chronic bronchitis (7). In contrast, although MUC5AC is present in chronic bronchitic sputum (10), there is a suggestion that it may be only a minor component of the secretions (9). Thus, although requiring confirmation in more samples, these preliminary observations indicate an intrinsic difference between the secretions from normal, asthmatic, and chronic bronchitic subjects, with MUC5AC more characteristic of normal and asthma and MUC5B more characteristic of chronic bronchitis. It also indicates that normal and asthmatic airway secretions are derived more from goblet cells, with COPD secretions derived more from glands (see above). If supported by further data, the above suggestions imply that specific MUC profiles may need to be generated in experimental models of asthma or chronic bronchitis if they are to have relevance to disease. Interestingly, MUC2 is not found in secretions from ''normals'' or bronchitics (6). However, in sputum from patients with cystic fibrosis (CF), although MUC5AC and MUC5B were the major mucin species found, small amounts of MUC2 were also present (12). This observation suggests a further MUC profile for a specific hypersecretory condition.

HUMAN RESPIRATORY MUCIN GENES IN HEALTH AND DISEASE

Currently, ten human mucin genes are recognized: MUC1–4, MUC5AC, MUC5B, and MUC6–9 (Fig. 1). Using currently available antisera, the MUC5AC and MUC5B gene products appear to be the major gel-forming moieties in airway secretions from healthy individuals and patients with asthma or chronic bronchitis (see previous section). A greater range of antisera is required to more precisely define the MUC content of secretions. Generation of molecular probes for MUC gene mRNA has superseded that of probes for MUC gene products, leading to information on the distribution of MUC genes in the airways.

The expression of the MUC2 gene in the airways is thought to depend upon the induction of mucus overproduction. For example, transcription of MUC2 is upregulated by *Pseudomonas aeruginosa* lipopolysaccharide, present in the lungs of CF patients (13), and by TNF-α in human airways in vitro (14). A separate study comparing expression of the MUC1, 2, and 5AC genes in noninflamed nasal epithelial cells from CF, allergic rhinitis, and control subjects found that in all groups the mRNA levels of MUC5AC were greater than those for MUC1 and 2. The expression of MUC2 in all subjects remained constant, whereas the ratio of MUC5AC to MUC2 gene expression decreased significantly in CF cells

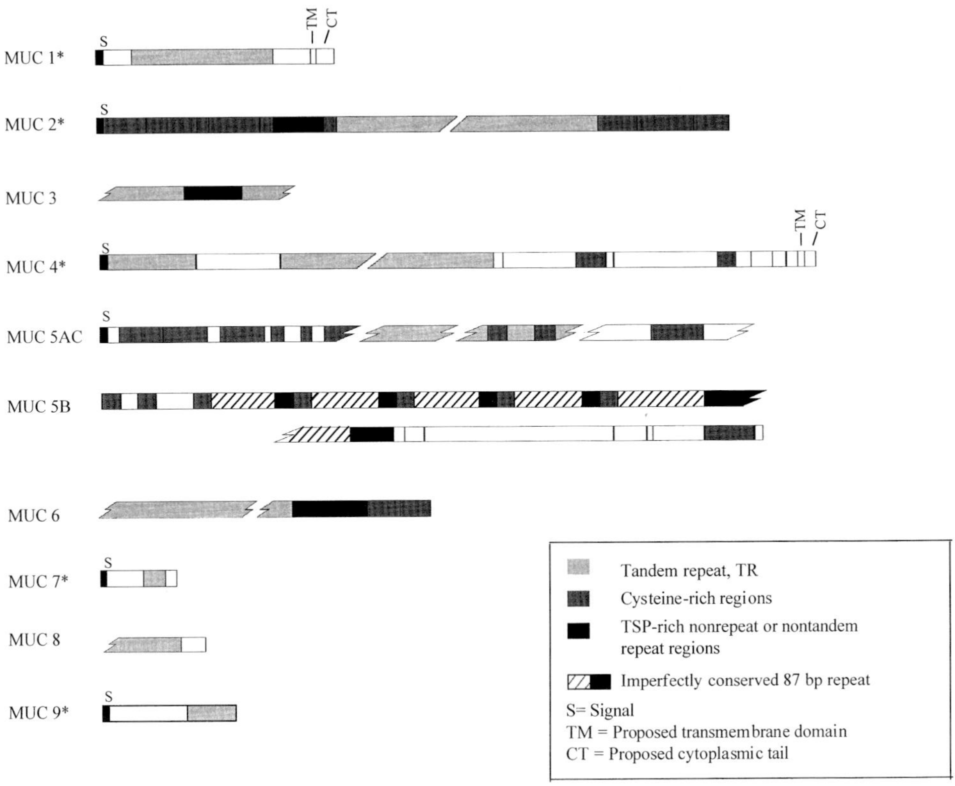

FIGURE 1 Models of the 10 currently described human MUC gene products. Structures deduced either from nucleotide sequences of complete *MUC* genes (*) or cDNA clones. Breaks = long tandem repeats; ragged ends = sequences awaiting elucidation.

when compared to normals (15). In contrast, levels of MUC5AC mRNA increase in cultured human airway epithelial cells exposed to airway bronchoalveolar lavage (BAL) fluid from asthmatic patients or allergen-challenged allergic dogs (16). Further work in vitro and in vivo in mice suggested the involvement of the Th2 cell cytokine IL-9, present in BAL and upregulated in asthma (17). This cytokine is thought to trigger a signal cascade, which in turn activates MUC5AC transcription by activating epithelial IL-9 receptors normally present on the airway epithelium.

There is increasing evidence to suggest that inflammatory mediators increase the expression of mucin genes. However, little is known about the molecular mechanisms involved in this upregulation. The inflammatory mediator neutrophil elastase, known to stimulate mucus secretion acutely, increases the mRNA and protein expression of MUC5AC in A549 human carcinoma cells and primary respiratory epithelial cells in culture (18). In A549 cells the increase was due to an increased stability of the mRNA. TNF-α also increases mucin biosynthesis by increasing MUC5AC mRNA half-life (19), indicating that postranscriptional regulation of mucins is an important mechanism controlling mRNA levels in disease states.

MUCIN GENES IN LABORATORY ANIMALS

As a result of the molecular cloning of human mucin genes, the identification of corresponding animal MUC genes, such as those of the rat and mouse, is now an active area of research. The cloning of animal genes allows for easier manipulation, permitting elucidation of their regulation, function, and abnormalities in experimental models of respiratory disease. Rodents are frequently used in these models due to their small size and, apart from a lack of submucosal glands, the similarity of their airway surface epithelia to that of human airways.

For the rat, homologs of human MUC1-4 and 5AC have been cloned (20–26). These rat mucin genes show a high degree of sequence similarity to their human counterparts. For example, the rat cDNA homolog of human MUC5AC has a sequence similarity of 73% at the amino acid level and 71% at the nucleotide level (25). The size and tissue distribution of rMUC5AC mRNA was found to be consistent with that of human MUC5AC. However, mRNA levels were found to be much lower in rat airways but became strongly expressed during mucous differentiation (25). The tissue distribution of rMUC2 is also similar to that of its human homolog in that it is strongly expressed in the small intestine and is dependent upon mucus hypersecretion (27).

At present, the only murine MUC homologs reported are those for MUC1, 2, 3, and 5AC (28–31). The murine mucins have not yet been extensively studied. However, murine MUC5AC is mainly confined to the stomach, with no expression in the trachea or lung, as for rMUC5AC (30). The expression of both murine MUC2 and MUC3 is absent in the airways of pathogen free mice. MUC2 is confined to the intestinal goblet cells (29), whereas MUC3 is mainly present in the caecum and small intestine (31).

GOBLET CELL HYPERPLASIA IN ASTHMA AND COPD

One of the cardinal pathophysiological features of chronic bronchitis is goblet cell hyperplasia (32). The hyperplasia is in larger airways, whereas metaplasia occurs in the bronchioles in which goblet cells are normally scarce or absent. This observation has been repeated by others and is confirmed by semi-quantitative analysis (33). Goblet cell hyperplasia has also been noted in patients dying of asthma (34). Quantitative analysis demonstrates marked goblet cell hyperplasia throughout the lower airways of patients dying of acute severe asthma but not of chronic asthma (35). These observations indicate that, at least for the latter group of patients, disproportionate goblet cell hyperplasia is associated with fatal attacks of asthma. From the above, goblet cell hyperplasia/metaplasia appear to be features of asthma and chronic bronchitis. However, there are exceptions. Goblet cell hyperplasia is not obvious in all patients with asthma or chronic bronchitis (36), and in a small group of Japanese patients with COPD quantitation did not demonstrate hyperplasia compared with controls (37).

GOBLET CELL HYPERPLASIA IN EXPERIMENTAL ANIMALS

Goblet cell hyperplasia and metaplasia can be readily induced in a variety of laboratory animals by a variety of experimental procedures (38,39) (Table 1). Most of the ''early'' (i.e., late 1960s, 1970s, and 1980s) models were primarily of chronic bronchitis, with goblet cell hyperplasia induced by inhaled gases, principally cigarette tobacco smoke, sulfur dioxide, nitrogen dioxide, or ozone. Elastase was also frequently used, although it tended to generate emphysematous changes rather than hypersecretory changes. The molecular and cellular mechanisms underlying the development of goblet cell hyperplasia in these COPD models were for the most part unexplored. More recently, models of allergy and asthma have been described, with mechanisms being delineated. These models are principally based on animals sensitized systemically to ovalbumin and subsequently challenged with inhaled ovalbumin, usually repeatedly every few days for a number of weeks. The allergy/asthma models demonstrate goblet cell hyperplasia associated with recruitment of inflammatory cells (40). Th2 cells may orchestrate the inflammatory response, including hypersecretion (41), in concert with the cytokines IL-4 and IL-13 (42–44). The latter cytokines probably both act through the α chain of the IL-4 receptor (45). Tumor necrosis factor α (TNF-α) alone has little effect on goblet cell number but potentiates goblet cell hyperplasia induced by platelet activating factor (PAF) (46). TNF-α also upregulates receptors for epidermal growth factor (EGRF-R) leading to associated goblet cell hyperplasia (47). The hyperplasia is prevented by an EGF-R tyrosine kinase inhibitor (BIBX 1522) in both allergic rats (47) and rats with agarose plugs instilled into their bronchi (48).

MUCIN GENE EXPRESSION IN RESPIRATORY DISEASE MODELS

It is thought that the pathophysiology of mucus hypersecretion involves abnormalities that lead to an increase in the production of mucin. These abnormalities

TABLE 1 Inducers of In Vivo Goblet Cell Hyperplasia and MUC Expression in the Lower Airways

Inducer	Hyperplasia	*MUC* mRNA	MUC protein
β-Adrenoceptor agonists	Yes	np	np
Acrolein	Yes	*MUC5AC*	MUC5AC
Agarose plugs	Yes	*MUC5AC*	np
Cigarette smoke	Yes	np	np
Cigar smoke	Yes	np	np
Cholinoceptor agonists	Yes	np	np
EGF (+ TNF-α)	Yes	*MUC5AC*	np
Elastase	Yes	np	np
Endotoxin	Limited effect	np	np
IL-4	Yes	*MUC5AC*	MUC5AC
IL-4Rα	Yes	*MUC5AC*	MUC5AC
IL-9	Yes	np	np
IL-13	Yes	*MUC5AC*	MUC5AC
Marijuana smoke	Yes	np	np
Neutrophil lysates	Yes	np	np
Nicotine	Yes	np	np
Nitrogen dioxide	Yes	np	np
Estrogen analogs	Yes	np	np
Ozone	Yes	np	np
PAF	Yes	*MUC5*	np
Proteinases	Yes	np	np
Sensitization followed by challenge	Yes	*MUC5AC*	MUC5AC
Sulfur dioxide	Yes	np	np
Sulfur dioxide + influenza infection	Yes	*MUC2*	np
Th2 cells	Yes	np	np
Th1 cells/IFN-γ	Inhibition	np	np
TNF-α + PAF	Yes	*MUC5*	np

Abbreviations: np, effect not published; EGF, epidermal growth factor; IL, interleukin; IL-4Rα, α chain of the IL-4 receptor; IFN-γ, interferon-γ, PAF, platelet-activating factor; TNF-α, tumor necrosis factor α.

may occur at the level of transcription or stabilization of mucin RNA transcripts or at the level of translation or protein stabilization. To help determine the association of mucin gene expression with specific airway diseases, mucus hypersecretion can be induced experimentally by the exposure of the respiratory tract to a variety of agents (Table 1).

Jany et al. (27) were the first to report that induced hypersecretion may be controlled in part at the level of mucin mRNA. Sprague-Dawley rats infected with Sendai virus and exposed subacutely to SO_2 to induce experimental chronic bronchitis had detectable levels of MUC2 mRNA in their airways, unlike control rats that were pathogen-free. Rats exposed to acrolein, an aldehyde found in high concentrations in tobacco smoke and that is absorbed in the upper respiratory tract, exhibit a pulmonary pathology similar to that of human bronchitis (49). The resulting mucus hypersecretion was in part due to an increase in the steady-state mRNA levels of MUC5AC in the trachea and lung and was accompanied by an increase in MUC5AC protein. In contrast, expression of MUC2 was not significantly different in the trachea or lung.

Models of allergy/asthma are currently receiving considerable attention. For example, brown Norway rats sensitized systemically to ovalbumin and the airways subsequently challenged by inhaled ovalbumin exhibit the bronchial hyperresponsiveness and extensive airway remodeling characteristic of asthma (50). There is also evidence of goblet cell hyperplasia (Fig. 2). We found that part of the remodeling includes marked increases in expression of mRNA for MUC5AC (Fig. 3), with associated decreases in MUC1 (51). Similarly, mice sensitized and

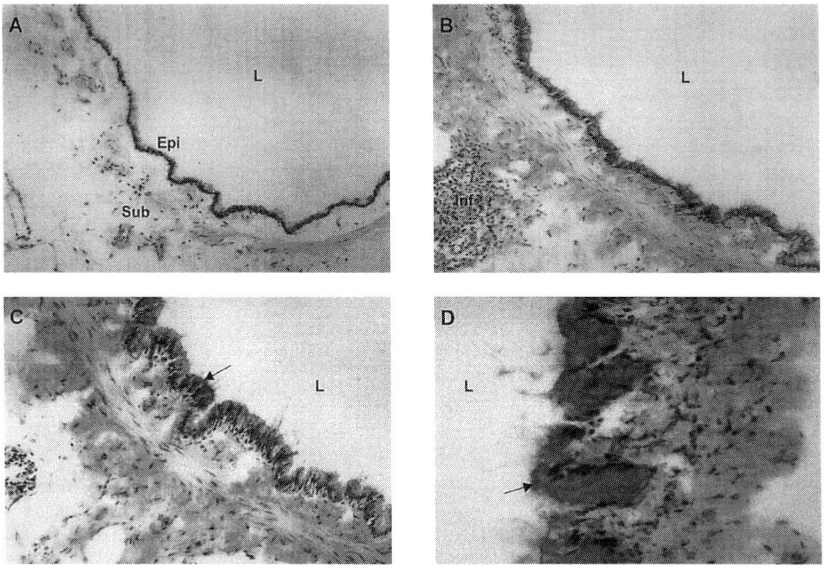

FIGURE 2 Effect of allergy on goblet cell number in airways of brown Norway rats. (A) Rat sensitized systemically to ovalbumin and challenged with inhaled ovalbumin 3 weeks later. Note shallow epithelium (Epi) and lack of inflammation in the submucosa (Sub). (B) Sensitized rat challenged with ovalbumin (same magnification as panel A). Note increased depth and prominence of epithelium and submucosal inflammation (Inf). (C) Higher magnification of panel B. Note prominent goblet cells (arrow). (D) Higher magnification of panel C. Note prominent intracellular mucus in goblet cells (arrow). L = lumen. Periodic acid-Schiff stain. (Courtesy of M. Salmon.)

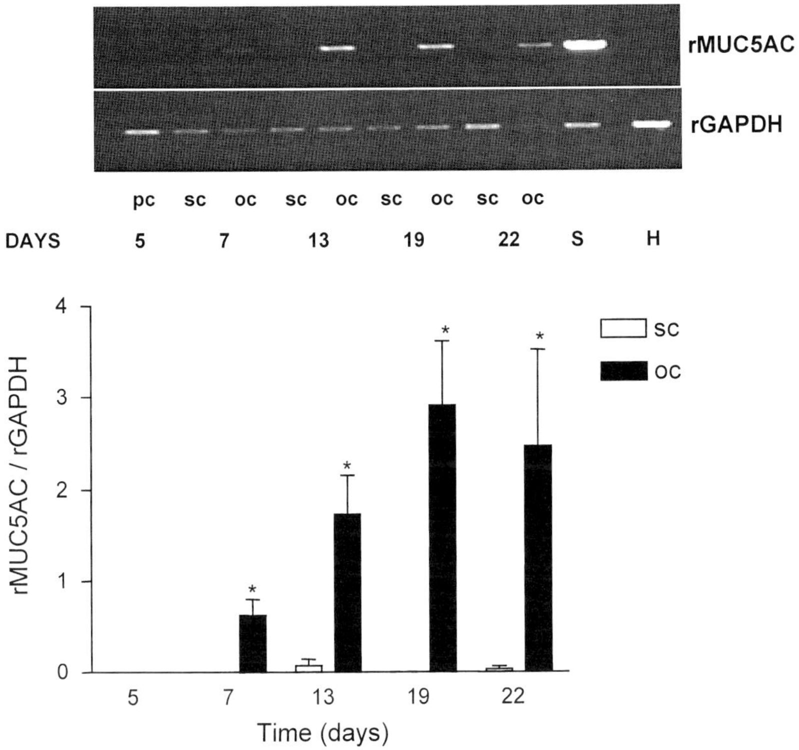

FIGURE 3 Effect of allergy on rat MUC5AC expression in tracheae of brown Norway rats. Rats were sensitized systemically to ovalbumin and challenged with inhaled ovalbumin on days 6, 12, 15, 18, and 21. Tracheae were sampled at day 5 (i.e., 1 day prechallenge, pc) and then 24 hours after a saline challenge (sc) or an ovalbumin challenge (oc). Upper panel: Representative RT-PCR of rMUC5AC and rGAPDH (S = stomach, positive control; H = heart, negative control). Lower panel: Mean data (vertical bars = one SE mean, $n = 4$ animals per group), $^{*}p < 0.05$ versus oc.

subsequently challenged with ovalbumin, or treated with the cytokine IL-13, demonstrate induction of MUC5AC mRNA and protein expression in the lungs (44). The ovalbumin murine model has pathophysiological changes similar to those seen in human allergic asthma, including goblet cell metaplasia (40), and IL-13, a Th2 cytokine, has been found to have a critical role in the expression of murine asthma (52). In addition to MUC5AC, the expression of MUC1, 2, and 3 mRNA was also investigated. Neither IL-13 nor ovalbumin treatment altered the levels of MUC1 or induced the expression of MUC2 or 3, which supports the concept that differential regulation of mucin genes occurs in the respiratory tract.

Interleukin-4 has been shown to produce similar effects to that of IL-13 in mice. For example, transgenic mice overexpressing Il-4 selectively within the lung exhibit goblet cell metaplasia and an increased expression of MUC5AC, but not MUC2 (42). IL-4 is predominantly produced by Th2 lymphocytes and is believed to play a role in animal models of asthma (53). Instillation of IL-4 in murine airways resulted in rapid expression of MUC5AC mRNA, the production of mucous glycoconjugates, and the induction of goblet cell metaplasia (43).

CONCLUSIONS

Goblet cell hyperplasia/metaplasia and expression of specific mucins (MUC gene products) characterize asthma and COPD. Early in vivo models of airway hypersecretion demonstrated that numerous agents, in particular the inhaled irritants associated with chronic bronchitis/COPD in humans, induced goblet cell hyperplasia in laboratory animals, particularly small rodents. These models were useful in terms of epidemiology but did not really explore mechanisms underlying development of the hypersecretory phenotype. More recently, in vivo models of allergy and asthma have been used to explore mechanisms of airway goblet cell hyperplasia. Th2 cells, IL-4Rα, and EGF-R are currently envisaged as important induction mechanisms underlying ''allergic'' goblet cell hyperplasia and expression of MUC5AC. Whether or not these same mechanisms apply to models of COPD needs to be assessed, particularly as chronic bronchitis may be associated more with expression of MUC5B than MUC5AC and, at least at present, there is no rodent homolog of human MUC5B. Further development of these models will increase our understanding of hypersecretory diseases and aid in rational design of therapeutic compounds.

ACKNOWLEDGMENT

We thank SmithKline Beecham, Philadelphia, Pennsylvania, for funding AKS.

REFERENCES

1. Liu YC, Khawaja AM, Rogers DF. Pathophysiology of airway mucus secretion in asthma. In: PJ Barnes, IW Rodger, NC Thomson, eds. Asthma: Basic Mechanisms and Clinical Management. 3rd ed. London: Academic Press, 1998, pp 205–227.
2. Rogers DF. Mucus pathophysiology in COPD: differences to asthma and pharmacotherapy. Monaldi Arch Chest Dis 55:324–332, 2000.
3. Rogers DF, Godfrey RWA, Castro K, Jeffery PK. Effects of a new compound (Zy 15850A) on cigarette smoke-induced bronchitis in the rat. Agents Actions 33:359–366, 1991.
4. Widdicombe JH, Widdicombe JG. Regulation of human airway surface liquid. Respir Physiol 99:3–12, 1994.
5. Thornton DJ, Davies JR, Carlstedt I, Sheehan JK. Structure and biochemistry of

human respiratory mucins. In: DF Rogers, MI Lethem, eds. Airway Mucus: Basic Mechanisms and Clinical Perspectives. Basel: Birkhäuser Verlag, 1997, pp 19–39.
6. Hovenberg HW, Davies JR, Carlstedt I. Different mucins are produced by the surface epithelium and the submucosa in human trachea: identification of MUC5AC as a major mucin from the goblet cells. Biochem J 318:319–324, 1996.
7. Wickström C, Davies JR, Eriksen GV, Veermans CI, Carlstedt I. MUC5B is a gel-forming, oligomeric mucin from human salivary gland, respiratory tract and endocervix: identification of glycoforms and C-terminal cleavage. Biochem J 334:685–693, 1998.
8. Thornton DJ, Davies JR, Kraayenbrink M, Richardson PS, Sheehan JK, Carlstedt I. Mucus glycoproteins from 'normal' human tracheobronchial secretion. Biochem J 265:179–186, 1990.
9. Thornton DJ, Carlstedt I, Howard M, Devine PL, Price MR, Sheehan JK. Respiratory mucins: identification of core proteins and glycoforms. Biochem J 316:967–975, 1996.
10. Hovenberg HW, Davies JR, Herrmann A, Linden C-J, Carlstedt I. MUC5AC, but not MUC2, is a prominent mucin in respiratory secretions. Glycoconjugate J 13: 839–847, 1996.
11. Sheehan JK, Howard M, Richardson PS, Longwill T, Thornton DJ. Physical characterization of a low-charge glycoform of the MUC5B mucin comprising the gel-phase of an asthmatic respiratory mucous plug. Biochem J 338:507–513, 1999.
12. Davies JR, Svitacheva N, Lannefors L, Kornfält R, Carlstedt I. Identification of MUC5B, MUC5AC and small amounts of MUC2 mucins in cystic fibrosis airway secretions. Biochem J 344:321–330, 1999.
13. Li J-D, Dohrman AF, Gallup M, Miyata S, Gum JR, Kim YS, Nadel JA, Prince A, Basbaum CB. Transcriptional activation of mucin by *Pseudomonas aeruginosa* lipopolysaccharide in the pathogenesis of cystic fibrosis lung disease. Proc Natl Acad Sci USA 94:967–972, 1997.
14. Levine SJ, Larivée P, Logun C, Angus CW, Ognibene FP, Shelhamer JH. Tumour necrosis factor-α induces mucin hypersecretion and MUC2 gene expression by human airway epithelial cells. Am J Respir Cell Mol Biol 12:196–204, 1995.
15. Voynow JA, Selby DM, Rose MC. Mucin gene expression (MUC1, MUC2 and MUC5/5AC) in nasal epithelial cells of cystic fibrosis, allergic rhinitis and normal individuals. Lung 176:345–354, 1998.
16. Longphre M, Li D, Gallup M, Drori E, Ordoñez CL, Redman T, Wenzel S, Bice DE, Fahy JV, Basbaum C. Allergen-induced IL-9 directly stimulates mucin transcription in respiratory epithelial cells. J Clin Invest 104:1375–1382, 1999.
17. Shimbara A, Christodoulopoulos P, Soussi-Gounni A, Olivenstein R, Nakamura Y, Levitt RC, Nicolaides NC, Holryd KJ, Tsicopoulos A, Lafitte JJ, Wallaert B, Hamid QA. IL-9 and its receptor in allergic and non-allergic lung disease: increased expression in asthma. J Allergy Clin Immuno 105:108–115, 2000.
18. Voynow JA, Rosenthal Young L, Wang Y, Horger T, Rose MC, Fischer BM. Neutrophil elastase increases MUC5AC mRNA and protein expression in respiratory epithelial cells. Am J Physiol 276:L835–L843, 1999.
19. Borchers MT, Carty MP, Leikauf GD. Regulation of human airway mucins by acrolein and inflammatory mediators. Am J Physiol 276:L549–L555, 1999.

20. DeSouza MM, Mani SK, Julian J, Carson DD. Reduction of mucin 1 expression during the receptive phase in the rat uterus. Biol Reprod 58:1503–1507, 1998.
21. Ohmori H, Dohrman AF, Gallup M, Tsuda T, Kai H, Gum JR Jr, Kim YS, Basbaum CB. Molecular cloning of the amino terminal region of a rat MUC2 mucin gene homologue. J Biol Chem 269:17833–17840, 1994.
22. Hansson GC, Baeckström D, Carlstedt I, Klinga-Levan K. Moecular cloning of a cDNA coding for a region of an apoprotein from the 'insoluble' mucin complex of rat small intestine. Biochem Biophys Res Commun 198:181–190, 1994.
23. Khatri IA, Forstner GG, Forstner JF. The carboxyl terminal sequence of rat intestinal mucin rMUC3 contains a putative transmembrane region and two EGF-like motifs. Biochim Biophys Acta 1326:7–11, 1997.
24. Moniaux N, Nollet S, Porchet N, Degand P, Laine A, Aubert J-P. Complete sequence of the human mucin MUC4: a putative cell membrane-associated mucin. Biochem J 338:325–333, 1999.
25. Guzman K, Bader T, Nettesheim P. Regulation of MUC5 and MUC1 gene expression: correlation with airway mucous differentiation. Am J Physiol 270:L846–L853, 1996.
26. Inatomi T, Tisdale AS, Zhan Q, Spurr-Michaud S, Gipson IK. Cloning of rat MUC5AC mucin gene: comparison of its structure and tissue distribution to that of human and mouse homologues. Biochem Biophys Res Commun 236:789–797, 1997.
27. Jany B, Gallup M, Tsuda T, Basbaum C. Mucin gene expression in rat airways following infection and irritation. Biochem Biophys Res Commun 181:1–8, 1991.
28. Spicer AP, Parry G, Patton S, Gendler SJ. Molecular cloning and analysis of the mouse homologue of the tumour-associated mucin MUC1 reveals conservation of potential O-glycosylation sites transmembrane and cytoplasmic domains and a loss of minisatellite-like polymorphism. J Biol Chem 266:15099–15109, 1991.
29. Van Klinken BJ-W, Einerhand AWC, Duits LA, Makkink NK, Twtgat KMAJ, Renes IB, Verburg M, Büller HA, Dekker J. Gastrointestinal expression and partial cDNA cloning of murine MUC2. Am J Physiol 276:G115–G124, 1999.
30. Shekels LL, Lyftogt C, Kieliszewski M, Filie JD, Kozak CA, Ho SB. Mouse gastric mucin: cloning and chromosomal localisation. Biochem J 311:775–785, 1995.
31. Shekels LL, Hunninghake DA, Tisdale AS, Gipsun IK, Kieliszewski M, Kozak CA, Ho SB. Cloning and characterisation of mouse intestinal MUC3 mucin: 3′ sequence contains epidermal growth factor-like domains. Biochem J 330:1301–1308, 1998.
32. Reid L. Pathology of chronic bronchitis. Lancet i:275–278, 1954.
33. Mullen JBM, Wright JL, Wiggs BR, Paré PD, Hogg JC. Structure of central airways in current smokers and ex-smokers with and without mucus hypersecretion: relationship to lung function. Thorax 42:843–848, 1987.
34. Dunnill MS. The pathology of asthma with special reference to changes in the bronchial mucosa. J Clin Pathol 13:27–33, 1960.
35. Aikawa T, Shimura S, Sasaki H, Ebina M, Takishima T. Marked goblet cell hyperplasia with accumulation in the airways of patients who died of severe acute asthma attack. Chest 101:916–921, 1992.
36. Glynn AA, Michaels L. Bronchial biopsy in chronic bronchitis and asthma. Thorax 15:142–153, 1960.

37. Aikawa T, Shimura S, Sasaki H, Takishima T, Taegashi H, Takahashi T. Morphometric analysis of intraluminal mucus in airways in chronic obstructive pulmonary disease. Am Rev Respir Dis 140:477–482, 1989.
38. Rogers, DF. Airway goblet cells: responsive and adaptable front line defenders. Eur Respir J 7:1690–1706, 1994.
39. Rogers, DF. In vivo test models for studying airway mucus secretion. Pulmonary Pharmacol. Therapeutics 10:121–128, 1997.
40. Blyth DI, Pedrick MS, Savage TJ, Hessel EM, Fattah D. Lung inflammation and epithelial changes in a murine model of atopic asthma. Am J Respir Cell Mol Biol 14:425–438, 1996.
41. Cohn L, Homer RJ, Niu N. T helper 1 cells and interferon γ regulate allergic airway inflammation and mucus production. J Exp Med 9:1309–1317, 1999.
42. Temann UA, Prasod B, Gallup MW, Basbaum C, Ho SB, Flavell RA, Rankin JA. A novel role for murine IL-4 in vivo: induction of MUC5AC gene expression and mucus hypersecretion. Am J Respir Cell Mol Biol 16:471–478, 1997.
43. Dabbagh K, Takeyama K, Lee H-M, Ueki IF, Lausier JA, Nadel JA. IL-4 induces mucin gene expression and goblet cell metaplasia in vitro and in vivo. J Immunol 162:6233–6237, 1999.
44. Alimam MZ, Piazza FM, Selby DM, Letwin N, Huang L, Rose MC. Muc-5/5AC mucin messenger RNA and protein expression is a marker of goblet cell metaplasia in murine airways. Am J Respir Cell Mol Biol 22:253–260, 2000.
45. Grünig G, Warnock M, Wakil AE, Venkayya R, Brombacher F, Rennick DM, Sheppard D, Mohrs M, Donaldson DD, Locksley RM, Corry DB. Requirement for IL-13 independently of IL-4 in experimental asthma. Science 282:2261–2263, 1998.
46. Lou Y-P, Takeyama K, Grattan KM, Lausier JA, Ueki IF, Agusti C, Nadel JA. Platelet-activating factor induces goblet cell hyperplasia and mucin gene expression in airways. Am J Respir Crit Care Med 157:1927–1934, 1998.
47. Takeyama K, Dabbagh K, Lee H-M, Agusti C, Lausier JA, Ueki IF, Grattan KM, Nadel JA. Epidermal growth factor system regulates mucin production in airways. Proc Natl Acad Sci USA 96:3081–3086, 1999.
48. Lee H-M, Takeyama K, Dabbagh K, Lausier JA, Ueki IF, Nadel JA. Agarose plug instillation causes goblet cell metaplasia by activating EGF receptors in rat airways. Am J Physiol 278:L185–L192, 2000.
49. Borchers MT, Wert SE, Leikauf GD. Acrolein-induced MUC5AC expression in rat airways. Am J Physiol 274:L573–L581, 1998.
50. Salmon M, Walsh DA, Koto H, Barnes PJ, Chung KF. Repeated allergen exposure of sensitized Brown-Norway rats induces airway cell DNA synthesis and remodelling. Eur Respir J 14:633–641, 1999.
51. Rush AK, Liu Y-C, Rogers DF. Differential expression of mucin (MUC) genes 1 and 5AC in airways of allergic rats. Respir Med 93:A.11, 1999.
52. Wills-Karp M, Luyimbazi J, Xu X, Schofield B, Neben TY, Carp CL, Donaldson DD. Interleukin 13: central mediator of allergic asthma. Science 282:2258–2261, 1998.
53. Gross A, Ben-Sasson SZ, Paul WE. Anti-IL-4 diminishes in vivo priming for antigen specific IL-4 production by T cells. J Immunol 150:2112–2120, 1993.

23

Interleukin-13–Induced Mucous Cell Hyperplasia in Airway Epithelium

Linda D. Martin, Brian W. Booth, Nancy J. Akley, and Kenneth B. Adler
North Carolina State University
Raleigh, North Carolina

Mariangela Macchione
North Carolina State University, Raleigh, North Carolina, and Universidade de São Paulo, São Paulo, Brazil

James C. Bonner
National Institutes of Health
Research Triangle Park, North Carolina

INTRODUCTION

Mucous cell hyperplasia is a prevalent lesion in chronic inflammatory airway diseases including chronic bronchitis (1), asthma (2), cystic fibrosis (3), and bronchiectasis (4). Described as a hypertrophy of mucus-secreting elements in the conducting airways, mucous cell hyperplasia is characterized by an increase in goblet cell number, cell size, and mucin content. Increased epithelial height and enlargement of submucosal glands are also observed throughout the airways (reviewed in Refs. 5 and 6). Clinical implications of the large amount of mucus secreted into the airways include increased bacterial infections and airway obstruction that, in some cases, can lead to death (7). Even with its toll of morbidity and mortality, molecular mechanisms within epithelial cells that govern mucous cell differentiation and proliferation during development of mucous cell hyperplasia have not been elucidated.

While detailed mechanisms are not currently understood, results of studies using animal models suggest that inflammation (8–13) and injury to epithelial cells (14–16) are important elements in the development of mucous cell hyperplasia. A diverse array of injurious insults including cigarette smoke (8), ozone (9), elastase (10), and sulfur dioxide (8) have been found to cause the lesion in experimental animal systems. An inflammatory component responding to this injury also appears to be important to the development of mucous cell hyperplasia, as development of the lesion can be enhanced by endotoxin (11) or previous viral infection (12). Conversely, the increase in goblet cell number induced by elastase (13), ozone, or cigarette smoke (14) can be attenuated with steroid treatment or nonsteroidal anti-inflammatory drugs (6,10).

IL-13

Recent studies indicate that within the inflammatory milieu, interleukin 13 (IL-13) is a central molecule in the development of mucous cell hyperplasia. IL-13 was found to be sufficient to increase goblet cell number in the airway epithelium when administered intranasally to wild-type mice (17). Similarly, a blockade of IL-13 inhibited the increase in mucus-containing cells observed in an ovalbumin-sensitized murine model of allergic asthma (18). Selective expression of IL-13 in the lungs of transgenic mice resulted in a phenotype characterized by airway epithelial cell hypertrophy, mucous cell hyperplasia, and increased neutral and acidic mucus (19). IL-13 was also found to induce expression of Muc5/5ac mucins in the murine allergic-asthma model, with or without ovalbumin challenge (20).

Increases in mucous cell number in the airways of IL-4 transgenic mice (21) have also been observed, along with induced mucin gene expression following stimulation with IL-4 in vivo and in vitro (22). Overexpression of IL-4, however, is predominantly found in the airways of allergic asthmatics, while elevated levels of IL-13 have been confirmed in both allergic and nonallergic asthma (23,24), as well as in pulmonary fibrosis (25). Thus, IL-13 may be a generalized effector molecule of the mucous cell hyperplasia observed in a wide variety of inflammatory airway diseases.

Little is currently known about intracellular mechanisms governing IL-13–induced mucous cell proliferation in airway epithelium. However, IL-13–induced proliferation has been extensively examined in activated B cells (26) and hematopoietic cell lines (27), providing information about downstream events in IL-13–

mediated signal transduction pathways. Upon binding IL-13, the IL-13 receptor is known to activate one or more members of the Janus kinase (JAK) family of tyrosine kinases (28). While IL-13 can induce phosphorylation of JAK1, JAK2, or tyk2, depending on the cell type, phosphorylation of JAK3 has not been observed. IL-13 induces rapid phosphorylation of JAK2 in a human colon carcinoma cell line, followed by JAK1 and tyk2 phosphorylation (29). Tyk2 phosphorylation is also augmented in response to IL-13 stimulation in an ovarian carcinoma cell line (30). IL-13 causes phosphorylation of JAK2, but not JAK1 or tyk2 in endothelial cells (31), but in many fibroblast cell lines JAK2 and tyk2, and less frequently JAK1, become phosphorylated (29). Taken together, these findings suggest that IL-13 is unlikely to stimulate JAK3 phosphorylation in airway epithelium, but may phosphorylate JAK2, tyk2, and (possibly) JAK1.

IL-13, and the related cytokine IL-4, require activation of an intracellular signaling molecule, the insulin receptor substrate (IRS-1 or IRS-2), in addition to JAKs, to provoke a proliferative response (27,28,32–34). IL-13 activates IRS-2 via phosphorylation in numerous cell lines and primary cells of lymphohemopoietic origin (27,35–38). This suggests that during development of mucous cell hyperplasia in the airway epithelium, the IL-13 receptor with its associated JAKs may use IRS-2 as an adapter molecule for intracellular signaling. As such, IRS-2 may interact with additional signaling molecules such as the p85 regulatory subunit of phosphatidylinositol 3′ kinase (PI 3′ kinase) (28,34), which has been shown to activate a number of cell cycle–regulating enzymes essential for cellular proliferation (39–42).

GROWTH FACTORS AND EPITHELIAL CELL INJURY

While a direct proliferative effect of IL-13 on epithelial cells has not been described previously, numerous studies have noted proliferation of airway epithelial cells in response to injury that triggers inflammation (43–45). Increased amounts of transforming growth factor alpha (TGF-α) and/or epidermal growth factor receptor (EGF-R) correlate with this proliferation in damaged respiratory tissue (43,46). Interestingly, recent studies using pathogen-free rats and an airway epithelial cell line (NCI-H292) also have implicated EGF-R in the development of airway mucous cell hyperplasia (47).

Proliferation of tumor cells (48,49), as well as epithelial cells in benign proliferative diseases ranging from benign prostatic hyperplasia (50) to gastroesophageal conditions (51,52) appears to involve the autocrine action of TGF-α on the EGF-R. One study even suggests that such a mechanism functions in the transformation of rat tracheal epithelial cells in vitro (53), giving precedence for TGF-α/EGF-R–activated proliferation of airway epithelium. Further support for such an autocrine/paracrine proliferative mechanism comes from observations that EGF-R and TGF-α are coincidentally localized in epithelial cells following naphthalene-induced bronchiolar injury (43). TGF-α participating in an autocrine/paracrine proliferative mechanism could function to activate EGF-R as an uncleaved TGF-α precursor (54), or as a soluble product cleaved by an elastase-like protease (55–57). Thus, there appears to be ample precedence, supported by data regarding proliferation of respiratory tissue, to suggest that airway epithelial cell proliferation may be controlled by a TGF-α/EGF-R autocrine/paracrine mechanism during the injury/repair process or during development of mucous cell hyperplasia.

PI 3′ KINASE

As discussed above, recent studies implicate IL-13 and/or EGF-R as important molecules in the development of mucous cell hyperplasia. While each of these molecules may contribute to development of the lesion under different circumstances, it is also possible that another mediator coordinates their actions during the pathogenic process. Both IL-13 (60,61) and EGF-R (62–64) can activate PI 3′ kinase, an enzyme shown to be activated in response to cytokines important in many proliferative responses (58,59). Thus, PI 3′ kinase may serve to integrate multiple intracellular signaling pathways to affect epithelial cell proliferation during the hyperplastic response.

PI 3′ kinase, which consists of a p85 regulatory subunit and a p110 catalytic subunit, phosphorylates phosphoinositides at the D-3 position of the inositol, creating multiple second messengers including PI-3-P, PI-3,4-P_2, and PI-3,4,5-P_3 (reviewed in Ref. 65). The precise mechanistic events leading to PI 3′ kinase activation currently are not well understood. Activation of PI 3′ kinase appears to involve phosphorylation of the p85 regulatory subunit (63), a process mediated, in part, by IRS-2 following IL-13 stimulation (21,27,33,36,43,58). In addition, direct physical interaction with cytoplasmic proteins, such as IRS-1/IRS-2 (28), also appears to play a role in PI 3′ kinase activation, although the proteins involved differ with cell type and stimulus. Activation of PI 3′ kinase in response to EGF-R stimulation, for example, may involve cytoplasmic proteins as diverse as c-Cbl, src, Gab1, p115, and p105 (66–69). Once the p85 subunit is phosphorylated and associated with additional cytoplasmic proteins, it may serve to target the p110 catalytic subunit to the plasma membrane, where it can access its substrate, ''inositol,'' or phosphatidylinositol. This translocation step appears to be required for downstream signaling via PI 3′ kinase (70,71). Thus, based on current knowledge regarding PI 3′ kinase activation, IL-13 stimulation is likely to result in phosphorylation of the p85 subunit and its physical association with IRS-2 and/or other cytoplasmic proteins, followed by translocation to the plasma membrane.

As a pivotal molecule coordinating IL-13–induced mucous cell hyperplasia, PI 3′ kinase would likely cause an increase in mucous cell proliferation by modulating cell cycle regulatory enzymes. PI 3′ kinase is known to upregulate two such enzymes, $p70^{S6k}$ (72,73) and cyclin-dependent kinase-2 (cdk2) (42), both of which are essential to the transition from G1 to S phase during cellular proliferation (74,75). Increased activity of $p70^{S6k}$ and cdk2 induced by PI 3′ kinase provokes proliferation in a variety of cell types including adipocytes, pancreatic cells, and aortic smooth muscle (39–42). Interestingly, cellular proliferation involving PI 3′ kinase activation of $p70^{S6k}$ appears to be mediated by IRS-1 or IRS-2, signaling molecules important to IL-13–induced activation of PI 3′ kinase (70,71).

FUTURE DIRECTION AND SIGNIFICANCE

Studies in our laboratory have begun to examine potential mechanisms by which IL-13 may induce mucous cell hyperplasia in human airway epithelium. To address this phenomenon, normal human bronchial epithelial (NHBE) cells grown in an air/liquid interface culture (76) are utilized. These NHBE cells, which begin growth in culture in an undifferentiated state, develop over time fully differentiated characteristics, including secretion of membrane-bound mucin-containing vesicles and fully developed cilia. Thus, experimentation can be carried out throughout the entire course of mucociliary differentiation, yielding a model of epithelial injury and repair. Such an in vitro model is of particular importance for studying effects of inflammatory mediators on airway epithelium, since focal points of injury are known to be interspersed with regions of normal epithelium in injured, inflamed airways (77). Thus, areas of undifferentiated and differentiated epithelium will respond differently to inflammatory mediators, a finding we have previously shown to be true when responses to such mediators (elastase, IL-13, TNF-α + IFN-γ + IL-1β) are examined using the NHBE cells in vitro early and late in their course of differentiation (78). In addition, such an in vitro model of mucous cell hyperplasia can be readily manipulated to allow elucidation of signaling pathways within the epithelial cells that regulate development of the lesion.

Preliminary studies from our laboratory suggest that continuous exposure of NHBE cells to IL-13 during the course of mucociliary differentiation in vitro results in a mucous phenotype with characteristics typical of mucous cell hyperplasia. In addition, IL-13 induces proliferation of differentiated NHBE cells in vitro. Thus, NHBE cells continuously exposed to IL-13 provide a good model of mucous cell hyperplasia, allowing mechanistic examination of two important phenomena leading to development of this lesion: differentiation and proliferation.

Any suggested in vitro mechanism must complement well what is known from in vivo models of airway mucous cell hyperplasia (8–11). For example, elastase and ozone, two agents known to provoke mucous cell hyperplasia in vivo, can directly interact with a proposed epithelial cell TGF-α/EGF-R autocrine/paracrine loop mechanism, as elastase can cleave TGF-α, while ozone may activate the EGF-R pathway via generation of oxidative species. In addition, any epithelial damage inflicted in vivo results in increased airway inflammation with inflammatory mediators such as IL-13 activating additional intracellular pathways such as those mediated by IRS-2. Of more general interest may be the role PI 3′ kinase plays in coordinating the mechanisms required for development of mucous cell hyperplasia. This enzyme, with its ability to be activated by several signaling pathways and to respond by producing multiple second messengers, is ideally suited to coordinate the seemingly opposing events of cellular differentiation and proliferation involved in this pathogenic process. Understanding the mechanistic details of this coordination is expected to provide a greater understanding of the ways epithelial cells regulate sequential processes to provoke permanent phenotypic changes.

ACKNOWLEDGMENTS

The authors would like to thank Dr. Frédéric Tournier from the Laboratoire Cytophysiologic et Toxicologie Cellulaire, Université Paris 7, for stimulating discussions, and Ms. Anne Crews for help in preparation of this manuscript. Dr. Macchione was funded by a fellowship from the Fundacão de Amparo a Pesquisa do Estado de São Paulo (Brazil).

REFERENCES

1. L Reid. Pathology of chronic bronchitis. Lancet i:275–278, 1954.
2. T Aikawa, S Shimura, H Sasaki, M Ebina, T Takishima. Marked goblet cell hyperplasia with mucus accumulation in the airways of patients who died of severe acute asthma attack. Chest 101:916–921, 1992.
3. Y Zhang, B Doranz, JR Yankaskas, JF Engelhardt. Genotypic analysis of respiratory mucous sulfation defects in cystic fibrosis. J Clin Invest 96:2997–3004, 1995.
4. JV Fahy, A Schuster, I Ueki, HA Boushey, JA Nadel. Mucus hypersecretion in bronchiectasis. The role of neutrophil proteases. Am Rev Respir Dis 146:1430–1433, 1992.
5. C Basbaum, B Jany. Plasticity in the airway epithelium. Am J Physiol 259:L38–L46, 1990.
6. DF Rogers. Airway goblet cells: responsive and adaptable front-line defenders. Eur Respir J 7:1690–1706, 1994.
7. RA Goldstein, WE Paul, DD Metcalfe, WW Busse, ER Reece. NIH conference. Asthma. Ann Intern Med 121:698–708, 1994.
8. MV Fanucchi, JA Hotchkiss, JR Harkema. Endotoxin potentiates ozone-induced mucous cell metaplasia in rat nasal epithelium. Toxicol Appl Pharmacol 152:1–9, 1998.
9. HY Cho, JA Hotchkiss, CB Bennett, JR Harkema. Effects of pre-existing rhinitis on ozone-induced mucous cell metaplasia in rat nasal epithelium. Toxicol Appl Pharmacol 158:92–102, 1999.
10. JD Lundgren, M Kaliner, C Logan, JH Shelhamer. Dexamethasone reduces rat tracheal goblet cell hyperplasia produced by human neutrophil products. Exp Lung Res 14:853–863, 1988.

11. JA Hotchkiss, R Hilaski, H Cho, K Regan, P Spencer, K Slack, JR Harkema. Fluticasone propionate attenuates ozone-induced rhinitis and mucous cell metaplasia in rat nasal airway epithelium. Am J Respir Cell Mol Biol 18:91–99, 1998.
12. N Greig, M Ayers, PK Jeffery. The effect of indomethacin on the response of bronchial epithelium to tobacco smoke. J Pathol 132:1–9, 1980.
13. DF Rogers, PK Jeffery. Inhibition of cigarette smoke-induced airway secretory cell hyperplasia by indomethacin, dexamethasone, prednisolone or hydrocortisone in the rat. Exp Lung Res 10:285–298, 1986.
14. D Lamb, L Reid. Mitotic rates, goblet cell increase and histochemical changes in mucus in rat bronchial epithelium during exposure to sulfur dioxide. J Pathol Bacteriol 96:97–111, 1968.
15. JA Hotchkiss, JR Harkema, RF Henderson. Effects of cumulative ozone exposure on ozone-induced nasal epithelial hyperplasia and secretory metaplasia in rats. Exp Lung Res 17:589–600, 1991.
16. TG Christensen, R Breuer, LJ Hornstra, EC Lucey, PI Stone, GL Snider. An ultrastructural study of response of hamster bronchial epithelium to human neutrophil elastase. Exp Lung Res 13:279–297, 1987.
17. G Grunig, M Warnock, AE Wakil, R Venkayya, F Brombacher, DM Rennick, D Sheppard, M Mohrs, DD Donaldson, RM Locksley, DB Corry. Requirement for IL-13 independently of IL-4 in experimental asthma. Science 282:2261–2263, 1998.
18. M Wills-Karp, J Luyimbazi, X Xu, B Schofield, TY Neben, CL Karp, DD Donaldson. Interleukin-13: central mediator of allergic asthma. Science 282:2258–2260, 1998.
19. Z Zhu, RJ Homer, Z Wang, Q Chen, GP Geba, J Wang, Y Zhang, JA Elias. Pulmonary expression of interleukin-13 causes inflammation, mucus hypersecretion, subepithelial fibrosis, physiologic abnormalities, and eotaxin production. J Clin Invest 103:779–788, 1999.
20. MZ Alimam, FM Piazza, DM Shelby, N Letwin, L Huang, MC Rose. Muc-5/5ac mucin messenger RNA and protein expression is a marker of goblet cell metaplasia in murine airways. Am J Respir Cell Mol Biol 22:253–260, 2000.
21. U-A Temann, B Prasad, MW Gallup, C Basbaum, SB Ho, RA Flavell, JA Rankin. A novel role for murine IL-4 in vivo: induction of MUC5AC gene expression and mucin hypersecretion. Am J Respir Cell Mol Biol 16:471–478, 1997.
22. K Dabbagh, K Takeyama, H-M Lee, IF Ueki, JA Lausier, JA Nadel. IL-4 induces mucin gene expression and goblet cell metaplasia in vitro and in vivo. J Immunol 162:6233–6237, 1999.
23. Z Jaffar, K Roberts, A Pandit, P Linsley, R Djukanovic, S Holgate. B7 costimulation is required for IL-5 and IL-13 secretion by bronchial biopsy tissue of atopic asthmatic subjects in response to allergen stimulation. Am J Respir Cell Mol Biol 20:153–162, 1999.
24. GJ McKenzie, A Bancroft, RK Grencis, AN McKenzie. A distinct role for interleukin-13 in Th2-cell-mediated immune responses. Curr Biol 8:339–342, 1998.
25. A Hancock, L Armstrong, R Gama, A Miller. Production of interleukin 13 by alveolar macrophages from normal and fibrotic lung. Am J Respir Cell Mol Biol 18:60–65, 1998.
26. ANJ McKenzie, X Li, DA Largaespada, A Sato, A Kaneda, SM Zurawski, EL Doyle, A Milatovich, U Franke, NG Copeland, NA Jenkins, G Zurawski. Structural compar-

ison and chromosomal localization of the human and mouse IL-13 genes. J Immunol 150:5436–5444, 1993.

27. LM Wang, P Michieli, WR Lie, F Liu, CC Lee, A Minty, XJ Sun, A Levine, MF White, JH Pierce. The insulin receptor substrate-1-related 4PS substrate but not the interleukin-2R gamma chain is involved in interleukin-13-mediated signal transduction. Blood 86:4218–4227, 1995.
28. C Carter-Su, LS Smit. Signaling via JAK tyrosine kinases: growth hormone receptor as a model system. Recent Prog Hormone Res 53:61–82, 1998.
29. T Murata, SR Husain, H Mohri, RJ Puri. Two different IL-13 receptor chains are expressed in normal human skin fibroblasts, and IL-4 and IL-13 mediate signal transduction through a common pathway. Int Immunol 10:1103–1110, 1998.
30. T Murata, NI Obiri, RK Puri. Human ovarian-carcinoma cell lines express IL-4 and IL-13 receptors: comparison between IL-4 and IL-13-induced signal transduction. Int J Cancer 70:230–240, 1997.
31. RL Palmer-Crocker, CCW Hughes, JS Pober. IL-4 and IL-13 activate the JAK2 tyrosine kinase and Stat6 in cultured human vascular endothelial cells through a common pathway that does not involve the γc chain. J Clin Invest 98:604–609, 1996.
32. J Punnonen, G Aversa, BG Cocks, AN McKenzie, S Menon, G Zurawski, R deWaal Malefyt, JE de Vries. Interleukin 13 induces interleukin 4-independent IgG and IgE synthesis and CD23 expression by human B cells. Proc Natl Acad Sci USA 90: 3730–3734, 1993.
33. L-M Wang, MG Myers, X-J Sun, SA Aaronson, M White, JH Pierce. IRS-1: essential for insulin- and IL-4-stimulated mitogenesis in hematopoietic cells. Science 261: 1591–1594, 1993.
34. MG Myers, Jr, MF White. Insulin signal transduction and the IRS proteins. Annu Rev Pharmacol Toxicol 36:615–658, 1996.
35. MJ Welham, L Learmonth, H Bone, JW Schrader. Interleukin-13 signal transduction in lymphohemopoietic cells: similarities and differences in signal transduction with interleukin-4 and insulin. J Biol Chem 270:12286–12296, 1995.
36. S Lefort, N Vita, R Reeb, D Caput, P Ferrara. IL-13 and IL-4 share signal transduction elements as well as receptor components in TF-1 cells. FEBS Lett 366:122–126, 1995.
37. AD Keegan, JA Johnston, PJ Tortolani, LJ McReynolds, C Kinzer, JJ O'Shea, WE Paul. Similarities and differences in signal transduction by interleukin 4 and interleukin 13: analysis of Janus kinase activation. Proc Natl Acad Sci USA 92: 7681–7685, 1995.
38. XJ Sun, L-M Wang, Y Zhang, L Yenush, MG Myers, Jr, E Glasheen, WS Lane, JH Pierce, MF White. Role of IRS-2 in insulin and cytokine signalling. Nature 377: 173–177, 1995.
39. PM Sharma, K Egawa, TA Gustafson, JL Martin, JM Olefsky. Adenovirus-mediated overexpression of IRS-1 interacting domains abolishes insulin-stimulated mitogenesis without affecting glucose transport in 3T3-L1 adipocytes. Mol Cell Biol 17: 7386–7397, 1997.
40. GT Schuppin, S Pons, S Hugle, LP Aiello, GL King, M White, CJ Rhodes. A specific increased expression of insulin receptor substrate 2 in pancreatic beta-cell lines is involved in mediating serum-stimulated beta-cell growth. Diabetes 47:1074–1085, 1998.

41. R Mendez, G Kollmorgen, MF White, RE Rhoads. Requirement of protein kinase C zeta for stimulation of protein synthesis by insulin. Mol Cell Biol 17:5184–5192, 1997.
42. D Bacqueville, F Casagrande, B Perret, H Chap, JM Darbon, M Berton-Douillon. Phosphatidylinositol 3-kinase inhibitors block aortic smooth muscle cell proliferation in mid-late G1 phase: effect on cyclin-dependent kinase 2 and the inhibitory protein p21KIP1. Biochem Biophys Res Comm 244:630–636, 1998.
43. LS Van Winkle, JM Isaac, CG Plopper. Distribution of epidermal growth factor receptor and ligands during bronchiolar epithelial repair from naphthalene-induced Clara cell injury in the mouse. Am J Pathol 151:443–459, 1997.
44. DK Madtes, HK Busby, TP Standjord, JG Clark. Expression of transforming growth factor-alpha and epidermal growth factor receptor is increased following bleomycin-induced lung injury in rats. Am J Respir Cell Mol Biol 11:540–551, 1994.
45. TP Standjord, JG Clark, DE Guralnick, DK Madtes. Immunolocalization of transforming growth factor-alpha, epidermal growth factor (EGF), and EGF-receptor in normal and injured developing human lung. Pediatr Res 38:851–856, 1995.
46. WD Hardie, PA Bejarano, MA Miller, JR Yankaskas, JH Ritter, JA Whitsett, TR Korfhagen. Immunolocalization of transforming growth factor alpha and epidermal growth factor receptor in lungs of patients with cystic fibrosis. Pediatr Dev Pathol 2:415–423, 1999.
47. K Takeyama, K Dabbagh, H.-M Lee, C Agusti, JA Lausier, IF Ueki, KM Grattan, JA Nadel. Epidermal growth factor system regulates mucin production in airways. Proc Natl Acad Sci USA 96:3081–3086, 1999.
48. DRF Haugen, LA Akslen, JE Vargaug, JR Lillehaug. Demonstration of a TGF-α-EGF-receptor autocrine loop and c-myc protein over-expression in papillary thyroid carcinomas. Int J Cancer 55:37–43, 1993.
49. K-I Morishige, H Kurachi, K Amemiya, Y Fujita, T Yamamoto, A Miyake, O Tanizawa. Evidence for the involvement of transforming growth factor α and epidermal growth factor receptor autocrine growth mechanism in primary human ovarian cancers in vitro. Cancer Res 51:5322–5328, 1991.
50. ER Sherwood, C Lee. Epidermal growth factor-related peptides and the epidermal growth factor receptor in normal and malignant prostate. World J Urol 13:290–296, 1995.
51. J Jankowski, D Hopwood, KG Wormsley. Expression of epidermal growth factor, transforming growth factor alpha and their receptor in gastro-oesophageal diseases. Dig Dis 11:1–11, 1993.
52. RJ Coffey, M Romano, WH Polk, PJ Dempsey. Roles for transforming growth factor-α in gastric physiology and pathophysiology. Yale J Biol Med 65:693–704, 1992.
53. PC Ferriola, HS Earp, RD Augustine, P Nettesheim. Role of TGFα and its receptor in proliferation of immortilized rat tracheal epithelial cells: studies with tyrphostin and TGFα antisera. J Cell Physiol 147:166–175, 1991.
54. Derynck, R. Transforming growth factor α. Cell 54:593–595, 1988.
55. ST Wong, LF Winchell, BK McCune, HS Earp, J Teixido, J Massague, B Herman, DC Lee. The TGF-α precursor expressed on the cell surface binds to the EGF receptor on adjacent cells, leading to signal transduction. Cell 56:495–506, 1989.
56. P Anklesaria, J Teixido, M Laiho, JM Pierce, JS Greenberger, J Massague. Cell-cell

adhesion mediated by binding of membrane-anchored transforming growth factor α to epidermal growth factor receptors promotes cell proliferation. Proc Natl Acad Sci USA 87:3289–3293, 1990.
57. A Pandiella, J Massague. Cleavage of the membrane precursor for transforming growth factor α is a regulated process. Proc Natl Acad Sci USA 88:1726–1730, 1991.
58. CL Carpenter, LC Cantley. Phosphoinositide 3-kinase and the regulation of cell growth. Biochem Biophys Acta 1288:M11–M16, 1996.
59. VP Krymskaya, RB Penn, MJ Orsini, PH Scott, RJ Plevin, TR Walker, AJ Eszterhas, Y Amrani, ER Chilners, RA Panettieri, Jr. Phosphatidylinositol 3-kinase mediates mitogen-induced human airway smooth muscle cell proliferation. Am J Physiol 277: L65–L78, 1999.
60. GR Dubois, RC Schweizer, C Versluis, CA Bruijnzeel-Koomen, PL Bruijnzeel. Human eosinophils constitutively express a functional interleukin-4 receptor: interleukin-4-induced priming of chemotactic responses and induction of PI-3 kinase activity. Am J Respir Cell Mol Biol 19:691–699, 1998.
61. K Wright, SG Ward, G Kolios, J Westwick. Activation of phosphatidylinositol 3-kinase by interleukin-13: an inhibitory signal for inducible nitric-oxide synthase expression in epithelial cell line HT-29. J Biol Chem 272:12626–12633, 1997.
62. S Roche, M Koegl, SA Courtneidge. The phosphatidylinositol 3-kinase α is required for DNA synthesis induced by some, but not all, growth factors. Proc Natl Acad Sci USA 389:725–730, 1994.
63. Y Zhang, RA Akhtar. Effect of epidermal growth factor on phosphatidylinositol 3-kinase activity in rabbit corneal epithelial cells. Exp Eye Res 63:265–275, 1996.
64. Y Zhang, RA Akhtar. Epidermal growth factor stimulation of phosphatidylinositol 3-kinase during wound closure in rabbit corneal epithelial cells. Invest Opthal Vis Sci 38:1139–1148, 1997.
65. CL Carpenter, LC Cantley. Phosphoinositide kinases. Biochemistry 29:11147–11156, 1990.
66. T Fukazawa, S Miyake, V Band, H Band. Tyrosine phosphorylation of Cbl upon epidermal growth factor (EGF) stimulation and its association with EGF receptor and downstream signaling proteins. J Biol Chem 271:14554–14559, 1996.
67. M Holgado-Madruga, DR Emlet, DK Moscatello, AK Godwin, AJ Wong. A Grb2-associated docking protein in EGF- and insulin-receptor signalling. Nature 379:560–564, 1996.
68. DR Stover, M Becker, J Liebetanz, NB Lydon. Src phosphorylation of the epidermal growth factor receptor at novel sites mediates receptor interaction with Src and p85α. J Biol Chem 270:15591–15597, 1995.
69. Y Takahashi, Y Akanuma, Y Yazaki, T Kadowaki. Formation of distinct signalling complexes involving phosphatidylinositol 3-kinase activity with stimulation of epidermal growth factor or insulin-like growth factor-1 in human skin fibroblasts. J Cell Physiol 178:69–75, 1999.
70. A Klippel, C Reinhard, WM Kavanaugh, G Apell, M-A Escobedo, LT Williams. Membrane localization of phosphatidylinositol 3-kinase is sufficient to activate multiple signal-transducing kinase pathways. Mol Cell Biol 16:4117–4127, 1996.
71. K Egawa, PM Sharma, N Nakashima, Y Huang, E Huver, GR Boss, JM Olefsky.

Membrane-targeted phosphatidylinositol 3-kinase mimics insulin actions and induces a state of cellular insulin resistance. J Biol Chem 274:14306–14314, 1999.
72. K Reif, BMT Burgering, DA Cantrell. Phosphatidylinositol 3-kinase links the interleukin-2 receptor to protein kinase B and p70 S6 kinase. J Biol Chem 272: 14426–14433, 1997.
73. M Myers, Jr, T Grammer, L-M Wang, X Sun, J Pierce, J Blenis, M White. Insulin receptor substrate-1 mediates phosphatidylinositol 3′-kinase and p70S6k signaling during insulin, insulin-like growth factor-1, interleukin-4 stimulation. J Biol Chem 269:28783–28789, 1994.
74. HA Lane, A Fernandez, NJC Lamb, G Thomas. $p70^{s6k}$ function is essential for G1 progression. Nature 363:170–172, 1993.
75. S Van den Heuvel, E Harlow. Distinct roles for cyclin-dependent kinases in cell cycle control. Science 262:2050–2054, 1993.
76. TM Krunkosky, BM Fischer, LD Martin, N Jones, NJ Akley, KB Adler. Effects of TNFα on expression of ICAM-1 in human airway epithelial cells in vitro: signaling pathways controlling surface and gene expression. Am J Respir Cell Mol Biol 22: 1–8, 2000.
77. LS Van Winkle, AR Buckpitt, SJ Nishio, JM Isaac, CG Plopper. Cellular response in naphthalene-induced Clara cell injury and bronchiolar epithelial repair in mice. Am J Physiol 269:L800–L818, 1995.
78. LD Martin, D Norford, JA Voynow, KB Adler. Response of human airway epithelium in vitro to inflammatory mediators: dependence on the state of cellular differentiation. Chest 117:267S, 2000.

24

Airway Mucins and Lung Cancer

Chong-Jen Yu and Pan-Chyr Yang
National Taiwan University Hospital
Taipei, Taiwan

MUCIN AND CANCER

Mucins are high molecular weight glycoproteins forming the mucus that coats the surface of human epithelial cells lining mammary glands, salivary glands, digestive tract, respiratory tract, and urogenital tract (1). The major function of mucins is to protect epithelial cells from dehydration, infection, and other physical or chemical injuries.

Expression of mucins often changes during the malignant transformation of epithelial cells (2,3). Mucin synthesis may increase markedly, resulting in loss of cellular polarity. Large amounts of mucus are shed or secreted by tumor cells. Furthermore, diverse patterns of glycosylation changes on mucoprotein peptide backbone usually occur, resulting in the expression of novel carbohydrate epitopes or the exposure of peptide backbone (4,5). Alteration of mucin glycoprotein may be involved in several stages of metastasis, including cell growth, motility during invasion, interactions between tumor cells, platelets, and the immune system, and adherence of tumor cells to endothelia and extracellular matrix (6). Aberrant expression of carbohydrate epitopes or mucin core proteins has been shown to negatively influence the outcome of several neoplasms, such as colorectal cancer (7–9).

The dysregulation of mucin expression in cancer includes aberrant glycosylation, underglycosylation, and overexpression of mucin peptides (10–12). In vitro studies have demonstrated that expression of mucin in cancer cells can decrease tumor cell aggregation, promote tumor cell invasion, and block lymphocyte targeting, thereby facilitating metastasis by escaping systems of immunosurveillance. Highly metastatic colon carcinoma cells express significantly more sialylated glycoproteins than poorly metastatic cells. The expression of sialylated antigens may ‘‘mask’’ recognition in tumor metastasis by T and natural killer cells in the immune system. Increased expression of sialyl Lewis x on cell surface glycoproteins can serve as better ligands for selectins present on endothelial cells and paltelets (13–16). Moreover, high mucin–producing cancer cells may secrete more proteolytic enzymes, such as collagenase, than low mucin–producing cells. Collagenase activity is further stimulated by purified mucin, thus promoting tumor invasion (17).

MUCIN GENES AND AIRWAY MUCINS

To date, at least 10 mucin genes have been completely or partially cloned, and they can be grouped into two types: membrane-bound and secreted. MUC1 and MUC4 are membrane-bound mucins (18,19), while MUC2, MUC5AC, MUC5B, and MUC7 are secreted mucins (20–23). The others, MUC3, MUC6, MUC8, and MUC9, remain unclassified (24–27).

MUC1, the gene encoding episialin, a membrane O-glycoprotein, has been located on chromosome 1q21-24. MUC2 and MUC3, genes coding for secreted intestinal mucins, were mapped to chromosomes 11q15 and 7q22, respectively. Tracheobronchial mucin cDNAs were used to identify three mucin genes, MUC4, MUC5AC, and MUC5B, which have been mapped to chromosome 3q29 (MUC4) and chromosome 11p15 (MUC5AC and MUC5B). MUC6 gene was cloned and identified in stomach and gall bladder and was mapped to chromosome 11p15. MUC7, expressed in salivary glands, has been localized to chromosome 4. MUC8, expressed in both goblet cells and submucosal glands of human trachea, was mapped to 12q24.

Of the 10 human mucin genes, seven (MUC1, MUC2, MUC4, MUC5AC, MUC5B, MUC7, MUC8) are expressed in human airway epithelium, as demonstrated by Northern blot analysis and in situ hybridization (Table 1). However, the exact role of each of the seven mucin gene remains unclear (28–32). MUC1 is expressed in the primary culture of human nasal and tracheobronchial epithelial cells but not in squamous metaplastic epithelia, implying its close relationship with mucus differentiation of epithelial cells (33). MUC2 is known to exist in several cell types of the respiratory tract. However, it has not been observed in cultured airway epithelial cells and is not considered to be a major mucin gene

TABLE 1 Expression of Mucin Genes in Normal Airway and Lung Tissues vs. Lung Cancer

	Normal airway and lung tissues				Lung cancer	
	Trachea	Bronchus	Bronchiole	Alveolus	Cell lines	Tissues
MUC1	E+/G−	E+/G−	E+	±	+++	+++
MUC2	E++/G+	E++/G−	E±	—	±	+
MUC3	—	—	—	—	+	+
MUC4	E+++/G−	E+++/G−	E++	—	+	++
MUC5AC	E+++/G−	E+++/G−	—	—	++	++
MUC5B	E+/G++	E+/G++	—	—	+	+++
MUC6	—	—	—	—	NA	NA
MUC7	E−/G+++	E−/G+++	—	—	NA	NA
MUC8	E+/G+	E+/G+	—	NA	NA	NA

E: Epithelium; G: submucosal gland; NA: not assessed.
Source: Refs. 26, 28, 30–33, 38–41, 43, 57.

in human respiratory mucosa (34). As previously stated, MUC4, MUC5AC, and MUC5B had been cloned from tracheobronchial mucin cDNA library. MUC4 is located in both ciliary and goblet cells and is considered to be the very first mucin gene expressed in the developmental process of human respiratory tract. In contrast, MUC5AC and MUC5B are expressed later in the developmental phase of the airway. Biochemical analysis has identified MUC5AC and MUC5B as major contributors to the mucin produced by human airway epithelium (35–37). Among three major gel-forming mucins in human sputum obtained from patients who died of asthmaticus, two are differently charged glycoforms of MUC5B, while the other is MUC5AC. And, in contrast to MUC5AC, which originates mainly in goblet cells (Fig. 1a), MUC5B is the major mucin product of mucous cells of submucosal glands. Although both MUC7 and MUC8 are produced by submucosal glands, they are unlikely to be the major secreted airway mucin (32).

AIRWAY MUCIN AND LUNG CANCER: EXPRESSION PATTERNS

So far, research investigating the expression patterns of mucin genes and their products in lung cancer has been limited. The expression patterns of mucin genes in cancer cell lines are different from those in cancer tissues. Berger et al. studied the expression patterns of five mucin genes (MUC1, MUC2, MUC4, MUC5AC, and MUC5B) in six lung cancer cell lines by Northern and Western blot analyses (38). They demonstrated variable expression patterns of mucin genes in different cell line. All cell lines expressed MUC1 mucin, but none expressed MUC2. Calu-3 and A549 cell lines expressed high to moderate amounts of MUC5AC mucin, whereas NCI-H292 cells did not express as much. On the other hand, H292 cells expressed the highest amount of MUC4 in all cell lines.

By studying lung cancer tissues, Nguyen et al. found that lung adenocarcinomas, especially well-differentiated cancers, exhibited increased MUC1, MUC3, and MUC4 mRNA levels. Yet, squamous-cell, adenosquamous, and large-cell carcinomas were characterized only by increased levels of MUC4 mucin (39).

Another study evaluating the expression patterns of mucins in lung cancer tissue samples was performed by Seregni et al. (40). Their results showed that the intensities of MUC1 and MUC4 expressions were always higher in cancer tissues than in normal cell lines. Similar to the study by Nguyen et al., the highest reactivity for MUC1 and MUC4 expression was observed mainly in the adenocarcinoma histotype, which is mucin secreting.

In one of our previous studies, we examined the expression patterns of mucin genes in 7 lung adenocarcinoma cell lines and 12 lung adenocarcinoma tissues. We performed Northern blot analysis with specific antisense oligonucleo-

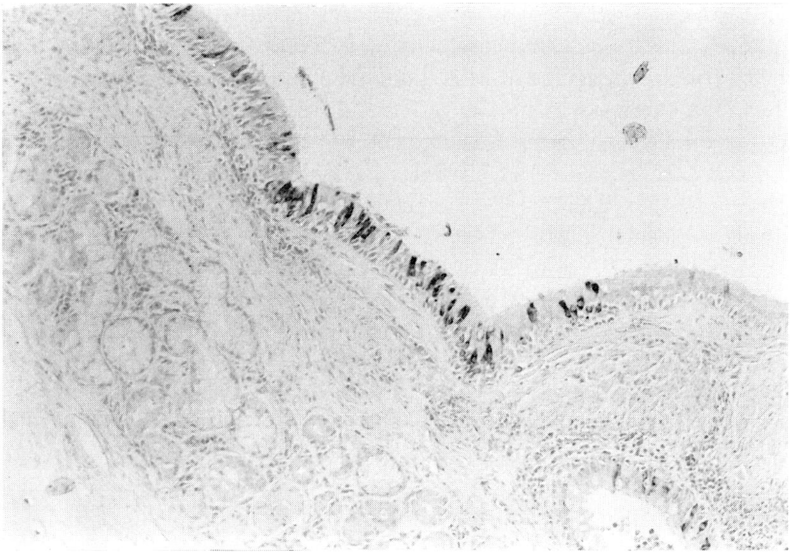

a

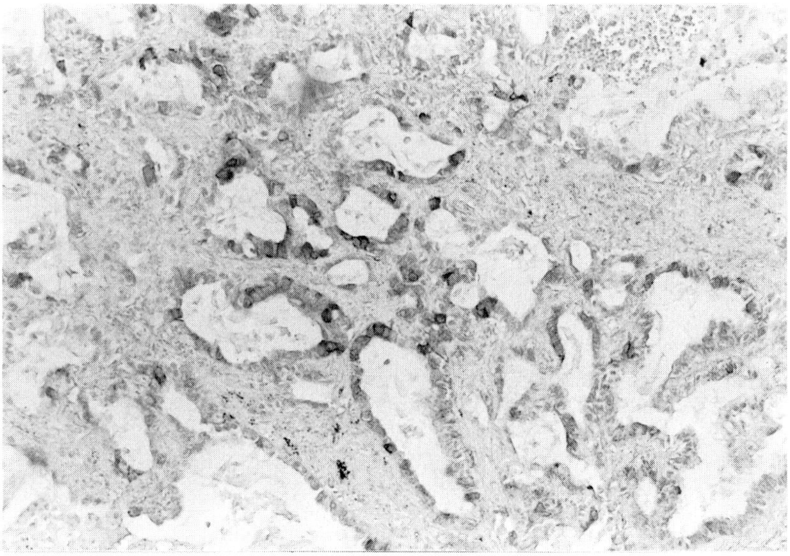

b

Figure 1 Immunohistochemistry using anti-MUC5AC monoclonal antibody for paraffin-embedded human specimens. (a) Bronchus. Positive staining appears in goblet cells and mucous epithelial cells, but not in submucosal glands. (b) Lung adenocarcinoma. Positive cytoplasmic staining appears in tumor cells. (Original magnification: ×100.)

tide probes that recognize mucin-specific tandem repeats of four mucin genes (MUC1, MUC2, MUC3, MUC4) and carried out RT-PCR to amplify the 3′ and 5′ nonrepetitive coding region of MUC1 and the 5′ nonrepetitive coding region of MUC2 (41). Of the seven cell lines, CL1 and PC13 are poorly differentiated with low mucin glycoprotein production (42), while the rest are well differentiated. All cell lines expressed MUC1, MUC2, MUC3, and MUC4 mRNAs, but in variable quantities. The poorly differentiated cell lines (CL1 and PC13) had a relatively low level of expression of MUC1-4. RT-PCR also revealed the presence of MUC1 and MUC2 mRNA in all the cell lines.

The expression pattern of mucin genes is consistent with that of mucin glycoproteins as studied using biochemical and immunological methods. Northern blotting and RT-PCR analyses in 12 lung adenocarcinoma tissues with various grades of differentiation (6 poorly differentiated adenocarcinoma and 6 moderately to well-differentiated adenocarcinomas) showed heterogeneous expressions of the four mucin genes in tissues without clear correlation with the differentiation grade.

We next studied mucin expression patterns in paired lung cancer and nontumorous lung portions (43). Sixty tissues pairs with varying types and stages were included. Slot blot analyses with specific antisense oligonucleotide probes derived from tandem repeat sequence of MUC1, MUC2, MUC3, MUC4, MUC5B, and MUC5AC were utilized to compare the amount of mucin gene mRNA in tumor samples with that of the nontumor counterparts. A ratio higher than 1.5 for each specific mucin mRNA amount was considered to indicate mucin gene overexpression in tumors. The study showed that mucin gene overexpressions frequently occurred in lung cancer (25 out of 60, 41.7%); MUC1 overexpression was found in 7 tumors (11.7%), MUC2 in 5 (8.3%), MUC3 in 12 (20%), MUC4 in 9 (15%), MUC5B in 10 (16.7%), and MUC5AC in 13 (21.7%). Immunohistochemical stain demonstrated the expression of mucin product in lung cancer tissue (Fig. 1b) However, there was no preferential expression of any particular or combination of mucin genes in lung tumors. Overexpression of mucin genes and mucin proteins appeared to have no correlation with tumor stage, nodal stage, histology, or pathological differentiation grade.

AIRWAY MUCIN AND LUNG CANCER: AS A PROGNOSTIC FACTOR

Lung cancer is one of the most commonly diagnosed malignancies throughout the world. The prognosis of lung cancer is generally poor. Only 20–30% of the patients are operable at diagnosis, and only half of these can be successfully resected. The 5-year survival rate of lung cancer is less than 15%. About 80% of the patients die of tumor recurrence following surgical resection (44,45).

Although many studies have focused on the carcinogenic and metastatic process of lung cancer, most have failed to make a significant breakthrough. To date, several studies have investigated the expression status of mucins in lung cancer. Most addressed the prognostic implications of mucin-associated carbohydrate antigens in lung cancer. The results indicated that the expression of some carbohydrate antigens, e.g., sialyl Le^x and sialyl Le^a antigens, in lung cancer may be correlated with shorter survival (46–48). Other evidence further supported the prognostic implications of sialylation or sialylated antigens for lung cancer. The amount of sialic acid present on the surface of malignant cells appeared to be correlated with the metastatic potential of malignancy (49). Lung cancer patients tended to have higher serum levels of total sialic acid and lipid-bound sialic acid (50). In human cancer cell lines, treatment with specific inhibitors of sialic acid could reduce the incidence of metastasis (51). Besides sialylation and sialylated antigens, mucin and mucin-associated antigens could also be prognostic factors for lung cancer (52,53). Alteration of mucin gene expression has also been demonstrated in various types of cancers. In particular, MUC1 protein expression has been proven to have prognostic significance in both colorectal cancer and breast cancer (54–56).

There were 117 patients with non-small cell lung cancer (NSCLC) in our study (57). Tumor specimens were stained immunohistochemically with monoclonal antibodies (mAbs) against mucin glycoprotein (17Q2, HMFG2, SM3). Samples were also stained histochemically with periodic acid Schiff/Alcian blue to differentiate neutral mucin from acid mucin and with high iron diamine/alcian blue to differentiate sialomucin from sulfomucin. The expression status of two established molecular prognostic factors, *p53* and *erb*B-2 oncoproteins, were evaluated with immunohistochemistry. Correlations were established between adenocarcinoma histotypes, *erb*B-2 overexpression, sialomucin expression, 17Q2, and HMFG2 immunohistochemical positivity ($p < 0.05$). Sialomucin expression was closely linked to *erb*B-2 overexpression ($p = 0.01$). Through our study, we found the significant univariate predictors ($p < 0.05$) of recurrence and cancer death to be surgical stage, *p53* expression, *erb*B-2 overexpression, and sialomucin expression. These four factors remained the independent predictors of early recurrence ($p < 0.05$) after multivariate analysis. We concluded that sialomucin expression is a poor prognostic indicator, which is associated with *erb*B-2 oncoprotein overexpression, early postoperative recurrence, and cancer death in NSCLC.

In our study investigating the overexpression of mucin genes in lung cancer, we found that tumors with increased expression of mucin genes tended to have postoperative relapse, especially when MUC5B and MUC5AC genes were overexpressed ($p = 0.015$ and 0.025, respectively). Our findings suggest that overexpression of major airway mucins may result in the increased likelihood of postoperative lung cancer recurrence or metastases.

AIRWAY MUCIN AND LUNG CANCER: AS A DIAGNOSTIC FACTOR

Mucin can also be used as a diagnostic adjunct for lung cancer. By assaying 135 pleural effusions with definite etiology (24 effusions were transudate, 45 nonmalignant exudate, and 66 malignant), we evaluated the diagnostic applicability of mucin-specific monoclonal antibody (17Q2)–derived enzyme-linked immunosorbent assay (ELISA) (58). Our results showed that the mean mucin concentration in malignant pleural effusions was significantly higher than that of benign exudates ($p < 0.01$). Mucin concentration in malignant pleural effusions of adenocarcinoma was also significantly higher than that in nonadenocarcinoma histotypes ($p < 0.01$). With the use of mean ± 2 SD of mucin concentration in benign exudates as a cut-off value, the sensitivity of this assay for diagnosis of malignant effusions was 66.7%, specificity 97.1%, and accuracy 82.2%.

We also established a modified quantitative competitive polymerase chain reaction (QC-PCR) analysis to evaluate the expression of specific mucin gene in clinical specimens (59). RNA internal standards of MUC1, MUC2, and MUC5AC nontandem repeat sequence were constructed, and known copy numbers of mucin RNA internal standards were introduced in reverse transcription-polymerase chain reaction (RT-PCR) for each mucin gene in order to compete with native mucin gene RNA during the reaction. The RNA of Gβ-like gene (a house-keeping gene) was used as internal control of RNA analysis. Twenty-five lung cancer tissues (13 adenocarcinomas and 12 squamous cell carcinomas), 15 paired nontumor lung tissues, and 3 nonpaired normal lung tissues were used for analysis. The results revealed that QC-PCR was more sensitive than slot blot analysis in evaluating the expressed quantity of mucin gene RNA. Mucin genes were overexpressed in 10 out of 15 paired tissues. Adenocarcinoma expressed higher quantities of MUC5AC gene than squamous cell carcinoma ($p = 0.03$). The expression of MUC5AC positively correlated with the expression status of sialomucin ($p = 0.012$). Because of its high sensitivity, QC-PCR can be applied in investigating many aspects of mucin gene expression, such as transcriptional regulation and the status of mucin composition in pleural fluid or biopsied specimens. We have recently applied mucin QC-PCR in order to assist the diagnosis of malignant pleural effusion and found that the test can reach acceptable sensitivity (60–70%) and specificity (80–90%) for clinical application (unpublished data).

In conclusion, recent studies investigating the expression of mucins in lung cancer have clarified the important role of mucins in the pathogenesis of tumor metastasis, as well as the clinical implications of measuring mucins in diagnostic assistance and prognostic stratification.

REFERENCES

1. B J-W Van Klinken, J Dekker, HA Buller, AWC Einerhand. Mucin gene structure and expression: protection vs. adhesion. Am J Physiol 269:G613–G627, 1995.
2. SJ Gendler, AP Spicer. Epithelial mucin genes. Annu Rev Physiol 57:607–634, 1995.
3. PL Devine, IFC McKenzie. Mucins: structure, function, and associations with malignancy. BioEssays 14:619–625, 1992.
4. YS Kim, JR Gum, I Brockhausen. Mucin glycoproteins in neoplasia. Glycoconjucate J 13:693–707, 1996.
5. M Fukuda. Possible role of tumor-associated carbohydrate antigens. Cancer Res 56: 2237–2244, 1996.
6. Y Niv. Mucin and colorectal cancer metastasis. Am J Gastroenterol 89:665–669, 1994.
7. YS Kim. Mucin glycoproteins in colonic neoplasia. Keio J Med 47:10–18, 1998.
8. S Yonezawa, T Tashikawa, S Shin, E Sato. Sialosyl-Tn antigen: Its distribution in normal human tissues and expression in adenocarcinomas. Am J Clin Pathol 98: 167–174, 1992.
9. SH Itzkowitz, EJ Bloom, WA Kokal, G Modin, S Hakomori, YS Kim. Sialosyl-Tn-A novel mucin antigen associated with prognosis in colorectal patients. Cancer 66:1960–1966, 1990.
10. CR Boland, CK Montgomery, YS Kim. Alterations in human colonic mucin occurring with cellular differentiation and malignant transformation. Proc Natl Acad Sci USA 79:2051–2055, 1982.
11. I Brockhausen, J-M Yang, J Burchell, C Whitehouse, J Taylor-Papadimitriou. Mechanisms underlying aberrant glycosylation of MUC1 mucin in breast cancer cells. Eur J Biochem 233:607–617, 1995.
12. C Hanski, K Drechsler, F-G Hanisch, J Sheehan, M Manske, D Ogorek, E Klussmann, M-L Hanski, M Blank, P-X Xing, IFC McKenzie, PL Devine, E-O Riecken. Altered glycosylation of the MUC-1 protein core contributes to the colon carcinoma-associated increase of mucin-bound sialyl Lewis^x expression. Cancer Res 53:4082–4088, 1993.
13. RS Bresalier, RW Rockwell, R Dahiya, QY Dun, YS Kim. Cell surface sialoprotein alterations in metastatic murine colon cancer cell lines selected in an animal model for colon cancer metastasis. Cancer Res 50:1299–1307, 1990.
14. RS Bresalier, SB Ho, HL Schoeppner, YS Kim, MH Sleisenger, P Brod, JC Byrd. Enhanced sialylation of mucin-associated carbohydrate structures in human colon cancer metastasis. Gastroenterology 110:1354–1367, 1996.
15. T Irimura, DA Carlson, J Price, T Yamori, R Giavazzi, DM Ota, KR Cleary. Differential expression of a sialoglycoprotein with an approximate molecular weight of 900,000 on metastatic human colon carcinoma cell growing in culture and in tumor tissues. Cancer Res 48:2353–2360, 1988.
16. O Saitoh, WC Wang, R Lotan, M Fukuda. Differential glycosylation and cell surface expression of lysosomal membrane glycoproteins in sublines of a human colon cancer exhibiting distinct metastatic potentials. J Biol Chem 267:5700–5711, 1992.

17. T Kanemoto, R Reich, L Royce, D Greatorex, SH Adler, N Shiraishi, GR Martin, Y Yamada, HK Kleinman. Identification of an amino acid sequence from laminin A chain that stimulates metastasis and collagenase IV production. Proc Natl Acad Sci USA 87:2279–2283, 1990.
18. S Gendler, O Lancaster, Taylor-Papadimitriou, T Duhig, N Peat, J Burchell, L Pemberton, EN Lalanl, D Wilson. Molecular cloning of human tumor-associated polymorphic epithelial mucin. J Biol Chem 265:15285–15293, 1990.
19. N Moniaux, S Nollet, N Prochet, P Degand, A Laine, J-P Aubert. Complete sequence of the human mucin MUC4: a putative cell membrane-associated mucin. Biochem J 338:325–333, 1999.
20. JR Gum, JW Hicks, NW Toribara, B Siddiki, YS Kim. Molecular cloning of human intestinal mucin (MUC2) cDNA: Identification of the amino terminus and overall sequence similarity to prepro-Von Willebrand factor. J Biol Chem 269:2440–2446, 1994.
21. D Li, M Gallup, N Fan, DE Szymkowski, CB Basbaum. Cloning of the amino-terminal and 5′-flanking region of the human MUC5AC mucin gene and transcriptional up-regulation by bacterial exoproducts. J Biol Chem 273:6812–6820, 1998.
22. J-L Desseyn, M-P Buisine, N Porchet, J-P Aubert, A Laine. Genomic organization of the human mucin gene MUC5B: cDNA and genomic sequences upstream of the large central exon. J Biol Chem 273:30157–30164, 1998.
23. LA Bobek, H Tsal, AR Biesbrock, MJ Levine. Molecular cloning, sequence, and specificity of expression of the mucin encoding the low molecular weight human salivary mucin (MUC7). J Biol Chem 268:20563–20569, 1993.
24. JR Gum, JW Hicks, DM Swallow, RL Lagace, JC Byrd, DTA Lamport, B Siddiki, YS Kim. Molecular cloning of cDNAs derived from a novel human intestinal mucin gene. Biochem Biophys Res Commun 171:407–415, 1990.
25. NW Toribara, AM Roberton, SB Ho, W-L Kuo, E Gum, JW Hicks, JR Gum, JC Byrd, B Siddik, YS Kim. Human gastric mucin: identification of a unique species by expression cloning. J Biol Chem 268:5879–5885, 1993.
26. V Shanker, MS Gilmore, RC Elkins, GP Sachdev. A novel human airway mucin cDNA encodes a protein with unique tandem-repeat organization. Biochem J 300: 295–298, 1994.
27. L Lapensee, Y Paquette, G Bleau. Allelic polymorphism and chromosomal localization of the human oviductin gene(MUC9). Fertil Steril 68:702–708, 1997.
28. J-P Audie, A Janin, N Porchet, MC Copin, B Gosselin, J-P Aubert. Expression of human mucin genes in respiratory, digestive and reproductive tracts ascertained by in situ hybridization. J Histochem Cytochem 41:1479–1485, 1993.
29. G Lamblin, J-P Aubert, JM Perinl, A Klein, N Porchet, P Degand, P Roussel. Human respiratory mucin. Eur Respir J 5:247–256, 1992.
30. M-P Buisine, L Devisime, M-C Copin, M Durand-Reville, P Gosselin, J-P Aubert, N Porchet. Developmental mucin gene expression in the human respiratory tract. Am J Respir Cell Mol Biol 20:209–218, 1999.
31. P Sharma, L Dudus, PA Nielsen, H Clausen, JR Yankaskas, MA Hollingsworth, JF Engelhardt. MUC5B and MUC7 are differentially expressed in mucous and serous cells of submucosal glands in human bronchial airways. Am J Respir Cell Mol Biol 19:30–37, 1998.

32. SH Bernacki, AL Nelsen, L Abdullah, JK Sheehan, A Harris, CW Davis, SH Rabdell. Mucin gene expression during differentiation of human airway epithelia in vitro: MUC4 and MUC5B are strongly induced. Am J Respir Cell Mol Biol 20: 595–604, 1999.
33. J-H Yoon, IY Park. Mucin gene expression and mucin secretion in human airway epithelium. Rhinology 36:146–152, 1998.
34. K Guzman, T Bader, P Nettesheim. Regulation of MUC5 and MUC1 gene expression: correlation with airway mucous differentiation. Am J Physiol 270:L846–L853, 1996.
35. JR Davies, HW Hovenberg, C-J Linden, R Howard, PS Richardson, JK Sheehan, I Carlstedt. Mucins in airway secretions from healthy and chronic bronchitic subjects. Biochem J 313:431–439, 1996.
36. DJ Thornton, I Carlstedt, M Howard, PL Devine, MR Prices, JK Sheehan. Respiratory mucins: identification of core proteins and glycoforms. Biochem J 316:967–975, 1996.
37. JK Sheehan, M Noward, PS Richardson, T Longwill, DJ Thornton. Physical characterization of a low-charge glycoform of the MUC5B mucin comprising the gel-phase of an asthmatic respiratory mucous plug. Biochem J 338:507–513, 1999.
38. JT Berger, JA Voynow, KW Peters, MC Rose. Respiratory carcinoma cell lines: MUC genes and glycoconjugates. Am J Respir Cell Mol Biol 20:500–510, 1999.
39. PL Nguyen, GA Niehans, DL Cherwitz, YS Kim, SB Ho. Membrane-bound (MUC1) and secretory (MUC2, MUC3, MUC4) mucin gene expression in human lung cancer. Tumor Biol 17:176–192, 1996.
40. E Seregni, C Botti, C Lombardo, A Cantonl, A Bogni, I Cataldo, E Bombardierl. Pattern of mucin gene expression in normal and neoplastic lung tissues. Anticancer Res 16:2209–2213, 1996.
41. CJ Yu, PC Yang, JY Shew, TM Hong, SC Yang, YC Lee, LN Lee, KT Luh, CW Wu. Mucin mRNA expression in lung adenocarcinoma cell lines and tissues. Oncology 53:118–126, 1996.
42. PC Yang, KT Luh, R Wu, CW Wu. Characterization of the mucin differentiation in human lung adenocarcinoma cell lines. Am J Respir Cell Mol Biol 7:161–171, 1992.
43. CJ Yu, PC Yang, CT Shun, YO Lee, SH Kuo, KT Luh. Overexpression of MUC5 genes is associated with early postoperative metastasis in non-small cell lung cancer. Int J Cancer 69:457–465, 1996.
44. RJ Ginsberg, EE Vokes, A Raben. Non-small cell lung cancer. In: VT Devita, S Hellman, SA Rosenberg, eds. Cancer: Principles and Practice of Oncology. 5th ed. Philadelphia: JB Lippincott, 1997, pp 858–911.
45. M Sagawa, Y Saito, S Takahashi, K Usuda, K Kamma, M Sato, S-I Ota, N Nagamoto, S Fujimura, T Nakada, K Hashimoto, H Suda, T Imai, H Saito. Clinical and prognostic assessment of patients with resected small peripheral lung cancer lesions. Cancer 66:2653–2657, 1990.
46. J-I Ogawa, A Sano, S Koide, A Shohtsu. Relation between recurrence and expression of proliferating cell nuclear antigen, sialyl Lewis^x, and sialyl Lewis ^{a} in lung cancer. J Thorac Cardiovas Surg 108:329–336, 1994.
47. J-I Ogawa, T Surumi, S Yamada, S Koide, A Shohtsu. Blood vessel invasion and

expression of sialyl Lewis^x and proliferating cell nuclear antigen in stage I non-small cell lung cancer: relationship to postoperative recurrence. Cancer 73:1177–1183, 1994.
48. J-I Ogawa, A Sano, H Inoue, S Koide. Expression of Lewis-related antigen and prognosis in stage I non-small cell lung cancer. Ann Thorac Surg 59:412–415, 1995.
49. G Yogeeswaran. Cell membrane glycolipids and glycoproteins in malignant transformation. Adv Cancer Res 38:289–350, 1982.
50. S Kakari, E Stringou, M Toumbis, AS Ferderigos, E Poulaki, K Chondros, A Dema, V Kotsovoulou, N Pavlidis. Five tumor markers in lung cancer: significance of total and ''lipid'' bound sialic acid. Anticancer Res 11:2107–2110, 1991.
51. AT Petrick, S Meterissian, G Steele Jr., P Thomas. Desialylation of human colorectal carcinoma cells facilitates binding to Kupffer cells. Clin Exp Metastasis 12:108–116, 1996.
52. PC Willsher, P-X Xing, CP Clarke, DWM Ho, IFC McKenzie. Mucin 1 antigens in the serum and bronchial lavage fluid of patients with lung cancer. Cancer 72: 2936–2942, 1993.
53. SI Graziano, AH Tatum, NR Newman, A Oler, LJ Kohman, LJ Vei, GP Gamble, MJ Coleman, S Barmada, S O'Lean. The prognostic significance of neuroendocrine markers and carcinoembryonic antigen in patients with resected stage I and II non-small cell lung cancer. Cancer Res 54:2908–2913, 1994.
54. MA McGuckin, MD Walsh, BG Hohn, BG Ward, RG Wright. Prognostic significance of MUC1 epithelial mucin expression in breast cancer. Hum Pathol 26:432–439, 1995.
55. S Nakamori, DM Ota, KR Clery, K Shirotani, T Irimura. MUC1 mucin expression as a marker of progression and metastasis of human colorectal carcinoma. Gastroenterology 106:353–361, 1994.
56. RA Graham, JM Burchell, J Taylor-Papadimitriou. The polymorphic epithelial mucin: potential as an immunogen for a cancer vaccine. Cancer Immunol Immunother 42:71–80, 1996.
57. CJ Yu, CT Shun, PC Yang, YC Lee, JY Shew, SH Kuo, KT Luh. Sialomucin expression is associated with *erb*B-2 oncoproein overexpression, early recurrence, and cancer death in non-small cell lung cancer. Am J Respir Crit Care Med 155:1419–1427, 1997.
58. PC Yang, KT Luh, SH Kuo, CW Wu. Immunocytochemistry and ELISA quantitation of mucin for diagnosis of malignant pleural effusions. Am Rev Respir Dis 146: 1571–1575, 1992.
59. CJ Yu, JY Shew, CT Shun, HZ Lin, SH Kuo, KT Luh, PC Yang. Quantitative analysis of mRNA encoding MUC1, MUC2, and MUC5AC genes: a correlation between specific mucin gene expression and sialomucin expression in non-small cell lung cancer. Am J Respir Cell Mol Biol 18:643–652, 1998.

25

Charged Oligosaccharides as Novel Mucolytic Therapies

Malcolm King, Eiichi Sudo,* and Martin M. Lee†
University of Alberta
Edmonton, Canada

INTRODUCTION

The structural crosslinking of mucus depends on several molecular factors (Fig. 1). These include intramolecular disulfide bonds that hold the mucin subunits together, entanglement crosslinks that occur because of the very high molecular weight, and various types of noncovalent bonds due to the partially charged oligosaccharide side chains characteristic of mucins (1,2). We have seen that agents that alter the ionic interactions or the hydrogen bonding of the mucous gel can produce potentially beneficial effects on mucus rheology and clearability. This was found to be the case with hypertonic saline (3–5) and with dextran (6,7), a compound that can also block bacterial adhesion (8) to epithelial cells.

As part of our ongoing search for improvements in mucoactive treatments, we have looked at mucoactive agents that might alter both the hydrogen bonds and the ionic interactions. We found that one such agent—charged, low molecular weight (LMW) heparin—had a greater mucolytic and mucokinetic capacity than the neutral saccharide polymer dextran. This was seen both in in vitro rheo-

* *Current affiliation*: Hospital of Printing Bureau of Ministry of Finance, Tokyo, Japan.
† *Current affiliation*: Brigham & Women's Hospital, Harvard Medical School, Boston, Massachusetts.

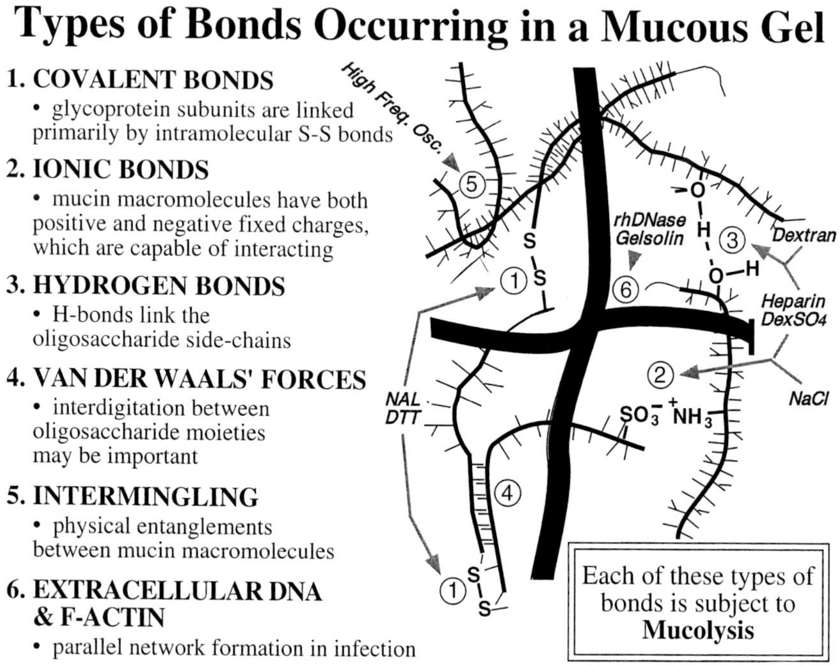

Figure 1 Diagram illustrating the types of bonding in a mucous gel. (Adapted from refs. 1 and 2.)

logical testing, where heparin decreased the viscoelasticity of mucus in vitro (9), and in excised frog palate clearance measurements, where topical heparin was found to increase the mucus clearance rate ex vivo (10). Heparin presumably exerts its mucokinetic action by interfering with intermolecular hydrogen bonding and/or ionic shielding effects in the mucous gel, thereby reducing the effective degree of crosslinking.

Having established these in vitro findings with heparin, we wished to see whether heparin had significant effects on mucus viscoelasticity in vivo, and whether administration of heparin by aerosol could increase the rate of mucociliary clearance. Further, we wished to see whether charged dextran, i.e., dextran sulfate, also had significant effects on mucus rheology and clearance in in vivo testing. If dextran sulfate were a more potent mucolytic agent than neutral dextran and if its mucokinetic effects were comparable to heparin, it would have a distinct advantage over LMW heparin and most other charged oligosaccharides from the viewpoint of cost; at current catalog prices (Sigma), the cost ratio of dextran: heparin is about 1:500.

In the present investigation, we therefore sought to determine the effect of LMW heparin and dextran sulfate on canine tracheal mucus velocity (TMV) and tracheal mucus viscoelasticity. In addition, since a charged macromolecule applied to the mucosal surface could alter epithelial ion currents, we also studied the changes in tracheal potential difference (PD) as an index of epithelial ion transport (11). We thus carried out a series of experiments involving administration of heparin and dextran sulfate by aerosol to anesthetized dogs, using protocols similar to those used in studying aerosolized neutral dextran. In addition to monitoring the effects of dextran sulfate aerosolization on tracheal mucus clearance, we also looked for evidence of adverse effects, such as airway hemorrhage, which could be associated with anticoagulant administration.

MATERIALS AND METHODS

Experimental Design

Heparin

Seven healthy mongrel dogs, weight 21–28 kg, were studied. The dogs were anesthetized with sodium pentobarbital (30 mg/kg i.v., supplemented as required) and intubated. Heparin sodium (MW 5000–6000) (Calbiochem 375097 or Sigma H-2149) was aerosolized via a Pari LC Star nebulizer for 30-minute intervals, separated by 30 minutes without aerosol. The nebulizer was loaded in each case with 10 mL of solution; the amount of solution remaining in the nebulizer after 30 minutes was recorded. A T-tube connection was provided to the endotracheal tube during aerosol delivery to ensure a minimal deadspace for ventilation and adequate delivery of the aerosol. Nonlactated Ringer's solution was aerosolized for 30 minutes as the initial control and used as the carrier for heparin at 2.0, 6.5, and 20.0 mg/mL. For comparison, a sham experiment involving four successive aerosolizations of Ringer's solution was carried out in the same seven dogs. Experiments in any given dog were separated by a period of at least 2 weeks.

After each aerosolization, tracheal transepithelial PD (using agar filled electrodes) and TMV (by charcoal marker particle transport) were measured under bronchoscopic control, and mucus for viscoelasticity analysis by magnetic rheometry was collected from the outside cuff of the endotracheal tube.

Dextran Sulfate

In a second series of experiments, aerosols of Ringer's solution or DexSO4 (Sigma, MW 5000) dissolved in Ringer's were generated by Pari LC STAR nebulizer and delivered during periods of spontaneous breathing. The dogs were administered Ringer aerosol, followed by 6.5, 20, and 65 mg/mL dextran sulfate by aerosol. As before, each aerosolization was of 30 minutes duration, with a 30-minute rest period before the next aerosol delivery. We performed $DexSO_4$

experiments in seven dogs; three of these seven dogs had also previously received heparin. PD and TMV measurements were made as follows, and mucus for physical and compositional analysis was collected from the endotracheal tube.

Transepithelial Potential Difference

PD (−mV) was measured with flexible agar-KCl microelectrodes, connected to calomel half cells (subcutaneous reference). The measurement probe was guided by bronchoscopy and carefully placed in contact with the epithelium ~2 cm above the main carina (11). PD was measured 5–10 minutes following each aerosol.

Tracheal Mucociliary Velocity

TMV (mm/min) was determined by observing the rate of charcoal marker particle transport in the lower trachea under bronchoscopic visualization (12). TMV was measured twice for each dose, i.e., during aerosolization, 15–25 minutes from the start of aerosol delivery (period 1), and 10–20 minutes after completion of aerosolization (period 2).

Mucus Collection Weight

Tracheal mucus was collected from the cuff of the endotracheal tube upon extubation (13). Mucus collections were performed twice after each aerosol, immediately following aerosol inhalation, and 30 minutes later.

Mucus Rheology

A ~100 μm steel ball was positioned in a 2–10 μL sample of mucus, and the motion of this sphere under the influence of an oscillating electromagnetic field gradient (1–100 rad/s) was used to determine the rheological properties of the mucus (14). The measured viscoelastic data were expressed as $\log G^*$ (rigidity factor) and $\tan \delta$ (recoil factor) and used to calculate a mucociliary clearability index and a cough clearability index from previously established relationships based on model studies (15).

Solids Content

The mucus samples, free of oil, were weighed on tared glass slides. The samples were then dried in a microwave oven (750 W for 30 minutes) and allowed to cool. These dried samples were then reweighed, and percent solids content (%SC) was calculated from the ratio of dry to wet weight (13).

RESULTS

The vehicle control experiments (Ringer's solution aerosolization) showed no significant differences in TMV, PD, or mucus rheology compared with pre-aerosol controls (16). With heparin aerosol administration, mucus viscoelasticity, G* at 10 rad/s, for 6.5 mg/mL was significantly decreased to 26% of control (p = 0.05) (Fig. 2). This change was accompanied by a nonsignificant, upward trend in TMV, to approximately 130% of control. There was a significant decrease in PD at 6.5 mg/mL to 55% of Ringer control (p = 0.02).

With dextran sulfate aerosolization, tracheal mucus viscoelasticity (average logG* over 1–100 rad/s) decreased progressively with increasing dose of $DexSO_4$ (Fig. 3). For the highest concentration (65 mg/mL), G* decreased by a factor of 2.61 (p = 0.021). The changes in mucus viscoelasticity predict increased clearability by ciliary and cough mechanisms (15). A modest increase in TMV was observed for the first dose of $DexSO_4$ (128% of baseline at 6.5 mg/mL, p =

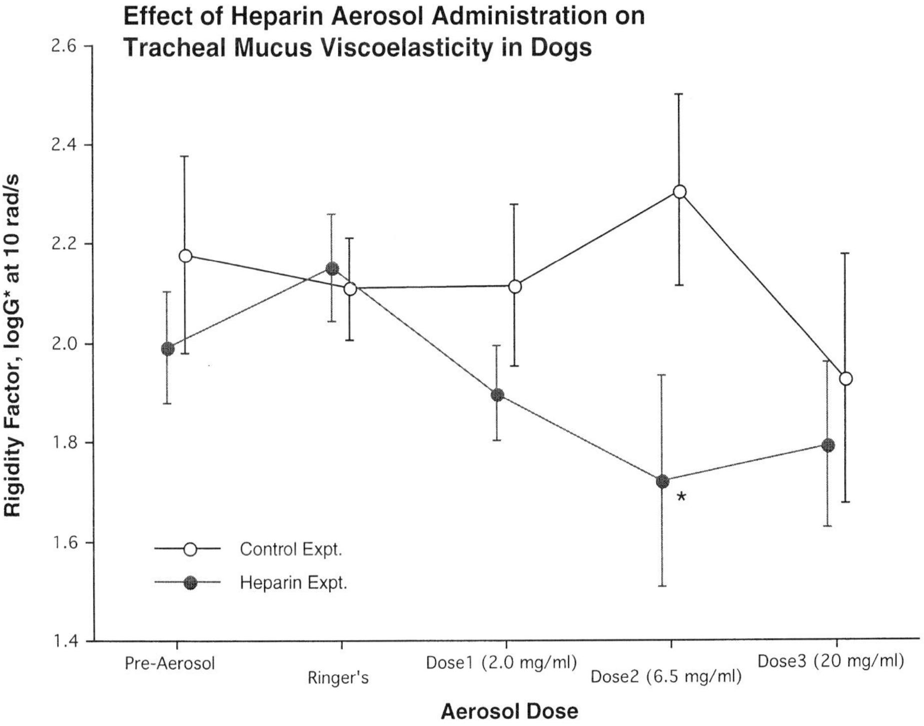

FIGURE 2 Tracheal mucus viscoelasticity as determined by magnetic microrheometry, expressed as rigidity factor logG* at 10 rad/s, in 7 dogs after delivery of Ringer aerosol, and with increasing concentrations of aerosolized LMW heparin sodium. For comparison, the variation of logG* during a sham experiment (delivery of four successive aerosols of Ringer's solution) is shown. There was a significant decrease in logG* following aerosolization of 6.5 mg/mL heparin.

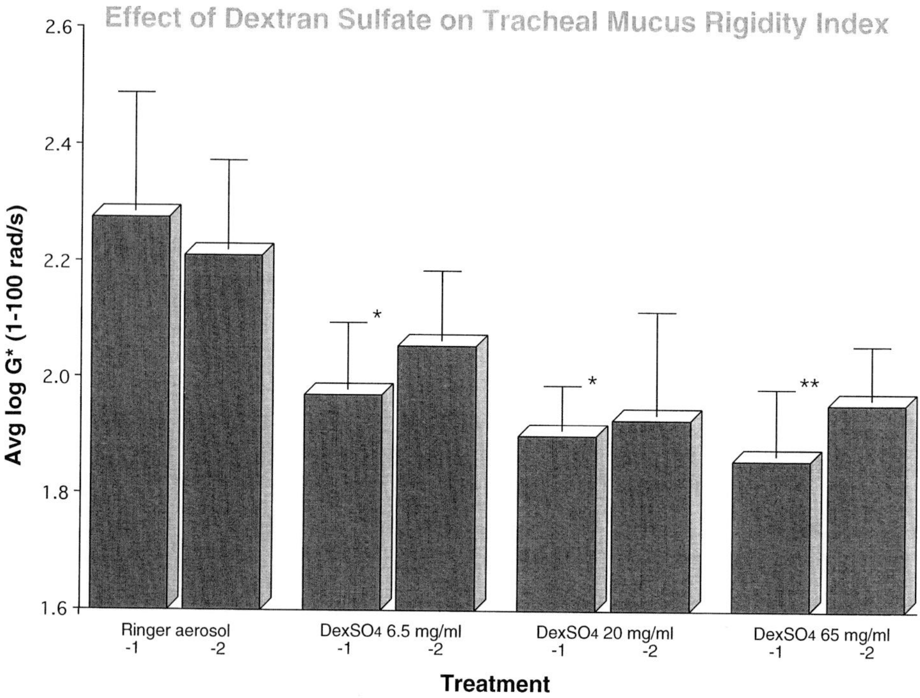

Figure 3 Tracheal mucus viscoelasticity as determined by magnetic microrheometry, expressed as average logG* over 1–100 rad/s in seven dogs during and after 30-minute Ringer aerosol, and with increasing concentrations of aerosolized $DexSO_4$. There was a significant decrease in average logG* during each $DexSO_4$ aerosolization.

0.066); thereafter TMV was quite stable (Fig. 4). PD increased significantly at each concentration of $DexSO_4$ compared with Ringer control; however, there was no additional change between the three groups (Fig. 5). The solids content of collected airway fluid (Fig. 6) gradually increased during successive 30-minute $DexSO_4$ aerosols, indicating a significant residence time for the dextran in the mucus, and correlating with the decrease in viscoelasticity (17).

DISCUSSION

Our results confirm the mucolytic action of LMW heparin in an in vivo model. At moderate concentration (6.5 mg/mL) it reduced the viscoelasticity of tracheal mucus, confirming the mucolytic effect seen in vitro (9). LMW dextran sulfate

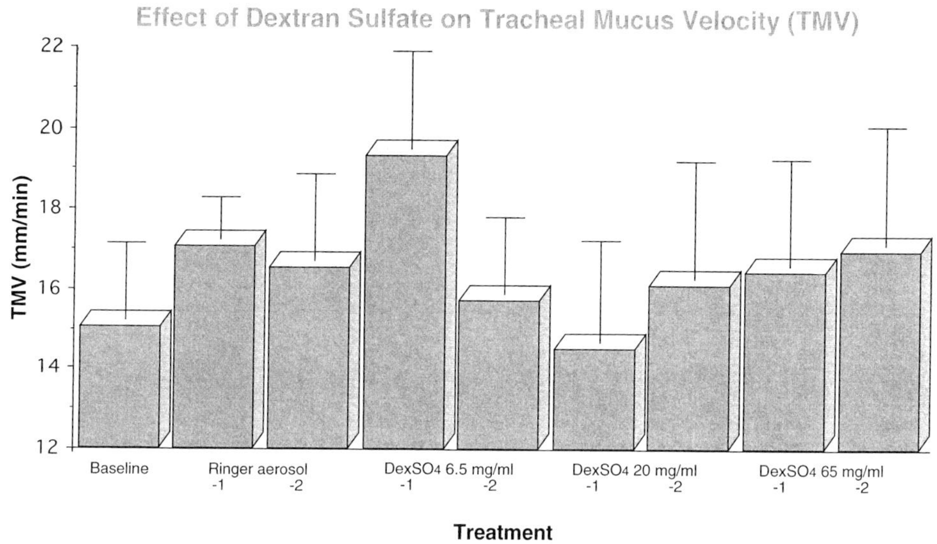

FIGURE 4 Tracheal mucociliary velocity (by charcoal marker displacement) in seven dogs during and after 30-minute Ringer aerosol and during and after increasing concentrations of aerosolized $DexSO_4$ in Ringer's. There were no significant differences in TMV with $DexSO_4$ administration.

applied in vivo also reduced the viscoelasticity of healthy dog mucus and increased predicted mucociliary and cough clearability. The mucolytic effect of $DexSO_4$ appears to be more potent than neutral dextran, since its effect on viscoelasticity is similar to that exhibited by dextran at a half-log higher concentration range (7). Heparin may be slightly more potent than $DexSO_4$, although this cannot be stated with certainty, since direct comparisons were only made in three dogs.

Tracheal PD was altered consistently and significantly to higher negative values after $DexSO_4$ administration. Changes in PD reflect changes in epithelial ion transport or altered integrity (11). An increase in negative PD value could indicate an increase in epithelial resistance, but in a healthy dog this is unlikely. It more likely indicates an increase in the driving potential for luminal ion transfer, either increased Cl^- on secretion or sodium absorption. Given the excess osmolarity of the fluid delivered to the luminal surface, the increase in tracheal PD seen with $DexSO_4$ suggests an increase in the driving potential for chloride-linked water transfer, indicating that an osmotic mechanism, as suggested by Winters and Yeates (18), was active. The elevation of osmolarity could increase cilia beat frequency and/or induce mucus secretion, thereby augmenting airway clearance. On the other hand, increasing the volume of airway surface fluid in a

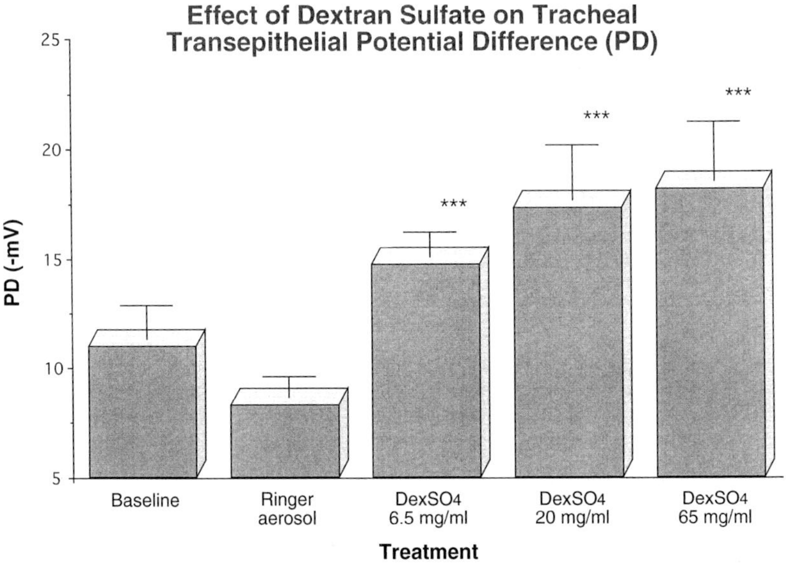

Figure 5 Tracheal transepithelial potential difference (by agar bridge technique) in seven dogs after 30-minute Ringer aerosol, and after increasing concentrations of aerosolized $DexSO_4$ in Ringer vehicle. PD was significantly more negative for all three concentrations of $DexSO_4$. Values are plotted as mean ± SEM.

normal system could impair the interaction between the cilia and the mucus and reduce the transport velocity. These potentially offsetting factors might account for the observed lack of net change in tracheal mucus velocity. The decrease in tracheal PD seen at low heparin concentration is not explained. The fact that no significant increase in TMV was seen in these healthy dogs, despite the improvement in mucus rheological properties, might also be ascribed to their healthy status and the probable fact that their TMV was already near optimal. More noticeable improvements in clearance associated with a similar change in mucus rheology could be revealed by studies in a pathological animal model.

Further research is needed to establish the optimal conditions for the use of charged dextran or heparin as mucokinetic agents. Our tentative conclusion is that charged oligosaccharides decrease mucus viscoelasticity by (1) interaction of its negative charge on the amino groups of the mucin molecule, thereby reducing its entanglement with neighboring mucin sulfate or sialic acid moieties, and/or (2) interfering with intermolecular hydrogen bonding due to their small molecular weights (similar to dextran), and/or (3) ionic shielding effects of mobile counterions (principally sodium) on the polyionic moieties of the mucin molecule

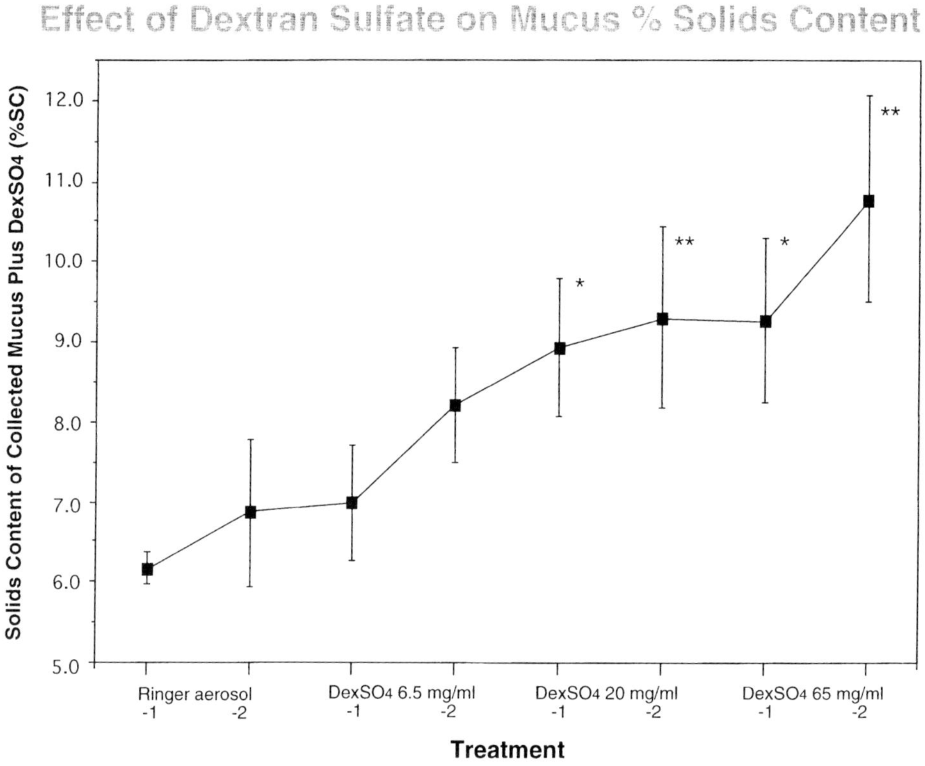

FIGURE 6 Solids content (%SC) of collected airway secretion in seven dogs during and after 30-minute Ringer aerosol and increasing concentrations of $DexSO_4$. Dextran sulfate induced a significant increase in %SC. $^{*}p < 0.05$ vs. Ringer-1; $^{**}p < 0.01$ vs. Ringer-2.

(9,17). Charged oligosaccharides may also stimulate the movement of ions across the epithelium, thereby increasing hydration of the airway surface fluid, similar to the action of chloride secretagogues (19).

Although tracheal mucociliary clearance did not significantly increase in these healthy dogs, it did not diminish, and there was no indication of any deleterious effect. The osmotic burden of the delivered aerosol was modest. Inhaled heparin as a potential asthma treatment has been shown to be safe (20). It is likely that the same safety considerations apply to dextran sulfate as apply to heparin, i.e., anticoagulant behavior is molecular weight dependent, and small oligomers, which are the ones of interest in terms of mucolytic effect, have little capacity to act as anticoagulants or to increase the risk of hemoptysis.

Based on the literature, we were fairly certain that dextran sulfate would turn out to be a potent mucolytic agent. In other laboratories, $DexSO_4$ has been used to induce experimental colitis in rodents (21); the mechanism probably involves erosion of the protective intestinal mucus layer. However, dextran sulfate (MW 7000) has also been administered orally to human subjects, with no apparent safety concerns (22). The potential of heparin or sulfated dextran as a mucokinetic therapy rather than an injurious agent is a question of dose, route of administration, molecular weight, and degree of sulfation.

In subsequent studies we intend to further explore the mechanisms involved in mucokinesis by charged oligosaccharides, including the role of the osmotic stimulus (18), in promoting secretion clearance. Further studies are also needed to establish the ideal oligomeric length and charge density for mucokinetic stimulation, as well as to optimize lung delivery by aerosolization (23).

ACKNOWLEDGMENTS

Studies were supported by the Canadian Cystic Fibrosis Foundation. Drs. E. Sudo and M. M. Lee were both recipients of MRC/CLA fellowships.

REFERENCES

1. B Dasgupta, M King. Molecular basis for mucolytic therapy. Can Respir J 2:223–230, 1995.
2. M King, BK Rubin. Mucus controlling agents: Past and present. In: JL Rau, ed. Aerosolized Drugs for the Respiratory Tract. Respir Care Clin North Am 575–594, 1999.
3. PJ Wills, RL Hall, WM Chan, PJ Cole. Sodium chloride increases the ciliary transportability of cystic fibrosis and bronchiectasis sputum on the mucus-depleted bovine trachea. J Clin Invest 99:9–13, 1997.
4. M King, B Dasgupta, RP Tomkiewicz, NE Brown. Rheology of cystic fibrosis sputum after in vitro treatment with hypertonic saline alone and in combination with rhDNase. Am J Respir Crit Care Med 156:173–177, 1997.
5. M Robinson, A Hemming, JA Regnis, DL Bailey, M King, W Feng, GJ Bautovich, PTP Bye. Improved mucociliary clearance following nebulisation with hypertonic saline in adults with cystic fibrosis. In: G Baum, ed. Cilia, Mucus and Mucociliary Interactions. New York: Marcel Dekker, 1998, pp 265–280.
6. W Feng, H Garrett, DP Speert, M King. Improved clearability of cystic fibrosis sputum with dextran treatment in vitro. Am J Respir Crit Care Med 157:710–714, 1998.
7. W Feng, S Nakamura, E Sudo, MM Lee, A Shao, M King. Effects of dextran on tracheal mucociliary velocity in dogs in vivo. Pulm Pharmacol Ther 12:35–41, 1999.
8. S Barghouthi, LM Guerdoud, DP Speert. Inhibition by dextran of Pseudomonas aeruginosa adherence to epithelial cells. Am J Respir Crit Care Med 154:1788–1793, 1996.

9. MM Lee, M King. Effect of low molecular weight heparin on the elasticity of dog mucus. Clin Invest Med 21:S102, 1998.
10. MM Lee, H Garrett, E Sudo, WA Boyd, M King. Mucociliary clearance increase due to low molecular weight heparin. Pediatr Pulmonol 386:S17, 1998.
11. EM App, JG Zayas, M King. Rheology of mucus and epithelial potential difference: Small airways vs. trachea. Eur Respir J 6:67–75, 1993.
12. M King, S Kelly, M Cosio. Alteration of airway reactivity by mucus. Respiration Physiol 62:47–59, 1985.
13. BK Rubin, O Ramirez, JG Zayas, B Finegan, M King. Collection and analysis of respiratory mucus from individuals without lung disease. Am Rev Respir Dis 141: 1040–1043, 1990.
14. M King. Magnetic microrheometer. In: PC Braga, L Allegra, eds. Methods in Bronchial Mucology. New York: Raven Press, 1988, pp 73–83.
15. M King. Role of mucus viscoelasticity in cough clearance. Biorheology 24:589–597, 1987.
16. M King, A Ghahary, R Franklin, M Hirji, D Malchenko, WA Boyd, H Garrett, MM Lee. Studies on aerosolized low mol. wt. heparin as a mucokinetic agent in dogs. Am J Respir Crit Care Med 159:A474, 1999.
17. E Sudo, WA Boyd, M King. Effects of dextran sulfate on tracheal mucociliary velocity in dogs. J Aerosol Med 13:87–96, 2000.
18. SL Winters, DB Yeates. Role of hydration, sodium, and chloride in regulation of canine mucociliary transport system. J Appl Physiol 83:1360–1369, 1997.
19. RP Tomkiewicz, EM App, GT De Sanctis, M Coffiner, P Maes, BK Rubin, M King. A comparison of a new mucolytic N-acetylcysteine L-lysinate with N-acetylcysteine: Airway epithelial function and mucus changes in dog. Pulm Pharmacol 8: 259–265, 1995.
20. T Ahmed, J Garrigo, I Danta. Preventing bronchoconstriction in exercise-induced asthma with inhaled heparin. N Engl J Med 329:90–95, 1993.
21. S Björk, E Jennische, A Dahlstrom, H Ahlman. Influence of topical rectal application of drugs on dextran sulfate-induced colitis in rats. Dig Dis Sci 42:824–832, 1997.
22. KJ Lorentsen, CW Hendrix, JM Collins, DM Kornhauser, BG Petty, RW Klecker, C Flexner, RH Eckel, PS Lietman. Dextran sulfate is poorly absorbed after oral administration. Ann Int Med 111:561–566, 1989.
23. WH Finlay, CF Lange, M King, DP Speert. Lung delivery of aerosolized dextran. Am J Respir Crit Care Med 161:91–97, 2000.

Part III

Mucociliary Interactions

Mucociliary interactions are ultimately responsible for removal of cellular debris and foreign materials from the airways. Recent advances in this field have shown that mucociliary clearance remains one of the most important host defenses of the respiratory system, and its dysfunction has been linked to morbidity and even mortality.

The chapters in this group all deal with the complex issues of mucociliary interactions. It is not surprising that this topic attracted chapters on quite diverse topics, ranging from the extreme of trying to measure the effectiveness of mucociliary interactions in the normal and diseased host (including the evaluation of drug therapy) to the other extreme of looking at backup mechanisms of failing mucociliary clearance. In addition, a first mechanistic look at how the normal epithelium can change into a diseased one is given here as well.

This part begins with a sophisticated mathematical modeling approach by Blake and Gaffney (Chapter 26). Such modeling will ultimately help us to understand mucociliary interactions better. Then, two chapters examine cell and molecular regulation mechanisms of clearance. The first, by Ribeiro et al. (Chapter 27), compares ion transport in the airway between normals and patients with cystic fibrosis; the second, by Nadel (Chapter 28), examines the molecular mechanisms that cause airway epithelial cell metaplasia. (Chapters 29–32) review the current state of the art of measuring mucociliary transport and assess factors that cause mucociliary dysfunction as well as drugs that increase clearance rates. The chapter by Foster and Wagner (Chapter 33) reminds us that beside the transport along the epithelium, soluble substances can actually be taken up into the vessels supplying the airway epithelium and are therefore transported away by a different mean. Finally, Chapters 34 and 35 assess cough clearance in vivo and in vitro.

This section reveals the immense advances that have been made in this field. However, they also remind us that much more needs to be learned about normal and abnormal function of mucociliary clearance. Clear progress has been made in understanding cellular and molecular changes that lead to abnormal ion transport or airway epithelial metaplasia and how this may be linked to abnormal function. On the other hand, therapy of mucociliary dysfunction is still suboptimal, and we hope that the better physiological understanding of clearance will identify mechanisms that can be targeted for new therapeutic approaches. Finally, we need to be able to assess mucus load in the airways directly. Current technology is unable to provide us this important information, but with all the advances described in this book, such measurements may be available in the near future.

Matthias Salathe

26

Modeling Aspects of Tracer Transport in Mucociliary Flows

J. R. Blake and E. A. Gaffney
University of Birmingham
Birmingham, United Kingdom

INTRODUCTION

On the epithelia of most of the respiratory tract, a dense mat of beating cilia interact with the overlying layer of airway surface liquid (ASL) to effect the removal of mucus and cellular debris from the lung. The understanding of this system, at a descriptive level, is well documented. However, the understanding of such mucociliary systems at a basic quantitative level is a fundamental field in respiratory mechanics; here we investigate one of many open topics of research in this area. After introducing techniques for the modeling mucociliary flows, we specifically consider recent measurements concerning the transport of an inert tracer within an in vitro model of the epithelial lung mucociliary system, based on human tracheobronchial epithelial cultures (1). In particular, it is found that these measurements are irreconcilable with previous modeling work. In this chapter we develop a simple modeling investigation of tracer transport within mucociliary flows to inform our understanding of this discrepancy, but must report in the negative that there is little further advancement in our understanding. This indicates that yet further modeling work is required to understand the physical basis of the experimental results, and possible research directions are suggested.

AN OVERVIEW OF MODELING MUCOCILIARY FLOWS

On the length scale of a metachronal wave, the epithelial substrate can be considered as a flat plane to an excellent approximation, and one typically assumes that the cilia are evenly spread throughout the substrate in a rectangular array and beating with the same period. Taking this into account, mucociliary fluid dynamics constitutes an example of a thin layer, low Reynolds number flow with an oscillatory forcing that exhibits both temporal and spatial periodicity. The first step in the modeling is to represent the cilia beat and the metachronal wave. Assuming the cilia beat in a planar fashion, this is conveniently done via Fourier series with the position of the point a distance s along the (i, j) cilium of the rectangular array at time t given by:

$$\boldsymbol{\xi}^{i,j}(s, t) = ia\hat{\mathbf{x}} + jb\hat{\mathbf{y}} + \sum_{m=1}^{M} (\mathbf{a}_{\mathrm{m}}(s)\cos(m\sigma(kia + \sigma t)) + \mathbf{b}_{\mathrm{m}}(s)\sin(m\sigma(kia + \sigma t))) \tag{1}$$

Here, a, b are the spacings between the cilia bases in the x and y directions, with $\hat{\mathbf{x}}, \hat{\mathbf{y}}$ representing unit Cartesian vectors across the plane of the epithelial substrate. $\sigma/2\pi$, $2\pi/k$ give the frequency and wavelength of the metachronal wave which is, without loss of generality, taken to be in the x direction. The coefficients $\mathbf{a}_{\mathrm{m}}(s)$, $\mathbf{b}_{\mathrm{m}}(s)$ are readily determined from experimental data by least-squares fitting.

The next step is to determine how the position of the beating cilia exerts a forcing on the ASL. The most convenient approximation is to assume first that the double layer stratification of the ASL can be modeled by two viscous, Newtonian fluids, one overlying the other, with the upper layer of substantially greater viscosity. This upper highly viscous layer is used as a crude representation of the mucus; the lower, less viscous, layer is reasonable representation of the periciliary layer (PCL). For simplicity, we only consider the situation in which the cilia do not penetrate the mucous layer.

The problem is greatly simplified by the linearity of the low Reynolds number fluid dynamical equations and associated boundary conditions.* This enables us to calculate the net fluid flow by summing over the fluid flows due to the resultant force exerted by each segment, of length δs, of each cilium on the fluid. The resultant force due to a segment, of length δs, of the cilium (i, j) can be written in the form

$$F_q(s, t)\delta s = P_{kq}\left(\frac{\partial \xi_q^{i,j}}{\partial t} - u_q\right)\delta s$$

---

* This does not take into account the fact that the boundary conditions on the cilia/fluid interface are inhomogeneous. This boundary condition is effectively ignored within the simplistic modeling framework presented above.

where $\xi_q^{i,j}$ is the qth component of the vector $\boldsymbol{\xi}^{i,j}$ defined in Eq. (1), P_{kq} is a constant tensor, and $(\partial \xi_q^{i,j}/\partial t - u_q)$ is the speed of the cilia relative to the speed of the surrounding fluid, moving at speed u_q. This relationship arises immediately from the linearity and homogeneity of the Stokes equations on the halfplane. The calculation of P_{kq} can be determined using slender body techniques, by generalizing the results of Gray and Hancock (5) and is essentially that of the calculation of drag coefficients for a cylinder moving in a low Reynolds number fluid flow. Such a calculation yields that

$$P_{kq} = C_T(\delta_{kq} - (\gamma - 1)(\partial \xi_k/\partial s)(\partial \xi_q/\partial s))$$

where C_T is the tangential resistance coefficient with γ the ratio of the normal to tangential resistance coefficients. The relationship between the forcing and velocity is given by summing over the individual solutions representing the forcing of the airway surface liquid by the segment of length δs, a distance s along the (i, j) cilium. Thus, we are summing over (i, j), and integrating over $s \in [0, L]$, where L is the length of cilia, for individual solutions satisfying, at time t, the Stokes equation:

$$\begin{aligned} \nabla p &= \mu \nabla^2 \mathbf{u} + (\mathbf{F}(s, t)\delta s)\delta(\mathbf{x} - \boldsymbol{\xi}^{i,j}(s, t)) && 0 \le z \le h \\ \nabla p &= \mu' \nabla^2 \mathbf{u} \qquad \mu \ll \mu' && h \le z \le H \end{aligned} \tag{2}$$

with the incompressibility condition $\nabla \cdot \mathbf{u} = 0$. The variables μ, μ' are viscosities in the periciliary layer and the mucous layer, respectively; h is the height of the pericilary layer, and H is the height of the airway surface liquid (assumed constant). The boundary conditions are $\mathbf{u} = 0$ at $h = 0$, $\hat{\mathbf{z}} \cdot \mathbf{u} = 0$ at $z = h$, $z = H$, with continuous velocity and normal stresses for the pericilary/mucous interface at $z = h$. Thus, the velocity within the periciliary layer is of the form

$$U_p = \sum_i \sum_j \int_0^L ds\, G^1_{pq}(\mathbf{x}, \boldsymbol{\xi}^{i,j}(s, t))F_q(s, t)$$

where U_p is the pth component of the ASL flow field, denoted $\mathbf{U}$ in vector form below. $G^1_{pq}(\mathbf{x}, \xi^{i,j}(s, t))$ is the (p, q) component of the Green's function matrix, i.e., the pth component of the solution of Eq. (2), denoted $\mathbf{u}$, for the special case where $\mathbf{F} = \hat{\mathbf{x}}^q$, with $\hat{\mathbf{x}}^q$ denoting the qth Cartesian unit vector. The solution in the mucous layer is basically the same, though the Green's function matrix is slightly different, due to the different viscosity and boundary conditions associated with the mucous layer. With this formalism, one can thus, in theory, calculate the velocity profile caused by the beating cilia. In practice, one typically calculates a time-averaged Green's function matrix, which takes a much simpler form than $G^1_{pq}(\mathbf{x}, \boldsymbol{\xi}^{i,j}(s, t)$; this is the basis of numerous papers modeling mucociliary fluid flow, which therefore calculate time-averaged properties of the flow only (2–4). A typical time averaged result for a mucociliary flow is displayed in Figure

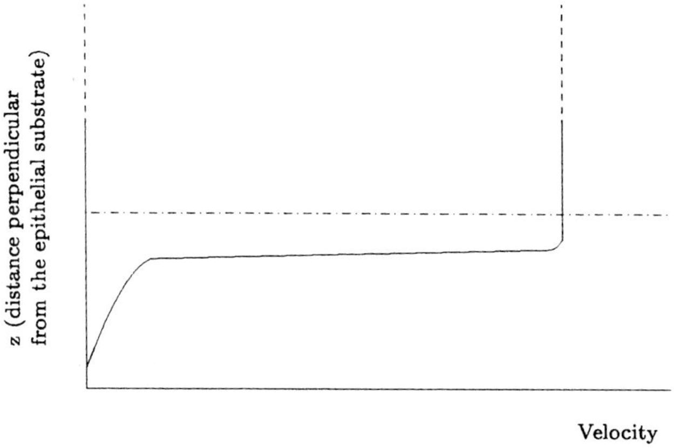

FIGURE 1 A typical time-averaged velocity profile representative of the typical results arising in the previous modeling of Blake et al. (3–5). The dot-dash horizontal line represents the mucous/periciliary layer interface.

1. Further details concerning such calculations may be found in Refs. 2–4. The key points to note about such results is that, first, there is essentially no time-averaged fluid flow within micrometers of the epithelial substrate, though oscillatory flow may occur at substantially smaller distances from the epithelial substrate. Second, from the assumption that the metachronal wave is planar and beats in the x direction, one can additionally deduce that the resulting fluid flow has no y dependence, which is a useful simplification below.

EXPERIMENTAL INVESTIGATIONS OF TRACER TRANSPORT

In a series of experiments, Matsui et al. (1) used conventional and confocal microscopy to investigate the transport of fluorescent microspheres and photoactivated fluorescent dyes within well-differentiated human tracheobronchial epithelial cell cultures exhibiting spontaneous, radial mucociliary transport. These studies resulted in a number of key observations, including:

The samples used were selected for the observation that the velocity of the fluorescent microspheres labeling the mucous layer exhibited a velocity profile with purely azimuthal motion of magnitude $v_{azi} = \omega r$, where r is the radial distance from a center of zero motion, and ω is constant.

Tracer transport was investigated at points midway along the radius of mu-

> cous rotation. Up to the limits of resolution, it was observed that the PCL was transported at approximately the same rate as the mucus (40 μms^{-1}) or, to be more precise, that the transport of tracer indicated that the PCL was transported at approximately the same rate as the mucus. Removing the mucous layer reduced PCL transport by $\approx$80%.

Clearly, such observations seem to refute the results from previous modeling; Figure 1 clearly shows that the periciliary layer is (on time averaging) stagnant. There are accordingly a number of interesting questions. The most important one, and the one we consider here, is: Does the tracer transport faithfully represent the fluid flow?

This requires careful consideration of whether tracer diffusion obscures the true motion of the periciliary layer, *especially* given the fact the fluid flow will have oscillatory components both parallel and perpendicular to the epithelial substrate. By definition, such flow contributions are not captured in the *time-averaged* calculations presented in the literature. Furthermore, the resultant mixing from the oscillatory components of the flow entails that the tracer is dispersed and mixed more than one would expect from considering just the time-averaged flow. The key question is therefore whether this additional mixing could explain the similar tracer transport rates observed in the mucous and periciliary layers, or whether one has to include additional mechanisms within the modeling to explain the experimental results.

MODELING TRACER TRANSPORT

To address the key question posed above we must model the transport of a tracer within a mucociliary fluid flow, denoted by $\mathbf{U}(x, z, t) = (u(x, z, t), v(x, z, t))$ below, where t represents time, and x, z are the coordinates perpendicular and parallel to the epithelial substrate, respectively. The tracer is dispersed by a combination of diffusion and convection (with neither mechanism completely dominating tracer transport). Consequently, the equation describing the concentration of tracer particles is

$$\frac{\partial c}{\partial t} = \underbrace{\nabla \cdot (D(z)\nabla c)}_{\text{Diffusion Term}} - \underbrace{\nabla \cdot (\mathbf{U}(x, z,t)c)}_{\text{Convection Term}} \quad (3)$$

where $D(z)$, $\mathbf{U}(x, z, t)$ are the diffusion and velocity profiles, respectively. The boundary conditions for this equation are that no tracer is lost through the epithelial substrate or through the air/mucous layer interface, and that $c \to 0$ as $x \to \pm\infty$. The initial conditions are plotted in Figure 2. The diffusion profile, $D(z)$, is plotted in Figure 3; it can be seen to decrease extremely rapidly at the boundary between the PCL and the mucus.

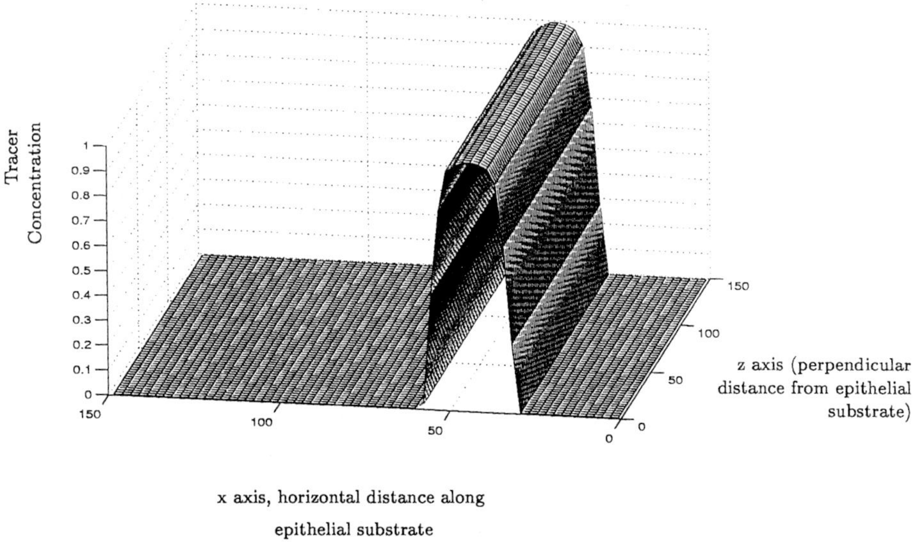

FIGURE 2 The initial distribution of the tracer.

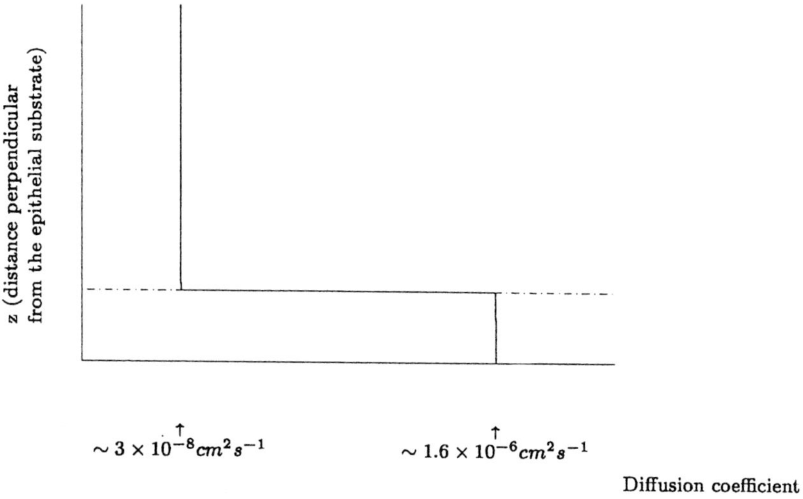

FIGURE 3 A diffusion profile appropriate for an airway surface liquid stratification. z is the height above the epithelial cell substrate. The dot-dash horizontal line represents the mucous/periciliary layer interface.

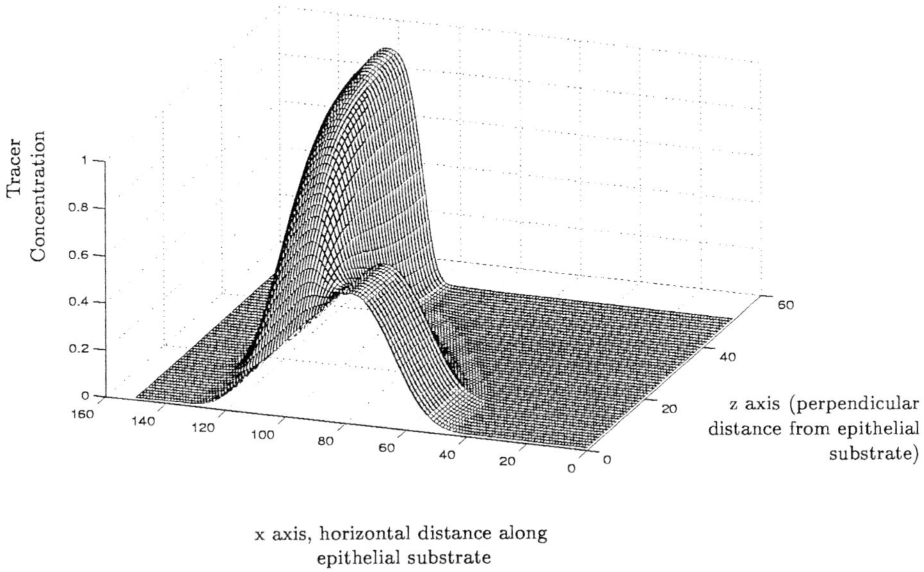

FIGURE 4 The dispersion of the tracer concentration (after 30 seconds) for the time-averaged velocity profile of Figure 1.

For a given velocity distribution, one can calculate the tracer transport computationally. [We do so in this paper using an extension of the alternating direction implicit (ADI) algorithm (7).] For the time-averaged velocity profile,

$$\mathbf{U}(x, z, t) = (u(x, z, t), v(x, z, t)) = (U(z),0)$$

where $U(z)$ is plotted in Figure 1. The resulting tracer distribution after 30 seconds, as calculated computationally, is depicted in Figure 4. This clearly shows that the tracer is predicted to disperse at different rates in the PCL as opposed to the mucous layer, which is in clear contradiction with experimental results (see, e.g., Fig. 5). However, this does not take into account the oscillatory components of the fluid flow that would be predicted by the theoretical model if one did not consider the simplifying step of time averaging, and these oscillatory components will increase mixing, thereby reducing the discrepancy between transport rates in the mucous and periciliary layers. The question is: Will the oscillatory components reduce this discrepancy enough?

Clearly, one should therefore extend the calculations presented in the literature to determine the velocity profile $\mathbf{U}(x, z, t) = (u(x, z, t), v(x, z, t))$, including its oscillatory components. This is an involved calculation, though it has been

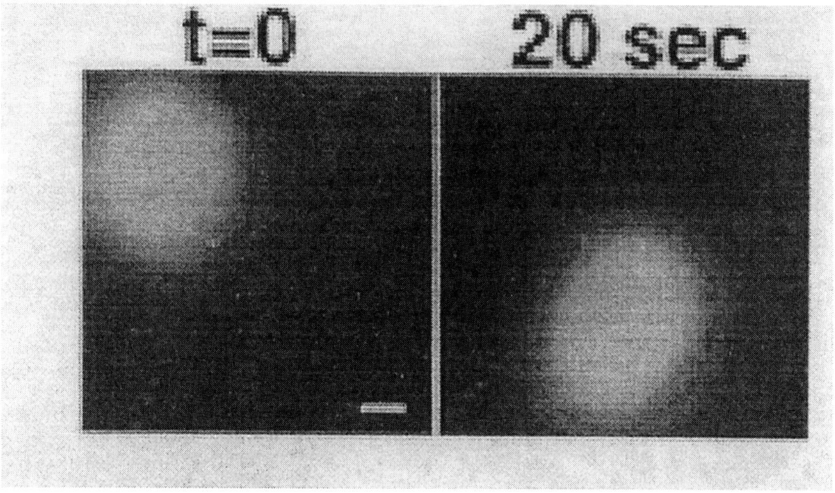

FIGURE 5 A representative result from the experimental studies of Matsui et al. The in vitro mucociliary system is viewed from above, with bright areas showing regions of high tracer concentration and dark areas showing regions with negligible tracer concentrations. The dispersion of tracer thus is observed to occur at similar rates in the mucus and periciliary layers, otherwise one would observe a streak in the picture on the right. (From Ref. 1.)

performed for some mucociliary systems in the past (e.g., Ref. 6). Such calculations are in progress for respiratory mucociliary systems, using simplifying, but accurate, approximations typically not considered in previous work, such as thin film techniques and lubrication theory, whereupon tracer transport will be investigated.

However, one can gain insight by considering oscillatory components with typical wavelengths and frequencies matching those of the metachronal wave and cilia beat, respectively, that are consistent with the boundary conditions required of the velocity field. For example, one could consider the velocity profile

$$\mathbf{U}(x, z, t) = (U(z), 0) + U_* H\left(\frac{z}{z_{\text{mucus}}}\right)\left\{\sqrt{\frac{4z}{z_{\text{mucus}}}\left(1 - \frac{z}{z_{\text{mucus}}}\right)}\right\} \times (\sin(kx - \sigma t), \cos(kx - \sigma t)) \tag{4}$$

This is the time-averaged velocity profile plus and an oscillatory components with wavelength k and frequency σ, weighted by the function

$$U_* H\left(\frac{z}{z_{\text{mucus}}}\right)\left\{\sqrt{\frac{4z}{z_{\text{mucus}}}\left(1 - \frac{z}{z_{\text{mucus}}}\right)}\right\}$$

TABLE 1 Typical Parameter Values

Parameter	Typical value(s)	Value used
Frequency of ciliary beat, σ	10 Hz	10 Hz
Metachronal wavelength, λ	30 μm	30 μm
Height of periciliary layer, *h*	6 μm	6 μm
Height of mucous and periciliary layers, *H*	20–100 μm	25 μm
Cilium length	6 μm	6 μm

Here, U_* is a measure of the oscillatory velocity, $H(z/z_{mucus})$ is unity for $z \leq z_{mucus}$ and zero for $z \geq z_{mucus}$, where z_{mucus} is the height of the mucous layer above the epithelial substrate. Thus, this velocity profile has no oscillatory components within the mucous layer. The function $\sqrt{4z/z_{mucus}(1 - z/z_{mucus})}$ entails the oscillatory components of the velocity are zero at the epithelial substrate, $z = 0$, and at the PCL/mucous layer interface, as required by the boundary conditions. Furthermore, this function rises from zero *very* rapidly on moving into the interior of the PCL from either the epithelial substrate or the mucous interface. In this sense, the oscillatory component in Eq. (4) is likely to match or surpass the mixing ability of the solution to the full equations, providing we choose the parameters U_*, k, σ appropriately.

The oscillatory components of the full solution will probably have a large contribution with the same wavelength and frequency as the forcing from the cilia, so we specifically consider this wavelength and frequency in the velocity profile [Eq. (4)]. From Table 1, this means we take

$$k = \frac{2\pi}{30}\ \mu\text{m}^{-1} \tag{5}$$

$$\sigma = \frac{2\pi}{0.1}\ s^{-1}$$

The oscillatory components of the full solution will be of a characteristic speed matching or less than that the speed of a ciliary tip, which we thus take to be U_* in Eq. (4); we estimate the ciliary tip speed considering the distance traveled by the ciliary tip in one oscillation as being very approximately given by the perimeter of a half circle of radius of a ciliary length. Thus, this approximate speed is

$$U_* \approx \frac{\sigma}{2\pi}(\pi \cdot 6\ \mu\text{m}) \approx 180\ \mu\text{ms}^{-1} \tag{6}$$

Before proceeding with solving Eq. (3), we note that Eq. (4) does not satisfy the incompressibility condition and hence does not represent a genuine fluid flow.

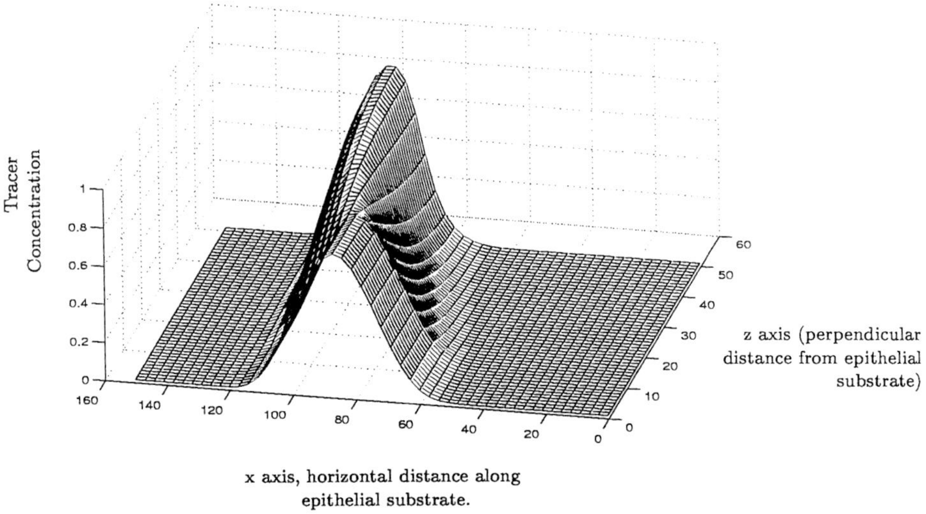

FIGURE 6 The dispersion of the tracer concentration (after 20 seconds) for the velocity profile containing oscillatory components [Eq. (2)].

Thus, one should consider Eq. (4) simply as an oscillatory functional form of **U** deliberately chosen to maximize mixing while satisfying a subset of the constraints required of the true velocity field. As such, the aim is to determine whether or not the current modeling framework, as it currently stands, is even likely to be capable of reproducing the experimental observations.

With the parameter values in Eqs. (5) and (6), solving Eq. (3) computationally results in Figure 6. We see that the inclusion of the above oscillatory terms reduces the difference in dispersion that occurs in the mucous layer compared to the PCL. However, it is still does not reduce this difference to negligible levels, and hence the modeling prediction is not in agreement with the experimental observations, as can be seen by comparing Figure 6 with Figure 5. We now proceed to consider the implications of this observation.

SUMMARY AND DISCUSSION

As described above, it has been suggested that including the oscillatory terms of Eq. (2) will increase general levels of mixing, thus reducing the differences in dispersion rates between mucous and periciliary layers that are observed in the *modeling simulations*. This has been observed (compare, e.g., Figs. 4 and 6). It was also suggested that the increase in levels of tracer transport induced by the fluid flow [Eq. (4)] would be likely to match or surpass transport due to the oscillatory terms that arise in the full solution of Eq. (2), as we have deliberately chosen velocity field that would be most likely to maximize mixing given a number of constraints of the physical system. We proceeded to note that the modeling simulations with the velocity field of Eq. (4) still exhibited differences in the tracer transport between the periciliary and mucous layers. Thus, it is extremely unlikely that the full solutions of Eq. (2) would yield modeling predictions that tracer transport occurs at approximately equivalent rates in the mucous and periciliary layers. In turn, it is therefore very unlikely that simply including the oscillatory components of fluid flow would result in modeling predictions that agreed with experiment. The confirmation of this, using the full solutions of Eq. (4), including the oscillatory components, is work in progress.

Nonetheless, we most likely have the scenario whereby more physics is required within the modeling to reconcile theory and experiment. The interesting question is: What mechanisms should be included within the modeling to reconcile theory and experiment? Two suggestions are:

Pressure gradients. Do extraneous pressure gradients exist in the experimental system? These might be mechanical pressure gradients or osmotic pressure gradients. How might these affect the flow field and the resulting tracer transport?

The interaction between the mucus and the periciliary layer, as mentioned in Ref. 1. The dramatic effects of the removal of the mucus suggest that there is interaction between the mucous and periciliary layers. The challenge is to pin down possible mechanisms for this interaction and to consequently assess whether such interaction significantly influences flow field. Systematically incorporating mucous/periciliary layer interaction into the modeling would be more challenging still.

This work is very much in progress and, with the full solutions of Eq. (2) in place, possible mechanisms such as those above will be investigated, with the aim of quantitatively understanding tracer transport in mucociliary systems; in particular, the similar transport rates in the mucus and the periciliary layers, the observation that the mucus effectively is undergoing rigid body rotation in numerous samples of the experimental system, and the reduction in transport rates on removal of the mucus.

REFERENCES

1. H Matsui, SH Randell, SW Peretti, C William Davis, RC Boucher. Coordinated clearance of periciliary liquid and mucous from airway surfaces. J Clin Invest 102:1125–1131, 1998.

2. GR Fulford, JR Blake, Mucociliary transport in the lung. J Theor Biol 121:381–402, 1986.
3. JR Blake, A model for the micro-structure in ciliated micro-organisms. J Fluid Mechanics 55:1–23, 1972.
4. MA Sleigh, JR Blake, N Liron, The propulsion of mucus by cilia, Am Rev Respir Dis 137:726–741, 1988.
5. J Gray, GJ Hancock, The propulsion of sea urchin spermatozoa. J Exp Biol 32:808–814, 1955.
6. SR Keller, Fluid mechanical investigations of ciliary propulsion. Ph.D. thesis. California Institute of Technology, Pasadena, 1975.
7. KW Morton, DF Mayers. Numerical solution of partial differential equations, Cambridge, UK: Cambridge University Press, 1996.

27

P2Y$_2$ Receptors and Ca^{2+}-Dependent Cl$^-$ Secretion in Normal and Cystic Fibrosis Human Airway Epithelia

Carla M. Pedrosa Ribeiro, Anthony M. Paradiso, Eduardo Lazarowski, and Richard C. Boucher
University of North Carolina at Chapel Hill
Chapel Hill, North Carolina

INTRODUCTION

Cystic fibrosis (CF) is a recessive disease resulting from mutations in the gene coding for the cystic fibrosis transmembrane conductance regulator (CFTR). In CF airways, the absence of the epithelial cyclic AMP (cAMP)–dependent, CFTR-mediated Cl$^-$ secretion coupled to an increased sodium absorption (1,2) results in production of an abnormal, viscous mucus and decreased airway clearance. Thus, persistent infection and inflammation underlay the chronic pulmonary disease involving the upper and lower airways, the most common cause of death in CF patients.

In healthy human subjects, CFTR-mediated Cl$^-$ transport is stimulated by agents that increase the intracellular levels of cAMP, e.g., adenosine and β-adrenergic receptor agonists. However, due to the lack of functional CFTR in CF patients, these agents are ineffective in stimulating Cl^- secretion. The discovery of an apical Cl^- conductance activated by rises in intracellular calcium ($[Ca^{2+}]_i$) in CF cells (3–6) led to the proposal that this alternative Cl^- channel may provide a compensatory mechanism to offset the absent cAMP-mediated Cl^- transport in CF airways. Evidence for the role of the Ca^{2+}-regulated Cl^- channel in protecting against CF lung disease came from CF knockout mice, which express a large endogenous Ca^{2+}-activated Cl^- conductance in the airway epithelia and are devoid of airway disease (7). Together with the known involvement of $[Ca^{2+}]_i$ in mediating ciliary beating and mucociliary clearance (8), these observations underscored the importance of studying Ca^{2+} signaling events in normal and CF human airway epithelial cells.

A considerable amount of research has been devoted to identifying plasma membrane receptors linked to the regulation of Ca^{2+}-dependent airway epithelial transport. In the past few years, several reports have described the expression of $P2Y_2$ purinergic receptors coupled to phospholipase Cβ (PLCβ) activation and $[Ca^{2+}]_i$ mobilization in normal and CF airway epithelia (see below). Since new therapies consisting of aerosol delivery of purinergic agonists aimed at improving the mucus clearance of CF patients as a result of Ca^{2+}-induced epithelial transport are underway (8–12), current research in our laboratory is aimed at advancing the understanding of Ca^{2+} signaling in airway epithelia by assessing the expression and function of the individual components of this system in normal and CF airways.

This chapter is, therefore, devoted to discussing the regulation of $[Ca^{2+}]_i$-dependent Cl^- transport triggered by $P2Y_2$ receptor activation in normal and CF human airway epithelia, with special attention given to apical $P2Y_2$ receptor–regulated Cl^- secretion.

$P2Y_2$ RECEPTORS AND THE Ca^{2+} SIGNALING SYSTEM

Airway epithelial cells are constantly challenged by extracellular signals, many of which reflect autocrine and/or paracrine events. Studies conducted in the past few years have unraveled the role of 5′ nucleotide triphosphates, released from endogenous (epithelial cells) or exogenous sources, in the regulation of Ca^{2+}-dependent chloride secretion in airways. The first study demonstrating the coupling between purinoceptors and PLCβ activity in human airway epithelia was performed in the cell line CF/T43 derived from a CF patient (13). In that investigation, extracellular ATP was as potent and efficacious as UTP in activating a common 5′-nucleotide or P_{2U}-purinergic ($P2Y_2$) receptor linked to PLC stimulation. Likewise, $P2Y_2$ receptors are functionally expressed in normal human bronchial epithelia, as illustrated by the similar dose responses for ATP- or UTP-induced PLC-dependent inositol phosphate accumulation in these cells (Fig. 1). The coupling between $P2Y_2$ receptor activation and stimulation of PLC occurs

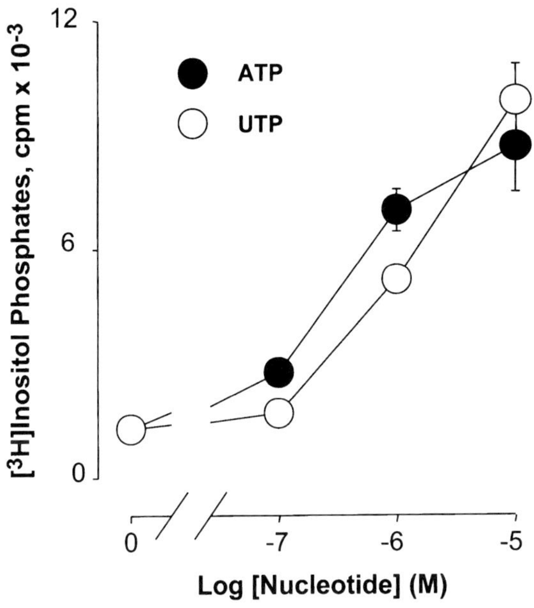

FIGURE 1 Activation of $P2Y_2$ receptors by ATP or UTP in human airway epithelia. Human bronchial airway epithelial (16HBE14o) cells, grown to confluence on 12 mm transwells, were labeled with [^{3}H]-inositol for 18 hours, followed by addition of 10 mM LiCl to inhibit inositol phosphate metabolism. ATP and UTP equipotently stimulated PLC-dependent inositol phosphate generation, suggesting the presence of $P2Y_2$ receptors in these epithelial cells.

in a heterotrimeric G protein–dependent manner (reviewed in Ref. 14), resulting in $[Ca^{2+}]_i$ mobilization through two related pathways (Fig. 2). At first, activation of PLC induces the breakdown of plasma membrane phosphatidylinositol 4,5-bisphosphate, resulting in generation of inositol 1,4,5-trisphosphate (IP_3); IP_3 then diffuses from its generation site and activates channel receptors in the endoplasmic reticulum (ER), the site of IP_3-sensitive Ca^{2+} stores, resulting in channel opening and release of stored calcium into the cytoplasm. Subsequently, depletion of IP_3-sensitive Ca^{2+} stores activates a Ca^{2+} influx pathway across the plasma membrane, which was originally termed ‘‘capacitative Ca^{2+} entry’’ (15,16) or, more recently, ‘‘store-operated calcium entry’’ (17). Both phases of Ca^{2+} mobilization can act in concert for the subsequent modulation of multiple plasma membrane transport processes, including the regulation of the alternative, Ca^{2+}-dependent Cl^- channel in airway epithelia. Although the contribution of each biochemical step underlying the coupling of $P2Y_2$ receptors to the modulation

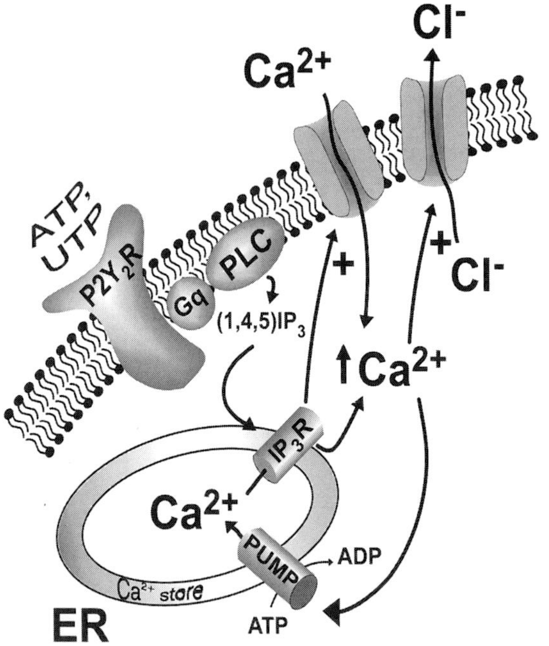

FIGURE 2 Schematic representation of the $P2Y_2$ receptor-activated Ca^{2+} signaling pathway in airway epithelia. $P2Y_2$ receptor activation by ATP or UTP is coupled to PLC-dependent generation of $(1,4,5)IP_3$, responsible for Ca^{2+} release from endoplasmic reticulum (ER) Ca^{2+} stores (first phase of $[Ca^{2+}]_i$ mobilization). Depletion of Ca^{2+} from ER stores activates capacitative Ca^{2+} entry, resulting in the second phase of $[Ca^{2+}]_i$ mobilization. The consequent rise in $[Ca^{2+}]_i$ subsequently modulates the activity of several membrane transporters, including the Ca^{2+}-activated Cl^- channel at the apical membrane of airway epithelia.

of the alternative Cl^- channel is far from clear, available techniques allow the study of expression and function of the individual components of the Ca^{2+} signaling system. Thus, the net responses of $P2Y_2$ receptor activation on $[Ca^{2+}]_i$ levels in normal and CF airway epithelia can be understood at the molecular level.

$P2Y_2$ RECEPTORS AND THE REGULATION OF Ca^{2+}-DEPENDENT Cl^- TRANSPORT IN NORMAL AND CF HUMAN AIRWAY EPITHELIA

The first evidence suggesting a link between $[Ca^{2+}]_i$ mobilization and Cl^- secretion in normal and CF human airway epithelia was obtained from studies with the Ca^{2+} ionophores A23187 and ionomycin (3). Through the use of double-barreled Cl^- microelectrodes, under conditions designed to inhibit Na^+ absorption and increase the driving force for Cl^- secretion, these studies demonstrated that both $[Ca^{2+}]_i$-mobilizing agents initiated Cl^- secretion via activation of an apical membrane Cl^- conductance in both normal and CF human nasal epithelia (3). Thus, despite the absence of cAMP-mediated Cl^- secretion in CF airway epithelia, Cl^- transport could still be attained by maximally raising $[Ca^{2+}]_i$ levels in presence of Ca^{2+} ionophores. A physiological role of $[Ca^{2+}]_i$ in airway epithelial Cl^- secretion was subsequently supported by studies that linked $P2Y_2$ receptor activation to changes in $[Ca^{2+}]_i$ and Cl^- transport in human nasal epithelia showing that apical application of ATP and UTP induced in a equipotent and equiefficacious manner increases in $[Ca^{2+}]_i$ and transepithelial Cl^- transport (18). The correlation between purinoceptor activation and the simultaneous increases in both $[Ca^{2+}]_i$ and Cl^- transport is illustrated in Figure 3, where similar dose responses for apical UTP-elicited $[Ca^{2+}]_i$ mobilization (Fig. 3A) and Cl^- secretion (Fig. 3B) are shown. These observations suggested that the 5′ nucleotide triphosphates were activating $P2Y_2$ receptors on the apical membrane to account for their effects on Ca^{2+} signaling and the Cl^- secretory response in human airways.

Several subsequent studies have described fundamental differences regarding Cl^- transport elicited by apical $P2Y_2$ receptor activation in normal and CF airway epithelia. For instance, in primary cultures of human nasal epithelia, UTP-stimulated Cl^- current was increased by twofold in CF compared to cultures derived from normal subjects (4). A following report, utilizing double-barreled

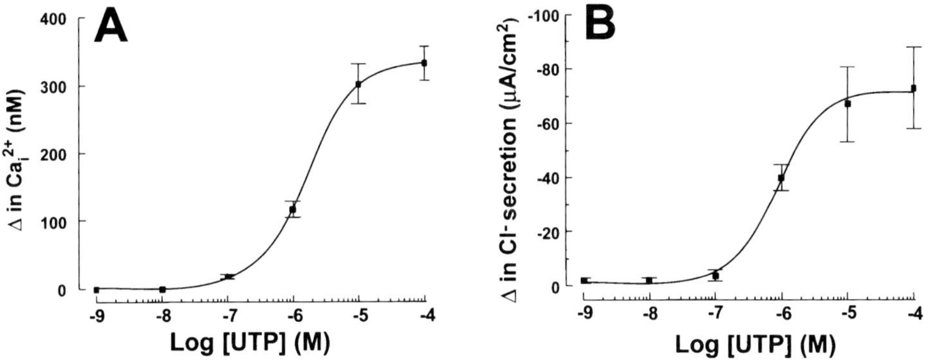

FIGURE 3 UTP-activated $P2Y_2$ receptors couples to intracellular calcium mobilization and Cl^- secretion in human airway epithelia. Seven-day-old primary cultures of polarized human nasal airway epithelia, grown on collagen-coated filters, were loaded with the Ca^{2+} fluorescent dye Fura-2 and mounted on a miniature Ussing chamber. Apical $P2Y_2$ receptor activation by UTP, in Na^+-free and low-Cl^- mucosal solution, promotes $[Ca^{2+}]_i$ mobilization and Cl^- secretion in a dose-dependent manner.

chloride-selective microelectrodes in normal and CF nasal epithelia, revealed that luminal ATP promoted a greater increase in Cl^- secretion in CF compared to normal tissue (19). Furthermore, consistent with these microelectrode data, in vivo measurements of ATP- or UTP-induced Cl^- secretion demonstrated that the Cl^- secretory response to these triphosphate nucleotides in CF patients was two times greater than that in normal subjects (20). Further evidence that the magnitude of the Ca^{2+}-dependent Cl^- secretory response is greater in CF than normal airway epithelia has been obtained by preliminary studies showing that apical application of ATP or UTP to polarized nasal epithelia promoted $[Ca^{2+}]_i$ mobilization and Cl^- secretion in normal and CF cells, but the elevations in $[Ca^{2+}]_i$ and the secretory Cl^- response were greater in CF (21).

These findings suggested that the $P2Y_2$ receptor–activated, Ca^{2+}-mediated apical Cl^- secretory pathway is unregulated in CF. Several possibilities may explain this upregulation, including an upregulation of the different components of the Ca^{2+} signaling pathway (e.g., increased $P2Y_2$ receptor number and/or PLC-generated IP_3 and/or increased expression of IP_3-sensitive Ca^{2+} stores) or an increased number of Ca^{2+}-activated Cl^- channels. Our laboratory is currently investigating the basis for the augmented $[Ca^{2+}]_i$ signals triggered by apical $P2Y_2$ receptor activation in CF compared to normal human airway epithelia.

Ca^{2+} STORE LOCALIZATION AND MEMBRANE-CONFINED $P2Y_2$ RECEPTOR-DEPENDENT Ca^{2+} SIGNALS IN HUMAN AIRWAY EPITHELIA

The ER, or a specialized portion of it, is considered to be the major Ca^{2+}-storing, buffering, and -signaling compartment within cells (22,23). These diverse ER functions result from the actions of (1) SERCA ATPases (sarcoplasmic-endoplasmic reticulum Ca^{2+}-ATPases) responsible for lumenal ER Ca^{2+} accumulation, (2) Ca^{2+}-binding proteins implicated in Ca^{2+} storage (e.g., calreticulin), and (3) Ca^{2+} channel receptors for IP_3 (and, in some cells, also for ryanodine) involved with Ca^{2+} release from stores. Thus, the activities of these transporting systems, together with the lumenal expression of Ca^{2+}-binding proteins, account for the homeostasis of ER Ca^{2+} (23). Furthermore, the cellular distribution of ER Ca^{2+} stores plays an important role on the spatial and temporal aspects of Ca^{2+} signaling (22,24–26) in response to external stimuli.

In the case of polarized airway epithelial cells, the localization of ER Ca^{2+} stores may determine the generation of localized Ca^{2+} signals, thereby restricting the spatial range for the Ca^{2+}-dependent modulation of apical or basolateral membrane transport processes. As depicted in Figure 4a, polarized human nasal epithelial cells form a simple columnar epithelium after 7–10 days in primary culture. Figure 4b illustrates the spatial distribution of ER in this polarized airway epithelium, where ribosome-decorated rough ER strands are seen in close proxim-

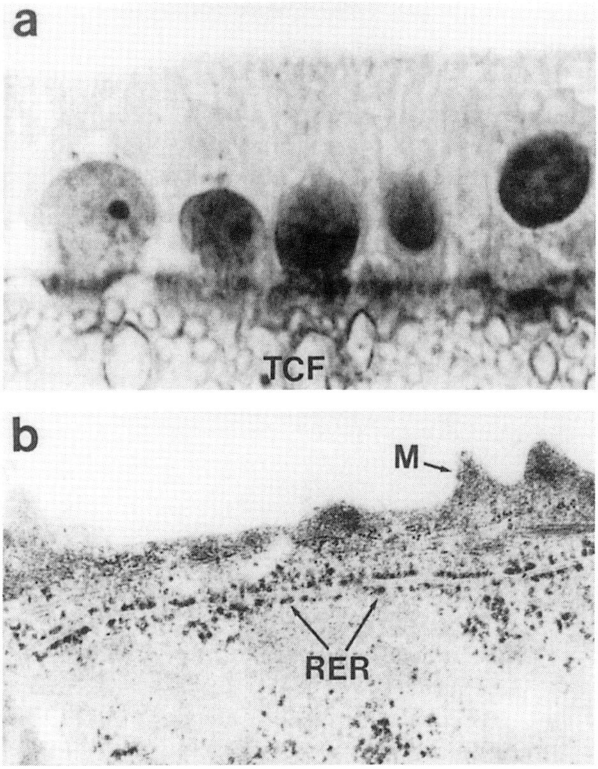

FIGURE 4 Morphology of a 7-day-old primary culture of polarized human airway epithelia. (a) Representative picture of a 7-day-old monolayer of human nasal columnar airway epithelium illustrating the polarizing aspect of the culture. (b) Electron micrograph from the same culture, depicting the close proximity of endoplasmic reticulum strands to the apical plasma membrane (magnification: 20,000×). TCF, Transwell collagen-coated filter; M, microvilli; RER, rough endoplasmic reticulum.

ity to the apical membrane. The proximity of the ER network and the apical membrane barrier may result in a more efficient coupling between apical $P2Y_2$ receptor activation and Ca^{2+}-induced Cl^- secretion. In support of this notion, our previous report in monolayers of human nasal epithelia suggested the existence of specific Ca^{2+} stores functionally associated with either the apical or basolateral poles of cells. Specifically, $P2Y_2$ receptor activation by mucosal or serosal application of ATP released Ca^{2+} from stores associated with the membrane domain ipsilateral but not contralateral to the activated receptors (27). Furthermore, activation of capacitative calcium entry by ATP was also limited to the membrane ipsilateral to receptor stimulation in human nasal epithelia (27). Similar to ATP, the expression of UTP-sensitive Ca^{2+} stores functionally associated with either the apical or the basolateral membrane is exemplified by the studies in primary cultures of human bronchial airway epithelia depicted in Figure 5. Like primary cultures of human nasal epithelial cells, $P2Y_2$ receptor activation by mucosal or serosal UTP released Ca^{2+} from stores associated with the membrane domain ipsilateral but not contralateral to activated receptors (Fig. 5, top panels). Furthermore, activation of capacitative Ca^{2+} entry by UTP is also restricted to the membrane ipsilateral to receptor activation (Fig. 5, bottom panels). The existence of specific Ca^{2+} stores functionally coupled to either the apical or basolateral poles was confirmed in a subsequent investigation in the human bronchial epithelial cell line 16HBE14o$^-$ (28). However, this latter study contrasts with our observations, since no capacitative Ca^{2+} entry was detected at the apical membrane. The reasons for these discrepancies are not clear, but they may result from the different origins of the investigated tissues, e.g., primary cultures of polarized human nasal (Fig. 4) or bronchial epithelia employed in our studies, versus a cell line of bronchial epithelial origin used in the studies of Kerstan et al. (28). Our results show that $P2Y_2$ receptor activation-induced $[Ca^{2+}]_i$ mobilization from Ca^{2+} store release and capacitative Ca^{2+} entry is confined to specific cellular domains in airway epithelia. It is, therefore, conceivable that localized $[Ca^{2+}]_i$ signals play a pivotal role on the modulation of membrane-restricted transport events, such as the apical Ca^{2+}-modulated Cl^- channel activity in airway epithelia. Consistent with this idea are the findings from a preliminary investigation in polarized nasal epithelial monolayers (21). In these studies, basolateral $P2Y_2$ receptor activation mobilized $[Ca^{2+}]_i$ in CF cells but did not result in Cl^- secretion. In contrast, apical $P2Y_2$ receptor activation coupled to elevation in $[Ca^{2+}]_i$ and Cl^- secretion. These results underscored the importance of localized, apical membrane–compartmentalized signal transduction by $P2Y_2$ receptor activation for the efficient modulation of Cl^- secretion in CF airways.

The findings discussed above highlight the importance of improving our understanding of the Ca^{2+} signaling pathways and Ca^{2+} store localization in normal and CF airway epithelia. Several possibilities may explain why apical $P2Y_2$ receptor activation–induced $[Ca^{2+}]_i$ mobilization is larger in CF cells, including a higher density of purinergic receptors at the apical membrane, a more efficient coupling between receptor activation and PLC-dependent IP_3 formation and/or a greater expression of Ca^{2+} stores. In addressing the mechanisms underlying the upregulation of apical $P2Y_2$ receptor–dependent Ca^{2+} signal in CF airways, we have obtained preliminary data suggesting that (1) agonist-sensitive Ca^{2+} stores are, primarily, distributed toward the apical domain in both normal and CF nasal and bronchial human airway epithelia and (2) their expression is increased in CF compared to normal epithelia (29). Thus, we speculate that the increased expression of Ca^{2+} stores may provide the basis for the higher apical $P2Y_2$ receptor activation–dependent $[Ca^{2+}]_i$ mobilization in CF airway epithelia.

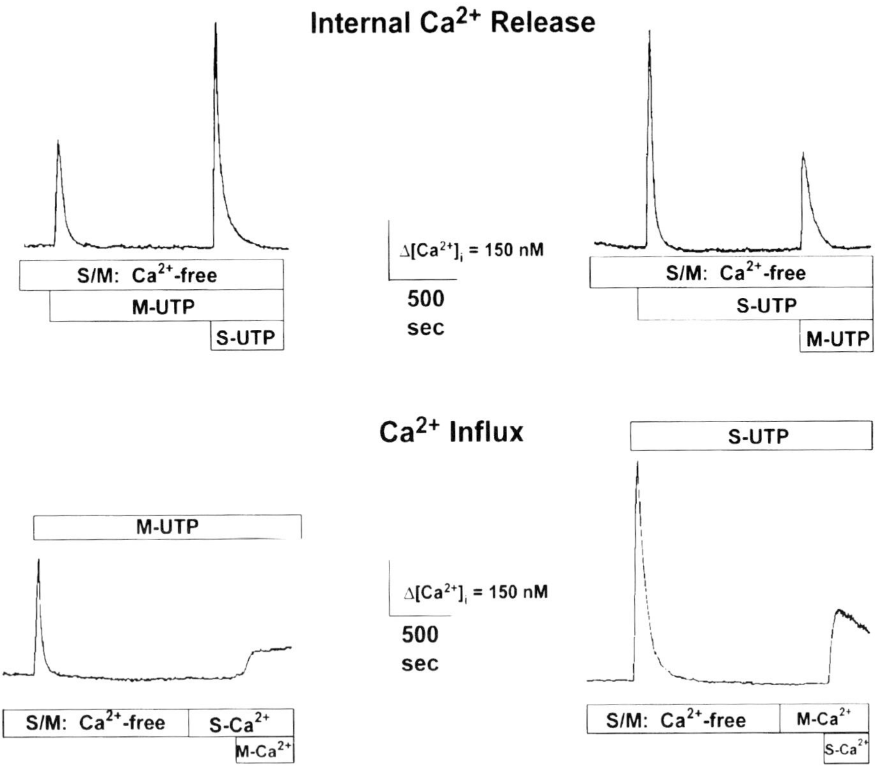

FIGURE 5 Ca^{2+} stores and capacitative Ca^{2+} entry are functionally associated with the apical or basolateral membranes in human airway epithelia. Seven-day-old primary cultures of polarized human bronchial airway epithelia, grown on collagen-coated filters, were loaded with the Ca^{2+} fluorescent dye Fura-2 and mounted on a miniature perfusion chamber. (Top) To address membrane-restricted, agonist-induced Ca^{2+} store release, polarized monolayers of human bronchial epithelial cells were bilaterally perfused with nominally Ca^{2+}-free NaCl Ringer and 100 μM UTP was added either to the mucosal (M; top left) or serosal (S; top right) compartment, followed by the sequential addition of the purinoceptor agonist to the contralateral side in both cases. Since the $[Ca^{2+}]_i$ signal elicited by mucosal or serosal UTP was of the same magnitude in either protocol, these data demonstrate that unilateral $P2Y_2$ receptor activation releases Ca^{2+} from stores associated with the plasma membrane ipsilateral but not contralateral to stimulated receptors. (Bottom) Polarized monolayers of human bronchial epithelial cells were bilaterally perfused with nominally Ca^{2+}-free NaCl Ringer and UTP-dependent Ca^{2+} store release was assessed by addition of the agonist to either the mucosal (M; bottom left) or serosal (S; bottom right) compartment. To address membrane-delimited, Ca^{2+} store release–activated capacitative calcium entry, the serosal and mucosal compartments were sequentially perfused with NaCl Ringer containing 1.3 mM Ca^{2+} (bottom left), while the perfusion sequence was reversed in parallel studies (bottom right). These observations demonstrate that UTP-activated capacitative calcium entry is also confined to the membrane ipsilateral to $P2Y_2$ receptor stimulation.

The hypothesis that Ca^{2+} signaling is altered in CF airway epithelia by virtue of its augmented Ca^{2+} storage is in agreement with a previous study showing increased levels of Ca^{2+} stores associated with a disease. This phenomenon has been demonstrated in a neuronal model system for Gaucher disease, where cells expressed an increase in ER and ryanodine receptors (ER Ca^{2+}-releasing channels equivalent to IP_3 receptors) and an augmented ER Ca^{2+} release in response to glutamate or caffeine stimulation (30). The elevated glutamate-induced Ca^{2+} signal was implicated in neuronal cell death, providing a molecular mechanism for the pathophysiology of Gaucher disease (30).

CLOSING REMARKS

Cl^- transport can be triggered by activation of the Ca^{2+} signaling pathway coupled to $P2Y_2$ receptors expressed at the apical membrane of airway epithelia. Activation of these receptors by 5′ nucleotide triphosphates released from epithelial or paracrine sources leads to a coordinated regulation of Cl^- secretion in the airways. This chapter dealt specifically with the regulation of the Ca^{2+}-dependent Cl^- channel by apical $P2Y_2$ receptor activation in normal and CF human airway epithelia. In normal airways, Cl^- transport resulting from cAMP- and Ca^{2+}-regulated pathways (CFTR and Ca^{2+}-activated Cl^- channel, respectively) plays a vital role in the composition of airway surface liquids, facilitating the maintenance of the mucociliary clearance and, thus, airway homeostasis. Conversely, in CF airways, which lack functional CFTR but express an upregulated Ca^{2+}-dependent Cl^- secretory pathway, the increased Ca^{2+}-mediated Cl^- secretion resulting from apical $P2Y_2$ receptor activation may serve to ameliorate the defective Cl^- transport characteristic of CF. The upregulation of the signaling pathway underlying $P2Y_2$ receptor activation–dependent Ca^{2+}-mediated Cl^- secretion in CF airway epithelia results, at least in part, from an increased $[Ca^{2+}]_i$ mobilization. Since preliminary studies suggest that agonist-sensitive Ca^{2+} store expression is increased in CF, these findings may provide the molecular basis for the increased coupling between $P2Y_2$ receptor activation and Ca^{2+}-mediated Cl^- secretion in CF airways. Thus, the understanding of $[Ca^{2+}]_i$ signaling events triggered by $P2Y_2$ receptor activation in airway epithelia should provide a strategy for therapeutic intervention in CF airway disease.

REFERENCES

1. MR Knowles, MJ Stutts, A Spock, N Fischer, JT Gatzy, RC Boucher. Abnormal ion permeation through cystic fibrosis respiratory epithelium. Science 221:1067–1070, 1983.
2. RC Boucher, CU Cotton, JT Gatzy, MR Knowles, JR Yankaskas. Evidence for reduced Cl^- and increased Na^+ permeability in cystic fibrosis human primary cell cultures. J Physiol (Lond) 405:77–103, 1988.

3. NJ Willumsen, RC Boucher. Activation of an apical Cl^- conductance by Ca^{2+} ionophores in cystic fibrosis airway epithelia. Am J Physiol 256:C226–C233, 1989.
4. MR Knowles, LL Clarke, RC Boucher. Activation by extracellular nucleotides of chloride secretion in the airway epithelia of patients with cystic fibrosis. N Engl J Med 325:533–538, 1991.
5. LL Clarke, BR Grubb, JR Yankaskas, CU Cotton, A McKenzie, RC Boucher. Relationship of a non-CFTR mediated chloride conductance to organ-level disease in *cftr*(−/−) mice. Proc Natl Acad Sci USA 91:479–483, 1994.
6. BR Grubb, RN Vick, RC Boucher. Hyperabsorption of Na^+ and raised Ca^{2+}-mediated Cl^- secretion in nasal epithelia of CF mice. Am J Physiol 266:C1478–C1483, 1994.
7. LL Clarke, BR Grubb, SE Gabriel, O Smithies, BH Koller, RC Boucher. Defective epithelial chloride transport in a gene targeted mouse model of cystic fibrosis. Science 257:1125–1128, 1992.
8. MR Knowles, PG Noone, WD Bennett, RC Boucher. Mucociliary and cough clearance: role of ion transport and the P_2Y_2 receptor-mediated system. In: GL Baum, Z Priel, Y Roth, N Liron, E Ostfeld, eds. Cilia, Mucus, and Mucociliary Interactions. New York: Marcel Dekker, Inc., 1998, pp 307–315.
9. RC Boucher. Drug therapy in the 1990s: What can we expect for cystic fibrosis. Drugs 43 (4):431–439, 1992.
10. RC Boucher. The genetics of cystic fibrosis: a paradigm for uncovering new drug targets. Curr Opin Biotechnol 5:639–642, 1994.
11. MR Knowles, K Olivier, P Noone, RC Boucher. Pharmacologic modulation of salt and water in the airway epithelium in cystic fibrosis. Am J Respir Crit Care Med 151:S65–S69, 1995.
12. SH Donaldson, RC Boucher. Therapeutic applications for nucleotides in lung disease. In: JT Turner, GA Weisman, JS Fedan, eds. The P2 Nucleotide Receptors. Totowa, NJ: Humana Press, 1998, pp 413–424.
13. HA Brown, ER Lazarowski, RC Boucher, TK Harden. Evidence that UTP and ATP regulate phospholipase C through a common extracellular 5′-nucleotide receptor in human airway epithelial cells. Mol Pharmacol 40:648–655, 1991.
14. JM Boeynaems, D Communi, R Janssens, S Motte, B Robaye, S Pirotton. Nucleotide receptors coupling to the phospholipase C signaling pathway. In: JT Turner, GA Weisman, JS Fedan, eds. The P2 Nucleotide Receptors. Totowa: Humana Press, 1998, pp 169–183.
15. JW Putney, Jr. A model for receptor-regulated calcium entry. Cell Calcium 7:1–12, 1986.
16. JW Putney, Jr. Capacitative calcium entry revisited. Cell Calcium 11:611–624, 1990.
17. DE Clapham. Intracellular calcium. Replenishing the stores. Nature 375:634–635, 1995.
18. SJ Mason, AM Paradiso, RC Boucher. Regulation of transepithelial ion transport and intracellular calcium by extracellular adenosine triphosphate in human normal and cystic fibrosis airway epithelium. Br J Pharmacol 103:1649–1656, 1991.
19. LL Clarke, RC Boucher. Chloride secretory response to extracellular ATP in normal and cystic fibrosis nasal epithelia. Am J Physiol 263:C348–C356, 1992.

20. MR Knowles, LL Clarke, RC Boucher. Extracellular ATP and UTP induce chloride secretion in nasal epithelia of CF patients and normal subjects in vivo. Chest 101(Suppl):60S–63S, 1992.
21. AM Paradiso, CMP Ribeiro, RC Boucher. Polarized signaling via purinoceptors in normal and cystic fibrosis airway epithelia. J Gen Physiol 117:53–67, 2001.
22. JW Putney, Jr. Capacitative Calcium Entry. Austin: R.G. Landes Company, 1997.
23. J Meldolesi, T Pozzan. The endoplasmic reticulum Ca^{2+} store: a view from the lumen. Trends Biochem Sci 23:10–14, 1998.
24. MJ Berridge, G Dupont. Spatial and temporal signalling by calcium. Curr Opin Cell Biol 6:267–274, 1994.
25. D Clapham. Calcium signaling. Cell 80:259–268, 1995.
26. MJ Berridge, MD Bootman, P Lipp. Calcium—a life and death signal. Nature 395: 645–648, 1998.
27. AM Paradiso, SJ Mason, ER Lazarowski, RC Boucher. Membrane-restricted regulation of Ca^{2+} release and influx in polarized epithelia. Nature 377:643–646, 1995.
28. D Kerstan, J Thomas, R Nitschke, J Leipziger. Basolateral store-operated Ca^{2+}-entry in polarized human bronchial and colonic epithelial cells. Cell Calcium 26:253–260, 1999.
29. CP Ribeiro, AM Paradiso, RC Boucher. Apical domains of inositol 1,4,5-triphosphate (IP_3)-sensitive Ca^{2+} stores in normal and cystic fibrosis (CF) human airway epithelia (abstr). Pediatr Pulmonol Suppl 19:253, 1999.
30. E Korkotian, A Schwarz, D Pelled, G Schwarzmann, M Segal, AH Futerman. Elevation of intracellular glucosylceramide levels results in an increase in endoplasmic reticulum density and in functional calcium stores in cultured neurons. J Biol Chem 274:21673–21678, 1999.

28

Role of Epidermal Growth Factor Receptor Cascade in Airway Hypersecretion and Proposal for Novel Therapy

Jay A. Nadel
University of California, San Francisco
San Francisco, California

INTRODUCTION

Mucus hypersecretion contributes to the morbidity and mortality in chronic obstructive pulmonary disease (COPD), cystic fibrosis, acute asthma, and bronchiectasis. At present, no effective therapy exists for hypersecretion in any disease. Because the physiology of secretion is presented elsewhere in this meeting, this work will focus on the first therapy of hypersecretion, which involves the epidermal growth factor receptor (EGFR) cascade. The methods used in the studies discussed here are described in detail in the original publications (see references).

RESULTS

Discovery that an Epidermal Growth Factor Cascade Regulates Mucin Production in Airways

EGFR is a 170 kDa membrane glycoprotein that is expressed in fetal airways and has been shown to be important in branching morphogenesis in airways (2).

The study of EGFR has been explored extensively in malignant tumors, where it has been implicated in cell multiplication (and thus tumor growth). Constitutive expression of EGFR is low in the airways of healthy adults (3) and in pathogen-free animals (1). Perhaps this absence of constitutive expression of EGFR made it more difficult to discover the role of the EGFR cascade in mucus cell production in the airways. There are limited studies of EGFR expression in inflammatory diseases. For example, Puddicombe et al. (4) have reported increased EGFR immunostaining in biopsies of asthmatics, and the authors implicated this pathway in bronchial epithelial repair and ''remodeling.'' Mucus cells were not implicated, and the mechanism of EGFR activation was not identified.

We hypothesized that a growth factor could be involved in airway secretory cell production. We found that EGFR expression is low in airways of pathogen-free rats (1), but we discovered that stimulation of the airways with tumor necrosis factor alpha (TNFα) induces EGFR in airway epithelial cells, but mucin production was not increased (Fig. 1). However, when we stimulated EGFR with its ligands (e.g., EGF, TGFα), mucin production occurred in the following sequence: in controls, the tracheal epithelium contained few goblet or pregoblet cells. The majority of the cells were ciliated, basal, and so-called nongranulated secretory cells. When TNFα was instilled in the airways, EGFR expression occurred, but the distribution of cell types was essentially unaltered. However, activation of EGFR by EGF or TGFα produced mucins and thereby profoundly changed the distribution of epithelial cells. There was an increase in pregoblet and goblet cells and a marked decrease in the number of nongranulated secretory cells, resulting in no change in the total number of epithelial cells (Table 1). We conclude that (1) mucin-containing goblet cells are derived from nongranulated secretory cells, and (2) goblet cells are formed by *differentiation* of precursor cells, rather than by the growth of new epithelial cells (*multiplication* of cells).

Most importantly, mucin production by TNFα followed by the EGFR ligand TGFα was inhibited dose-dependently by pretreatment with a selective inhibitor of EGFR tyrosine kinase (Fig. 2A). From these studies we conclude that the activation of EGFR causes mucin gene and protein expression. The findings were the first to provide a mechanism and a strategy for therapy in hypersecretory diseases.

Role of EGFR Cascade in Asthmatic Hypersecretion

Mucus hypersecretion contributes to asthma mortality (7–9). When we discovered that EGFR activation induces mucin production in airways, we hypothesized that this pathway could be involved in hypersecretion in asthma, so we investigated its role in experimental asthma induced in pathogen-free rats. Intraperitoneal injection of ovalbumin (OVA) alone did not increase the number of goblet cells, but when this was followed by airway instillation of OVA (''active sensitization''), the number of goblet cells was increased markedly (Fig. 1B); the num-

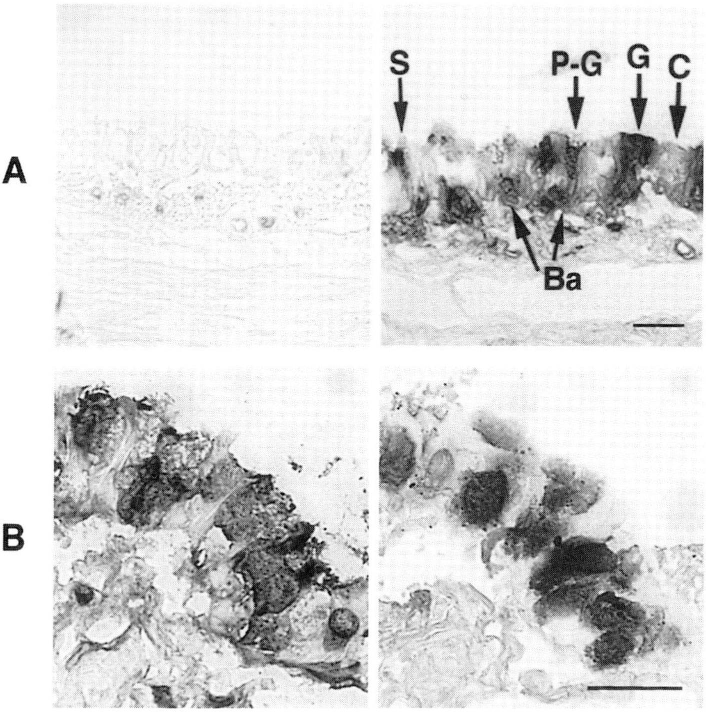

FIGURE 1 Immunohistochemical analysis of EGF-R with an anti-EGF-R antibody in pathogen-free rats. (A) TNFα-treated rats. Control animals showed little EGF-R staining (*left*); 24 hours after intratracheal instillation of TNFα (200 ng, 100μL), EGF-R–positive staining was present in goblet cells (G), pregoblet cells (P-G), nongranulated secretory cells (S), and basal cells (Ba), but not in ciliated cells. (Bar = 50 μm.) (B) Ovalbumin sensitization. After three intratracheal installations of ovalbumin (0.1%, 100 μL), EGF-R immunoreactivity was strongly expressed in goblet and pregoblet cells (*left*), the same cells that stained positively with AB-PAS (*right*). (Bar = 50 μm.)

ber of ciliated and basal cells was unchanged (Table 1). Immunohistochemical studies with an antibody to EGFR showed no staining in control tracheas, but actively sensitized rats showed EGFR staining in the cells that stained positively with Alcian blue/PAS (a stain for mucus glyconconjugates). Because OVA induced EGFR expression, we studied the effect of pretreatment with a selective EGFR inhibitor, BIBX1522. The inhibitor completely prevented OVA-induced goblet cell growth (1). From these results we conclude that inhibition of EGFR prevents goblet cell production due to allergen in vivo.

TABLE 1 Effects of Mediators and Ovalbumin Sensitization on Tracheal Epithelial Cells in Rats

				OVA sensitization	
Cell type	Control	TGFα	TNFα/TGFα	i.p. only	i.p. + i.t.
Goblet	2.8 ± 0.7	5.8 ± 1.2	28.8 ± 3.4*	5.4 ± 1.5	38.2 ± 6.3*
Pregoblet	7.8 ± 1.3	12.8 ± 1.6	44.8 ± 3.6*	13.8 ± 1.4	36.0 ± 6.3*
Secretory	82.0 ± 2.0	72.2 ± 4.0	40.8 ± 2.4*	67.6 ± 7.0	49.8 ± 4.2*
Ciliated	49.6 ± 2.0	54.6 ± 2.3	53.2 ± 1.8	56.4 ± 3.8	52.4 ± 7.1
Basal	57.8 ± 2.6	56.8 ± 2.3	43.0 ± 3.5*	60.2 ± 3.4	59.8 ± 2.9
Indeterminate	1.4 ± 0.5	2.0 ± 0.4	0.8 ± 0.4	1.4 ± 0.2	2.6 ± 0.5
Total	201.4 ± 2.2	204.2 ± 3.3	211.4 ± 4.8	204.8 ± 6.6	238.8 ± 4.4*

[a] Cells were analyzed as described in Methods section; five rats were in each group. Control airways and airways stimulated by TGFα (250 ng) alone contained few goblet and pregoblet cells; TNFα (200 ng) followed by TGFα (250 ng) resulted in increased numbers of goblet ($p <$ 0.001) and pregoblet ($p < 0.0001$) cells, accompanied by a decrease of nongranulated secretory cells ($p < 0.0001$) and basal cells ($p < 0.05$). Sensitization of rats with ovalbumin (OVA) intraperitoneally (i.p.) had no effect on cell distribution, but when OVA was given i.p. followed by intratracheal (i.t.) instillation, the total number of epithelial cells increased significantly ($p < 0.001$), and there were increased numbers of goblet ($p < 0.001$) and pregoblet ($p < 0.001$) cells. Nongranulated secretory cells decreased significantly ($p < 0.01$). In all conditions, the number of ciliated cells was unchanged.

Source: Ref. 1.

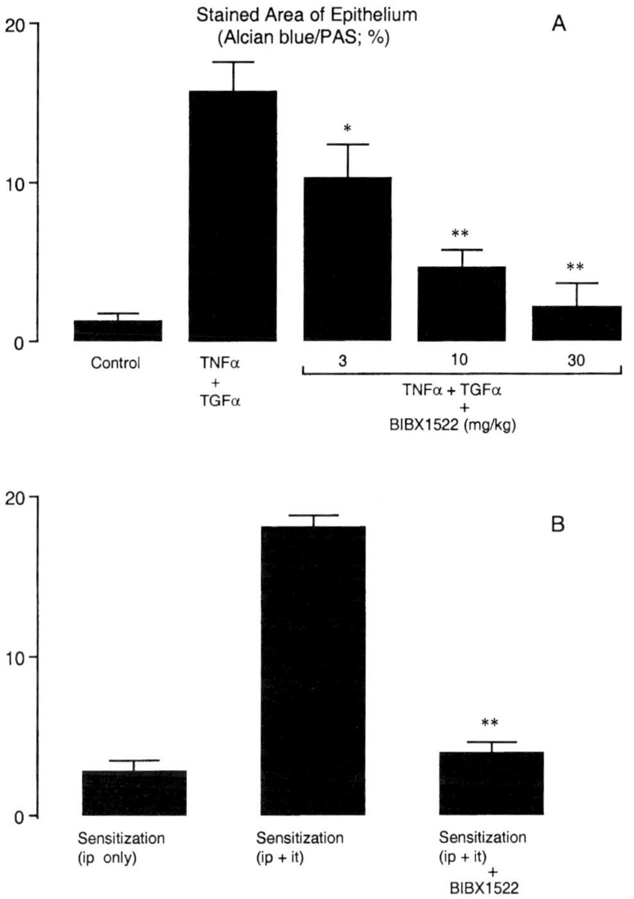

FIGURE 2 Effect of EGF-R tyrosine kinase inhibitor (BIBX1522) on production of goblet cells (expressed as % stained area of airway epithelium occupied by AB-PAS–positive stained cells). (A) Stimulation with TNFα (200 ng, 100 μL). Tracheal instillation of TNFα, followed by the EGF-R ligand TGFα, increased goblet cell production significantly ($n = 5$; $^{*}p < 0.0001$), an effect that was inhibited by pretreatment with BIBX1522 (3–30 mg/kg, i.p.) dose dependently ($n = 5$; p compared with TNFα followed by TGFα: $^{*}p = 0.003$; $^{**}p < 0.0001$). (B) Ovalbumin sensitization. Animals given ovalbumin intraperitoneally (i.p.) only showed little AB-PAS–positive staining in bronchial epithelium. Animals first sensitized with ovalbumin (OVA) i.p., followed by three intratracheal (i.t.) installations of OVA, showed a marked increase in AB-PAS–positive staining ($n = 5$; $^{*}p < 0.0001$). Pretreatment with BIBX1522 (10 mg/kg, i.p.) inhibited OVA-induced production of goblet cells ($n = 5$; $^{**}p < 0.0001$). *Source*: Ref. 1.

Th2 cytokines (IL-4 and IL-13) have been implicated in allergen-induced goblet cell metaplasia in sensitized mice (10–13). The involvement of Th2 cytokines in allergen-induced hypersecretion was proven by blocking the OVA-induced metaplasia by pretreatment with a neutralizing antibody to the appropriate receptor (14). In addition, airway instillation of IL-4 or IL-13 in nonallergic animals is reported to cause goblet cell metaplasia (11,15,16), implicating these cytokines in mucin production. Because OVA-induced goblet cell metaplasia is prevented by pretreatment with EGFR inhibitors, and because Th2 cytokines are implicated in goblet cell metaplasia, we hypothesized that Th2 cytokines cause goblet cell production indirectly by activating an EGFR cascade. Because previous studies suggested that IL-13 plays an important role in OVA-induced airway effects (12), we studied the role of EGFR activation in IL-13–induced goblet cell metaplasia in pathogen-free rats (5). Instillation of IL-13 induced goblet cell metaplasia, an effect that was prevented by pretreatment with a selective inhibitor of EGFR (Figs. 3,4), implicating this growth factor cascade in IL-13–induced goblet cell metaplasia. Next, we examined the mechanism of IL-13–induced EGFR expression and activation. Control airway epithelium contained few leukocytes, but intratracheal instillation of IL-13 resulted in leukocyte recruitment and induced IL-8 expression in the airway epithelium (Fig. 5). TNFα protein expression was prominent in the neutrophils that were recruited into the airways by IL-13 and are the probable mechanism of IL-13–induced EGFR expression. Pretreatment with an inhibitor of leukocytes in the bone marrow (cyclophosphamide) or with a blocking antibody to IL-8 prevented both IL-13–induced leukocyte recruitment and goblet cell metaplasia. From these studies we recognize the importance of recruited leukocytes and their activation in IL-13–induced EGFR-dependent goblet cell metaplasia. (For a discussion of the role of neutrophil activation in EGFR transactivation, see section on Oxidative Stress and Neutrophil Activation.)

In spite of the convincing studies of experimental asthma in animals, it is difficult to confirm that EGFR activation plays a role in mucin synthesis in human asthma because of the present inability to use EGFR tyrosine kinase inhibitors to prevent hypersecretion in humans. However, we have initiated studies to examine whether EGFR is upregulated in asthmatic airways. We performed in situ hybridization and immunohistochemical analysis for EGFR and MUC5AC (as a marker of airway goblet cell mucin expression in sections obtained from bronchial biopsies in healthy and asthmatic airways) (6). Healthy airways showed little epithelial expression of EGFR mRNA; asthmatic airways showed greater expression. MUC5AC mRNA expression was also greater in asthmatic airways, and mucus glycoconjugates (stained with Alcian blue/PAS) were also increased in asthmatic airways. Ciliated cells were negative for EGFR and MUC5AC both in asthmatic and in healthy airways at both mRNA and protein levels. There was a significant correlation between EGFR immunoreactivity and the area of MUC5AC-positive staining. These findings suggest the possible role of EGFR activation in mucin

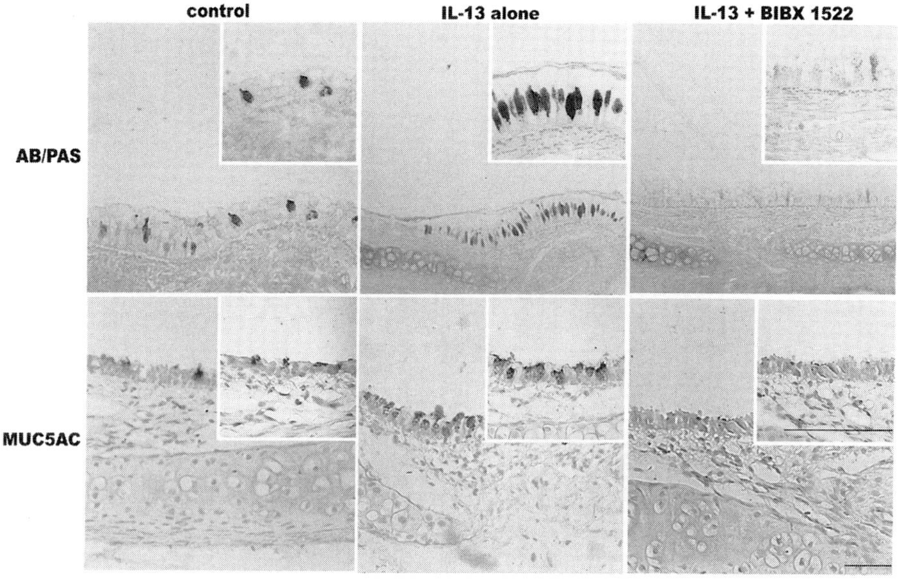

FIGURE 3 Photomicrographs of rat airway epithelium stained with Alcian blue (AB)/PAS or with MUC5AC protein. In the control state, AB/PAS and MUC5AC staining were minimal (*left columns*). IL-13 increased AB/PAS and MUC5AC staining (*middle columns*). Pretreatment with a selective EGFR tyrosine kinase inhibitor (BIBX 1522; 30 mg/kg/day, i.p.) prevented IL-13–induced effects (*right columns*). Results are typical of studies in five rats. Photomicrographs are shown at 20× magnification; inserts show magnification, 40× bar = 100 μm. *Source*: Ref. 5.

synthesis in asthmatic airways. Proof of concept in humans will require testing of EGFR inhibitors in patients with asthma and with other hypersecretory diseases.

Oxidative Stress and Neutrophil Activation Cause Mucin Synthesis Via EGFR Transactivation

Oxidative stress has been implicated in the pathogenesis of inflammatory airway diseases such as COPD and asthma (18). In relation to EGFR activation, oxidative stress has important implications. EGFR tyrosine kinase can be activated by two separate processes: the binding of EGFR ligands (e.g., TGFα, EGF) to EGFR activates the receptor tyrosine kinase and induces tyrosine phosphorylation (19). Alternatively, EGFR tyrosine phosphorylation can be activated by a ligand-independent mechanism (''transactivation'') such as by stimulation with oxygen free radicals (20). We examined the hypothesis that oxidative stress causes mucin

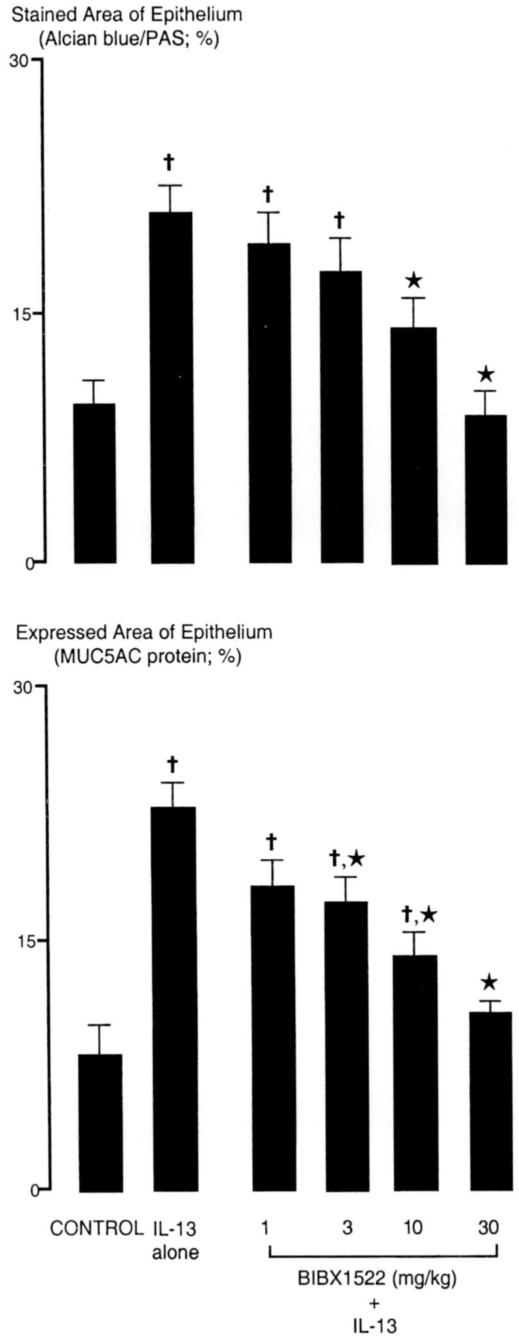

Stained Area of Epithelium
(Alcian blue/PAS; %)
30
15
0
†
†
†
★
★
Expressed Area of Epithelium
(MUC5AC protein; %)
30
15
0
†
†
†,★
†,★
★
CONTROL
IL-13
alone
1
3
10
30
BIBX1522 (mg/kg)
+
IL-13

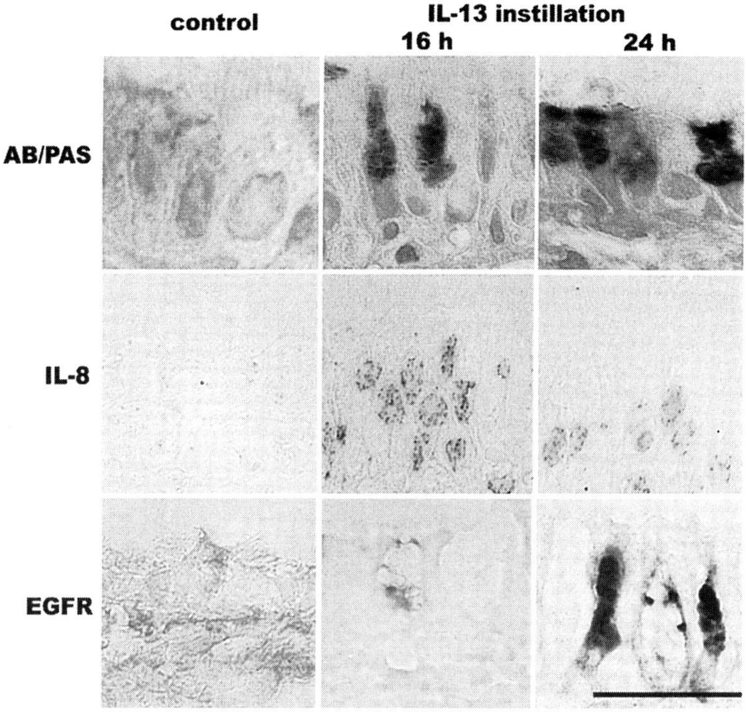

FIGURE 5 Photomicrographs of rat airway epithelium stained with Alcian blue (AB)/PAS, with an antibody to interleukin (IL)-8, or with an antibody to epidermal growth factor receptor (EGFR). In the control state, there was only sparse AB/PAS, IL-8, and EGFR staining. Instillation of IL-13 caused a time-dependent increase in AB/PAS, IL-8, and EGFR staining. Results are typical of studies in five rats. Photomicrographs are shown at 100 × magnification; bar = 50 μm. *Source*: Ref. 5.

FIGURE 4 Dose-dependent inhibition of IL-13–induced staining of mucous glycoconjugates with Alcian blue/PAS (*upper columns*) and MUC5AC protein (*lower columns*) by a selective EGFR tyrosine kinase inhibitor (BIBX 1522, 1–30 mg/kg/day, i.p., given 1 day before instillation of IL-13 and daily thereafter) in rats. Pretreatment with BIBX 1522 prevented IL-13–induced staining dose dependently. Values are expressed as means ± SEM ($n = 5$; $^{\dagger}p < 0.05$ compared with control rats; $^{*}p < 0.05$ compared with rats instilled with IL-13 alone). *Source*: Ref. 5.

production by EGFR transactivation by incubating NCl-H292 cells (a line of human epithelial cells that produce mucins) (1). H_2O_2 increased mucin synthesis dose-dependently. Next, we examined the effect of activated neutrophils on mucin production (17). When we activated neutrophils with IL-8, FMLP, or TNFα and then incubated the supernatant with NCl-H292 cells, mucin synthesis increased dose-dependently. A transduction pathway known to be downstream of EGFR activation, $p22/42^{mapk}$, was also stimulated by activated neutrophil supernatant–stimulated NCl-H292 cells. Pretreatment with selective EGFR tyrosine kinase inhibitors prevented activated neutrophil supernatant–induced mucin production, implicating EGFR activation in the response. Other tyrosine kinase inhibitors were without effect. Neutrophil supernatant–induced EGFR tyrosine phosphorylation, activation of $p44/42^{mapk}$, and mucin synthesis were inhibited by antioxidants (*N*-acetyl-L-cysteine, DMSO, dimethyl thiourea, or superoxide dismutase). These results show for the first time that oxidative stress (in particular neutrophil-derived oxidant stress) induces mucin synthesis by causing ligand-independent EGFR transactivation.

Many chronic airway diseases associated with mucus hypersecretion [e.g., chronic obstructive pulmonary disease (21), cystic fibrosis (22,23), and bronchiectasis (24)] are also characterized by the accumulation of activated neutrophils. In addition to oxidative damage to DNA by activated neutrophils, which leads to airway epithelial cell damage (25), oxidative stress mediates a variety of adaptive responses. The present studies add oxidative stress–induced mucin production in airways to these adaptive responses. Inhaled cigarette smoke, an important source of oxidative stress, induces mucus hypersecretion by a similar process (see following section].

Cigarette Smoke Causes Mucin Synthesis via EGFR Activation

Mucus hypersecretion from hyperplastic airway mucin-secreting cells is a landmark of COPD. Exposure to cigarette smoke is known to induce goblet call hyperplasia and mucus production in several species (27,28), but because the cellular mechanisms were unknown, treatment of airway hypersecretion has not been possible. We hypothesized that cigarette smoke activates in EGFR cascade, resulting in mucin synthesis (26). First, we showed that cigarette smoke upregulates EGFR gene expression, activates EGFR tyrosine phosphorylation, and increases mucin gene expression in mucin-producing airway epithelial cells in vitro. Pretreatment with selective EGFR tyrosine kinase inhibitors prevented cigarette smoke–induced mucin production, implicating an EGFR cascade in the response. The cigarette smoke–induced effects were also inhibited significantly by pretreatment with free radical scavengers, suggesting that smoke-induced mucin production was due at least in part to oxidative stress from cigarette smoke. Next, we examined the effect of inhalation of cigarette smoke in pathogen-free rats in vivo. Smoke inhalation increased the production of mucins in the airway epithelium markedly, and these effects were prevented dose-dependently by a selective EGFR inhibitor, implicating EGFR activation in the response. Cigarette smoke is known to contain oxygen free radicals, which explains the inhibition by antioxidants in the in vitro studies [where the only cells involved were NCI-H292 (epithelial) cells]. However, cigarette smoke also causes neutrophil migration into the airways (29), and activated (migrated) neutrophils cause EGFR activation and mucin production by relcasing free radicals (see Section on Oxidative Stress and Neutrophil Activation). Thus, mucin synthesis in vivo might be caused both by the direct stimulation by smoke and by indirect stimulation resulting from neutrophil recruitment, activation and release of oxygen free radicals.

Cigarette smoke–induced mucin production in vitro was inhibited entirely by EGFR inhibitors, indicating that the EGFR cascade was fully responsible for the smoke-induced response. However, only half of the EGFR activation by cigarette smoke in vitro was explained by oxidative stress, so other stimuli must be involved. Cigarette smoke–induced goblet cell metaplasia in rats is reported to be prevented by pretreatment with indomethacin, an inhibitor of cyclooxygenase products (30), and arachidonic acid induces EGFR tyrosine phosphorylation and its association with Shc, resulting in MAPK activation in renal epithelial cells (31). Another possible stimulus is acrolein. This molecule, an ingredient of cigarette smoke, is reported to produce cyclo-oxygenase products in airway epithelial cells (32), so acrolein-induced prostaglandin synthesis could play a role in smoke-induced, EGFR-dependent mucin synthesis.

Our animal studies focus on goblet cells, because these species do not contain significant numbers of airway submucosal glands. In humans, cigarette smoking may cause hypersecretion in glands, which are located in the large conducting airways, resulting in chronic bronchitis (manifested by cough and sputum production). Smoking also has been incriminated in goblet cell hyperplasia and mucus plugging in peripheral airways (33). We suggest that the hypersecretion in glands that occurs in chronic bronchitis and in peripheral goblet cell hyperplasia are both due to activation of an EGFR cascade, both treatable with EGFR antagonists.

Agar Plugs in Airways Cause Goblet Cell Metaplasia via EGFR Cascade: Implications for Chronic Airway Intubation

Mechanical wounding of the epithelium leads to repair processes. Thus, a denuding mechanical injury to hamster tracheal epithelium produces an epithelium largely composed of cells with secretory characteristics (35). Similarly, mechanical injury with a cotton swab is reported to cause excess mucus production (36). Likewise, orotracheal intubation in horses is reported to result in abundant mucus secretion (37). These findings led us to hypothesize that mechanical irritation of the airways could lead to EGFR activation and mucus hypersecretion. Plugs with an irregular surface can be produced using agarose. When appropriately sized plugs are instilled in rats, they lodge in airways. Their irregularity prevents total obstruction and atelectasis; breathing probably causes some movement of the plugs, exaggerating the irritation. Instilled agarose plugs lodged in the bronchi of pathogen-free rats and caused an increase in mucin production in the area of the plugs (Fig. 6A), which increased strikingly over the 3-day period of observation (34). Areas without agarose plugs contained few goblet cells. Areas with plugs showed a time-dependent increase in the number of pregoblet and goblet cells and a proportionate decrease in the number of nongranulated secretory cells. The number of ciliated cells was unchanged. The total number of epithelial cells in the epithelium was unchanged. These results suggest that agarose plugs cause the conversion of nongranulated secretory cells to mucin-containing cells via a process of epithelial cell differentiation, and not by the generation of newly generated goblet cells (cell multiplication). Control bronchi showed sparse staining for EGFR protein, but plugged bronchi showed intense EGFR staining in the epithelium. Pretreatment with an EGFR tyrosine kinase inhibitor prevented staining for mucus glycoconjugates (AB/PAS) and mucin gene expression (Fig. 7), implicating an EGFR cascade in agarose plug–induced goblet cell metaplasia. Next, we examined possible mechanisms of EGFR expression in the epithelium adjacent to agarose plugs. Because Takeyama et al. (1) showed that TNFα induces EGFR expression, we performed immunocytochemical studies for the localization of TNFα. We found that plugs caused the appearance of TNFα, especially in neutrophils in airways with agarose plugs (Fig. 8). Pretreatment with a TNFα-neutralizing antibody prevented agarose-induced EGFR expression and goblet cell metaplasia (Fig. 9).

We noted that agarose plugs caused epithelial damage and inflammatory cell infiltration. In particular, bronchoalveolar lavage in animals with plugs showed a marked increase in neutrophils and affected airways contained striking infiltrates containing neutrophils (Fig. 6B). Pretreatment with cyclophosphamide prevented neutrophil recruitment, EGFR expression, and goblet cell metaplasia (Figs. 8,9). These results implicate neutrophils in agarose plug–induced goblet cell metaplasia.

The discovery that mechanical irritation induced by agarose plugs results in EGFR cascade–dependent goblet cell metaplasia has several potential implications. First, in chronically intubated patients, mucus accumulation is generally assumed to be due to mucus production in the lower airways. However, while the orotracheal tube is fixed at the outlet, the trachea moves into the chest during each inspiration, resulting in mechanical irritation with each breath. If mucus is produced at the tracheal end of the tube, mucus could easily be aspirated into the lungs, especially during sleep. It is suggested that this potential mechanism be evaluated in patients in intensive care units.

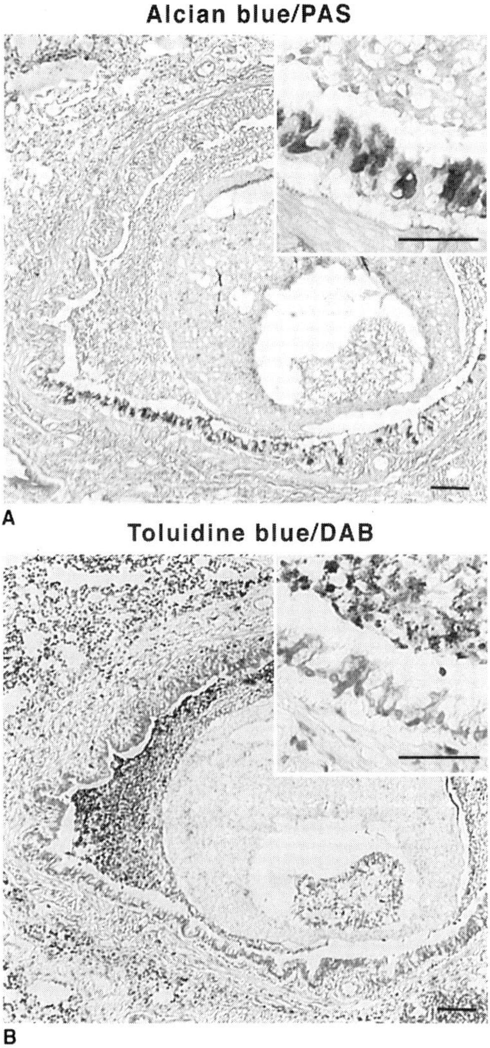

FIGURE 6 Local effect of intrabronchial instillation of agarose plugs on epithelial production of mucus glycoconjugates (stained with Alcian blue-PAS) (*A*) and on infiltration of inflammatory cells (stained with diaminobenzidine, or DAB) (*B*). Low-power fields show cross sections of entire plugged bronchus. Original magnification ×5; bars, 100 μm. Insets: magnified fields showing goblet (*A*) and inflammatory (*B*) cells. Original magnification ×20; bars = 50 μm. *Source*: Ref. 34.

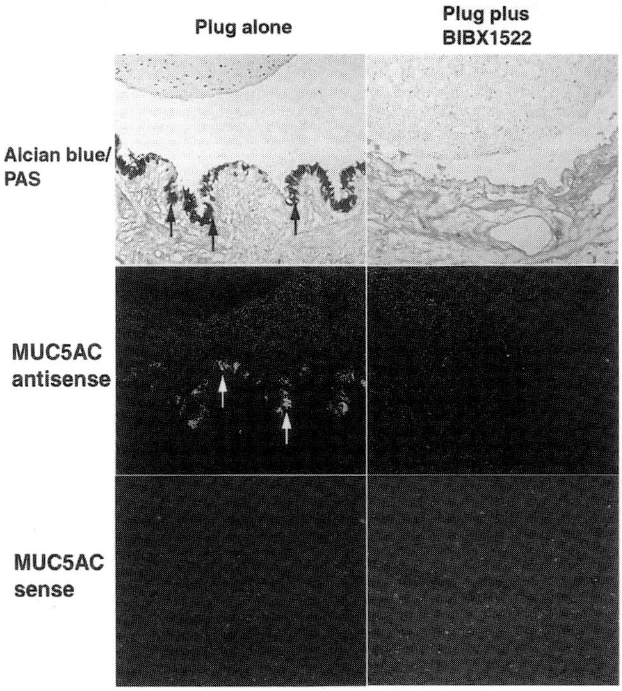

FIGURE 7 Effect of EGFR tyrosine kinase inhibitor BIBX1522 (80 mg kg^{-1} day^{-1} ip) on agarose plug–induced mucus glycoconjugate production (expressed as Alcian blue-PAS–positive staining) (*top*) and MUC5AC mRNA expression (*middle and bottom*) at 72 hours in pathogen-free rats. Agarose plugs alone (*left*) caused Alcian blue-PAS–positive staining in epithelium and MUC5AC mRNA expression (MUC5AC antisense). Pretreatment with EGFR tyrosine kinase inhibitor BIBX1522 prevented Alcian blue-PAS staining and MUC5AC expression in plugged bronchi (*right*). MUC5AC sense probe showed no expression (*bottom*). Arrows, location of epithelial basal lamina. Results were confirmed in four animals. *Source*: Ref. 34.

DISCUSSION

EGFR and Epithelial Cell Regulation

Here I present a novel cascade, which we have shown to be involved in the production of mucins in the airways. The discovery involves the following features:

1. The airways of pathogen-free animals and healthy humans contain only sparse goblet cells in the airway epithelium.

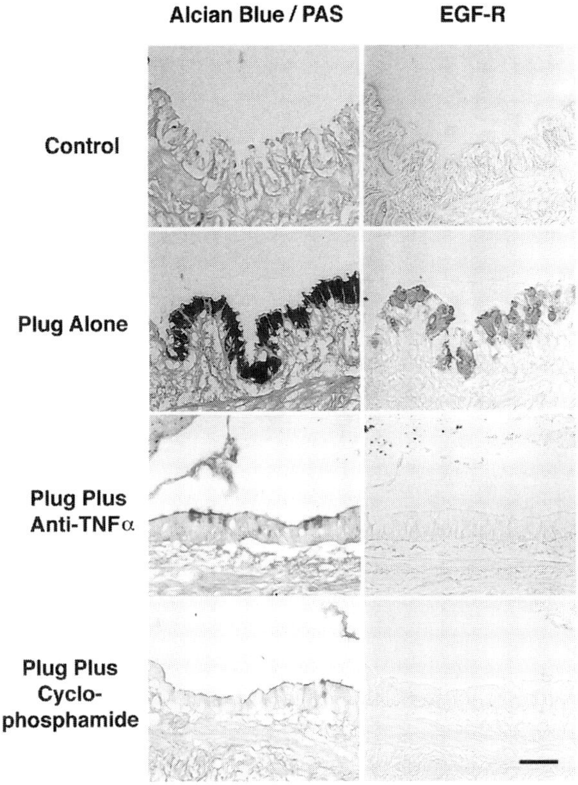

Figure 8 Effect of intrabronchial instillation of agarose plugs on staining for mucus glycoconjugates (*left*) and on expression of EGFR protein (*right*) in pathogen-free rats. In control airways, staining with Alcian blue-PAS and anti-EGFR antibody was negative (control), but adjacent to agarose plugs Alcian blue-PAS and EGFR staining were positive in epithelium (plug alone). After pretreatment with a TNFα-neutralizing antibody (plug plus anti-TNFα) or cyclophosphamide (plus plus cyclophosphamide), agarose plug–induced Alcian blue-PAS and EGFR staining were inhibited. Results were confirmed in five animals. Bar = 50 μm. *Source*: Ref. 34.

2. Similarly, in healthy animals and humans, the airway epithelium contains few EGFR.
3. Stimulation of the epithelium with TNFα causes EGFR gene and protein expression, but expression of EGFR alone does not cause mucin production.
4. Activation of EGFR by ligands (e.g., EGF, TGFα) causes EGFR tyrosine kinase phosphorylation and other downstream signaling, leading

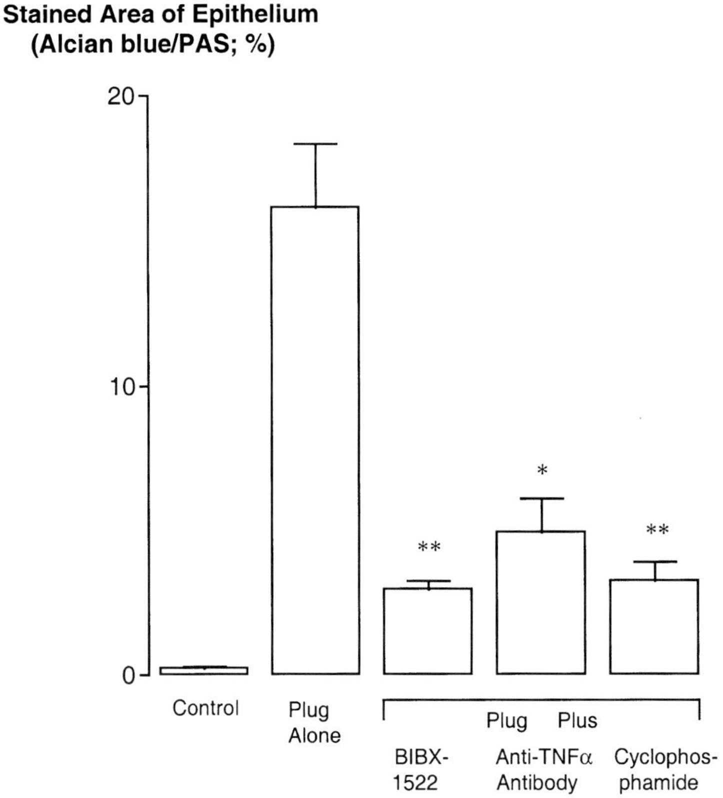

FIGURE 9 Percentage of Alcian blue-PAS–stained area of epithelium in bronchi of pathogen-free rats treated with various drugs. Agarose plugs were instilled intraronchially, and rats were euthanized 72 hours later. Values are means ± SE. In animals pretreated with EGFR tyrosine kinase inhibitor BIBX1522 (80 mg kg^{-1} day^{-1}), TNFα-neuturalizing antibody, or cyclophosphasmide, Alcian blue-PAS–stained area of epithelium decreased significantly compared with that in animals with plug alone. Values are means ± SE for five rats per group. Significantly different from plug alone: *$p < 0.05$; **$p < 0.01$. *Source*: Ref. 34.

to mucin gene and protein expression. At this point, the cells differentiate first into pregoblet and then into goblet cells.

5. Activation of EGFR may also occur in the absence of ligand binding (''ligand-independent'' EGFR activation), such as by stimulation by oxygen free radicals (''oxidative stress'').

The discovery of this pathway was made more difficult because EGFR are not well expressed in noninflamed airways. Overexpression of EGFR in ''clean'' animals would also not be expected to result in goblet cell hypersecretion in the absence of EGFR activation. EGFR is an interesting molecule: It plays an important role in branching morphogenesis in airways (2). With development, EGFR is normally downregulated. Epidermal growth Factor was discovered by Cohen, who received the Nobel Prize for the discovery (38). This was followed by studies that concentrated on biology of malignant cells and their behavior. Among tumors that express EGFR is lung cancer (non–small-cell cancer or adenocarcinoma). The discovery in my laboratory proved that EGFR expression and activation in normal (noncancer) airways leads to the development of goblet cells containing mucins involving a different process: stimulation of EGFR leads to the growth of goblet cells without evidence of cell multiplication. The results of our studies suggest the following evolution of epithelial cell differentiation leading to goblet cell production:

1. When airways are stimulated with TNFα, basal cells elongate and EGFR expression occurs on the luminal side of the cells.
2. The cells further elongate to become nongranulated secretory cells.
3. Activation of EGFR in nongranulated secretory cells induces the production of mucins. As mucin granules accumulate, the cells become pregoblet and then goblet cells.

If stimulation of epithelial cells leads to goblet cell production under one circumstance and to tumor growth in another circumstance, what are the critical differences? In tumor growth the cyclin pathway is stimulated, leading to cell multiplication. In the case of goblet cell growth, cell multiplication does not appear to be stimulated, but preformed cells differentiate into goblet cells. Studies of mechanisms regulating cell signaling and cell cycle regulation are likely to provide important information concerning this issue.

Role of EGFR Activation in Goblet Cell Metaplasia in Disease

The studies reviewed here show that EGFR expression and activation regulate mucin production in response to a large number of stimuli that are related to different hypersecretory diseases. Thus, allergens, cigarette smoke, mechanical irritation, neutrophilic inflammation, and certain cytokines all appear to depend on EGFR activation for induction of mucin hypersecretion. In each disease state, the precise mechanism of induction of the EGFR cascade appears to vary, but the cascade remains the same.

Role of Neutrophils in EGFR-Mediated Mucus Hypersecretion

The roles of these inflammatory cells is often underestimated. In mucus secretion, the following roles of activated neutrophils are noted:

1. Neutrophil elastase is a potent secretagogue for mucus (39). When neutrophils are recruited into the airways, elastase moves from azurophilic granules to the surface. Contact between neutrophils and goblet cells results in adhesive interactions between the cells, and localized release of elastase at the cell interface, causing mucus secretion.
2. Activated neutrophils also release TNFα, which can stimulate EGFR expression (34).
3. Activated neutrophils release oxygen free radicals, which activate EGFR, inducing mucin expression (17).

Thus, neutrophils can play a major role in the production and secretion of mucins in many hypersecretory diseases.

There may be multiple reasons why investigators have minimized possible neutrophil effects, and here I will use experimental asthma as an example: When an antigen is instilled in an airway of a sensitized animal, recruitment of leukocytes occurs. Neutrophil recruitment occurs early after antigen exposure and may reach a maximum within 1–2 hours and may disappear within a few hours (40).

Because of the short half-life of neutrophils, they may disappear rapidly from airway tissue. On the other hand, eosinophil recruitment occurs more slowly after exposure to antigen, but eosinophils may remain in the airways for weeks. During the short residence of neutrophils in the airway epithelium, the release of TNFα could cause the induction of the EGFR gene. Neutrophils could subsequently disappear from the airway epithelium, but the gene and protein induction would continue, resulting in mucin production hours or days later. Thus, the relationship between the presence of mucus cells and the physical presence of neutrophils may not provide the most important correlative information. Furthermore, neutrophils may provide only one part of the EGFR cascade, and other cells and mediators may also contribute. Another issue is the possible overlap in function of different cells. Thus, both leukocytes and epithelium may provide TNFα for the induction of EGFR (34) in one situation and activated eosinophils could provide TNFα in another pathological state such as nasal polyposis (41).

Studies utilizing knockout of selected genes have provided some insights concerning the roles of specific leukocytes. Thus, Cohn et al. (16) studied allergic mice whose IL-5 gene was absent and showed that allergic stimulation still caused goblet cell metaplasia in the absence of eosinophilia. Further, the authors showed that mast cells are not required for allergen-induced goblet cell metaplasia. The authors commented that "IL-13 and/or IL-4 stimulate mucus production either by direct effects on mucus secreting goblet cells or indirectly through mediators activated locally in the lung." Shim et al. (5) showed that IL-13 causes IL-8 expression in airway epithelial cells, which in turn recruit neutrophils into the airways. Activated neutrophils cause EGFR activation and mucin production (17), an effect that is prevented by EGFR inhibitors. Thus, the effect of IL-13 on airway goblet cell growth appears to be indirect and mediated via neutrophil recruitment and activation.

As mentioned, the accumulation of activated neutrophils occurs in many hypersecretory diseases, and activated neutrophils cause goblet cell metaplasia (17). In COPD, cigarette smoke may be the stimulus for the neutrophil response. In cystic fibrosis, *Pseudomonas aeruginosa* products may be the culprit, and in asthma Th2 cytokines appear to be important. Viral infections could also cause goblet cell production and degranulation via neutrophil recruitment.

Secretory Cell Production and Degranulation: Submucosal Glands Versus Goblet Cells

Mucins in airways are produced by two types of cells: submucosal glands and surface goblet cells. Glands are located selectively in large conducting airways, and they produce mucins constitutively in mucous tubules. The gland duct openings are located adjacent to cough receptors, which probably explains why glandular hypersecretion is associated with cough. Cough causes the shearing of accumulated mucus and is a normal mechanism for clearance in conducting airways.

The second type of mucin-producing cells are goblet cells. These are sparse in healthy individuals, especially in peripheral airways. However, in hypersecretory diseases goblet cell metaplasia occurs both in central and in peripheral airways. Because of the geometry of the airway wall, degranulation of goblet cells in peripheral airways may lead to plugging of the airways. Obstruction in peripheral airways may be "silent" clinically because of the geometry of the peripheral airways (42). Thus, mucus plugging can be advanced before it is detected clinically or by pulmonary function testing.

Of course, peripheral airway plugging by mucins is probably not possible in healthy subjects, because goblet cells are so sparse. However, goblet cell growth can occur rapidly and degranulation can then cause airway plugging. In pathogen-free animals, goblet cell mucin synthesis occurs within 24 hours and may reach a maximum within 2–3 days (34). Thus, peripheral airways could change from normal to totally obstructed (with mucus) in a short time.

The second aspect of mucus hypersecretion is the degranulation process, which occurs within minutes to hours (17). Thus, an individual with intact goblet cells could acquire a viral infection, recruit neutrophils, and develop peripheral airway plugging within hours! The dynamics of goblet cell growth and the relationship of mucin production to degranulation requires further study.

SUMMARY

Airway hypersecretion is an important pathophysiological feature in many chronic airway diseases. Mucins are derived from glands in large conducting airways and from goblet cells, especially in the periphery. At present, there is no effective therapy for hypersecretion, but it is believed to cause clinical deterioration and death in COPD, cystic fibrosis, and acute asthma.

My laboratory has recently discovered a novel cascade (EGFR) that appears to be responsible for mucin production in airways. A wide variety of stimuli, including cigarette smoke, allergens, mechanical irritation, bacteria, and probably viruses, cause mucin production via the EGFR pathway. Neutrophilic inflammation appears to be an important stimulus to activation of the EGFR cascade. Treatment with selective inhibitors of EGFR tyrosine kinase may provide effective therapy in airway hypersecretory diseases. Clinical testing with such drugs in patients with these diseases will determine their effectiveness in eliminating the deleterious effects of hypersecretion.

ACKNOWLEDGMENTS

I thank K. Takeyama for original work on the EGFR cascade, for studies on oxidative stress, and for studies on clinical asthma. I thank H.-M. Lee for ingenious studies with agarose plugs, J. J. Shim and K. Dabbagh for novel studies on Th2 cytokines, and P.-P. Burgel for excellent clinical studies of nasal polyposis, together with E. Escudier, A. Coste, and A. H. Murr. I thank C. Agustí for creative studies of neutrophils and allergen. I acknowledge I. Ueki for her superb knowledge and technical skills in helping to organize all of the studies. I thank K. Takeyama and B. Jung for the cooperative study on cigarette smoke. I also gratefully acknowledge assistance from J. Lausier, K. M. Grattan, T. Dao-Pick, D. C.-W. Tam, U. Protin, and P. Kroschel, who assisted in these studies. I thank Boehringer Ingelheim KG for providing BIBX1522.

REFERENCES

1. K Takeyama, K Dabbagh, H-M Lee, C Agustí, JA Lausier, IF Ueki, KM Grattan, JA Nadel. Epidermal growth factor system regulates mucin production in airways. Proc Natl Acad Sci USA. 96:3081–3086, 1999.
2. S Ruocco, A Lallemand, J Tournier, D Gaillard. Expression and localization of epi-

dermal growth factor, transforming growth factor-alpha, and localization of their common receptor in fetal human lung development. Pediatr Res 39:448–455, 1996.
3. K Takeyama, JV Fahy, JA Nadel. Relationship of epidermal growth factor receptors to airway goblet-cell production. Am J Respir Crit Care Med (in press, 2000).
4. SM Puddicombe, R Polosa, A Richter, MT Krishna, PH Howarth, ST Holgate, DE Davis. Involvement of the epidermal growth factor receptor in epithelial repair in asthma. FASEB J 14:1362–1374, 2000.
5. JJ Shim, K Dabbagh, IF Ueki, T Dao-Pick, P-R Burgel, K Takeyama, D C-W Tam, JA Nadel. Interleukin-13 induces mucin production by stimulating epidermal growth factor receptors and by activating neutrophils. Am J Physiol Lung Cell Mol Physiol 280:L134–L140, 2001.
6. K Takeyama, JV Fahy, JA Nadel. Relationship of epidermal growth factor receptors to goblet cell production. Am J Respir Crit Care Med 163:511–516, 2001.
7. T Aikawa, S Shimura, H Sasaki, M Ebina, T Takishima. Marked goblet cell hyperplasia with mucus accumulation in the airways of patients who died of severe acute asthma attack. Chest 101:916–921, 1992.
8. MA Saetta, A DiStefano, C Rosino, G Thiene, LM Fabbri. Quantitative structural analysis of peripheral airways and arteries in sudden fatal asthma. Am J Respir Dis 143:138–143, 1991.
9. BS Cardell, RSB Pearson. Death in asthmatics. Thorax 14:341–352, 1959.
10. U-A Temann, B Prasad, MW Gallup, C Basbaum, SB Ho, RA Flavell, JA Rankin. A novel role for murine IL-4 in vivo: induction of MUC5AC gene expression and mucin hypersecretion. Am J Respir Cell Mol Biol 16:471–478, 1997.
11. K Dabbagh, K Takeyama, H-M Lee, IF Ueki, JA Lausier, JA Nadel. IL-4 induces mucin gene expression and goblet cell metaplasia in vitro and in vivo. J Immunol 162:6233–6237, 1999.
12. G Grunig, M Warnock, A Wakil, R Venkayya, F Brombacher, DM Rennick, D Sheppard, M Mohrs, DD Donaldson, RM Locksley, DB Corry. Requirement for IL-13 independently of IL-4 in experimental asthma. Science 282:2261–2263, 1999.
13. M Wills-Karp, J Luyimbazi, X Xu, B. Scholfield, TY Neben, CL Karp, DD Donaldson. Interleukin-13: central mediator of allergic asthma. Science 282:2258–2260, 1999.
14. SH Gavett, DJ O'Hearn, CL Karp, EA Patel, BH Schofield, FD Finkelman, M Willis-Karp. Interleukin-4 receptor blockade presents airway responses induced by antigen challenge in mice. Am J Physiol 272:L253–L256, 1997.
15. DI Blyth, MS Pedrick, TJ Savage, EM Hessel, D Fattah. Lung inflammation and epithelial change in murine model of atopic asthma. Am J Respir Cell Mol Biol 14: 425–438, 1996.
16. L Cohn, R Homer, H MacLeod, M Mohrs, F Brombacher, K Bottomly. Th2-induced airway mucus production is dependent on IL-4Ralpha, but not on eosinophils. J Immunol 162:6178–6183, 1999.
17. K Takeyama, K Dabbagh, JJ Shim, T Dao-Pick, IF Ueki, JA Nadel. Oxidative stress causes mucin synthesis via transactivation of epidermal growth factor receptor: role of neutrophils. J Immunol 164:1546–1552, 2000.
18. I Rahman, D Morrison, K Donaldson, W Macnee. Systemic oxidative stress in asthma, COPD, and smokers. Am J Respir Crit Care Med 154:1055–1060, 1996.

19. T Hunter, J Cooper. Epidermal growth factor induces rapid tyrosine phosphorylation of proteins in A431 human tumor cells. Cell 24:741–752, 1981.
20. T Goldkorn, N Balaban, K Matsukuma, V Chea, R Gould, J Last, C Chan, C Chavez. EGF-Receptor phosphorylation and signaling are targeted by H_2O_2 redox stress. Am J Respir Cell Mol Biol 19:786–798, 1998.
21. JE Repine, A Base, I Lankhorst. Oxidative stress in chronic obstructive pulmonary disease: oxidative stress study group. Am J Respir Crit Care Med 156:341–357, 1997
22. V Witko-Sarsat, C Delacourt, D Rabier, J Bardet, AT Nguyen, B Descamps-Latscha. Neutrophil-derived long-lived oxidants in cystic fibrosis sputum. Am J Respir Crit Care Med 152:1910–1916, 1995
23. RK Brown, H Wyatt, JF Price, FJ Kelly. Pulmonary dysfunction in cystic fibrosis is associated with oxidative stress. Eur Respir J 9:334–339, 1996
24. S Loukides, SI Horvath, T Wodehouse, PJ Cole, PJ Barnes. Elevated levels of expired breath hydrogen peroxide in bronchiectasis. Am J Respir Crit Care Med 158: 991–994, 1998
25. M Yamaya, K Sekizawa, T Masuda, M Morikawa, T Sawai, H Sasaki. Oxidants affect permeability and repair of the cultured human tracheal epithelium. Am J Physiol 268:L284–L293, 1995.
26. K Takeyama, B Jung, JJ Shim, P-R Burgel, T Dao-Pick, IF Ueki, U Protin, P Kroschel, JA Nadel. Activation of epidermal growth factor receptors is responsible for mucin synthesis induced by cigarette smoke. Am J Physiol Lung Cell Mol Physiol 280:L165–L172, 2001.
27. SJ Coles, LR Levine, L Reid. Hypersecretion of mucus glycoproteins in rat airways induced by tobacco smoke. Am J Pathol 94:459–471, 1979.
28. D Lamb, L Reid. Goblet cell increase in rat bronchial epithelium after exposure to cigarette and cigar tobacco smoke. Br Med J 1:33–35, 1969.
29. NM Nishikawa, N Kakemizu, T Ito, M Kudo, T Kaneko, M Suzuki, N Udaka, H Ikeda, T Okubo. Superoxide mediates cigarette smoke-induced infiltration of neutrophils into the airways through nuclear factor-kappaB activation and IL-8 mRNA expression in guinea pigs In vivo. Am J Respir Cell Mol Biol 20:189–198, 1999.
30. DF Rogers, PK Jeffery. Inhibition of cigarette smoke-induced airway secretory cell hyperplasia by indomethacin, dexamethasone, prednisolone, or hydrocortisone in the rat. Exp Lung Res 10:285–298, 1986.
31. NO Dulin, A Sorokin, JG Douglas. Arachidonate-induced tyrosine phosphorylation of epidermal growth factor receptor and Shc-Grb2-Sos association. Hypertension 32:1089–1093, 1998.
32. CA Doupnlk, GD Leikauf. Acrolein stimulates eicosanoid release from bovine airway epithelial cells. Am J Physiol 259:L222–L229, 1990.
33. MG Cosio, KA Hale, DE Niewoehner. Morphologic and mophometric effects of prolonged cigarette smoking on the small airways. Am Rev Respir Dis 122:265–321, 1980.
34. H-M Lee, K Takeyama, K Dabbagh, JA Lausier, IF Ueki, JA Nadel. Agarose plug instillation causes goblet cell metaplasia by activating EGF receptors in rat airways. Am J Physiol Lung Cell Mol Physiol 278:L185–L192, 2000.
35. KP Keenan, TS Wilson, EM McDowell. Regeneration of hamster tracheal epithe-

lium after mechanical injury. IV, Histochemical, immunocytochemical and ultrastructural studies. Virchows Arch B Cell Pathol Incl Mol Pathol 43:213–240, 1983.
36. AC Hilding. Regeneration of respiratory epithelium after minimal surface trauma. Ann Otol Rhinol Laryngol 74:903–914, 1965.
37. RB Heath, EP Steffey, JC Thurmon, EM Wertz, DM Meagher, T Hyyppa, GL Van Slyke. Laryngotracheal lesions following routine orotracheal intubation in the horse. Equine Vet J 21:434–437, 1989.
38. S Cohen. Epidermal growth factor. Bioscience Reports 6:1017–1028, 1986.
39. CP Sommerhoff, JA Nadel, CB Basbaum, GH Caughey. Neutrophil elastase and cathepsin G stimulate secretion from cultured bovine airway gland serous cells. J Clin Invest 85:682–689, 1990.
40. C Agustí, K Takeyama, LO Cardell, I Ueki, J Lausier, Y-P Lou, JA Nadel. Goblet cell degranulation after antigen challenge in sensitized guinea pigs: role of neutrophils. Am J Crit Care Med 158:1253–1258, 1998.
41. P-R Burgel, E Escudier, A Coste, T Dao-Pick, IF Ueki, K Takeyama, JJ Shim, AH Murr, JA Nadel. Relation of EGF receptor expression to goblet cell hyperplasia in nasal polyps. J Allergy Clin Immunol 106:705–712, 2000.
42. WM Gold. Pulmonary Function Testing. In: JF Murray, JA Nadel, eds. Textbook of Respiratory Medicine. 3rd ed. Philadelphia: WB Saunders Company, 2000, pp 781–871.

29

In Vivo Measurement of Mucociliary Clearance: The Use of Gamma Camera Imaging

Joy Conway
Southampton General Hospital
Southampton, United Kingdom

INTRODUCTION

The measurement of mucus transport in the lungs can be achieved by a variety of methods, but when there is a need to establish mucociliary clearance in vivo, the most widely used method is the inhaled radioaerosol technique (1–3). This technique, in its most simple form, involves the inhalation of an aerosol labeled with a gamma-emitting isotope and the subsequent imaging of the lungs using one or more gamma cameras. The rate of removal of the radiolabeled aerosol can then be determined by serial gamma camera images (Fig. 1).

THE RADIOLABELED AEROSOL

The most common isotope used for this type of work is technetium (^{99m}Tc), which with a half-life of 6 hours is ideal for medical imaging. The technetium does, however, need to be linked to a larger molecular structure to ensure that it stays within the lung and is not rapidly absorbed across the respiratory epithelium. The

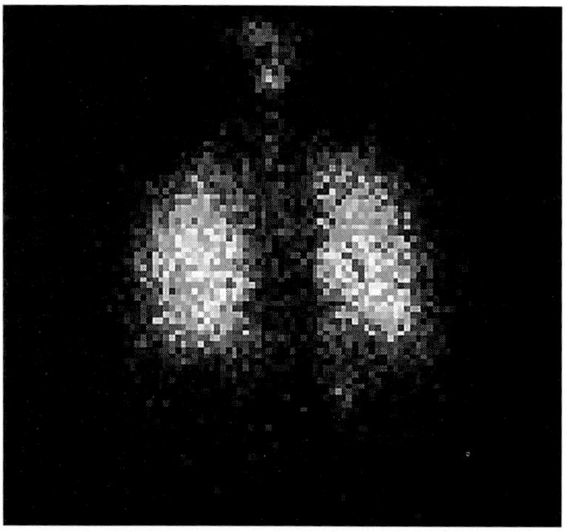

Figure 1 Example of a gamma camera image of an inhaled radio-aerosol. In this case the example is that of an adult male with mild asthma who has inhaled a technetium-labeled nebulized aerosol. This anterior view of the thorax shows deposition in both lungs with some deposition in the mouth.

most popular ''mixes'' are technetium associated with human serum albumin (4,5), iron oxide (Fe_2O_3) (6), and sulfur colloid (7). All of these radioaerosols satisfy the criteria of being safe to inhale and are relatively inert in that they will not be transported across the respiratory epithelium or degrade in any way during the imaging period.

The methods of delivery for these aerosols vary. Human serum albumin and sulfur colloid can be administered as a nebulized suspension. Iron oxide particles are produced by a spinning disc generator, which uses centrifugal force to create a spray of droplets from a liquid feed of iron oxide and solvents (8). Whatever the delivery system, the resulting aerosol must contain particles small enough to penetrate through to the periphery of the lung. Radioaerosol needs to be deposited on all of the ciliated airway generations. The radioaerosol cloud should ideally be made up of particles less than 5 μm to achieve this (4).

All radioaerosols must demonstrate that the technetium is tightly bound to the inert, carrier particle, otherwise losses due to absorption across the epithelium will be attributed to losses due to mucociliary clearance. Dissociation of technetium can be assessed both in vitro and in vivo via simple bench studies such as paper chromatography and via the immediate presence of activity circulating in the heart and the appearance of activity in the urine 24 hours after inhalation (8).

RADIATION DOSIMETRY

The 6-hour half-life of technetium is ideal in that it is short enough to minimize retention of radioactivity but long enough to conduct clinical studies. A starting lung activity of 5–10 MBq is sufficient to acquire good quality images for several hours. This level of activity equates to an effective dose of less than 0.1 mSv.

IMAGING MUCOCILIARY CLEARANCE USING A GAMMA CAMERA

This method views the lungs two-dimensionally via the use of static gamma camera(s). An anterior or posterior view can be used but a geometric mean of both views is preferred so as to reduce the effect of attenuation by body tissues on the count rate. All images need to be corrected for background activity and decay over time. A typical acquisition time of 30 seconds would be used for the initial image.

Once an image such as the one in Figure 1 has been obtained, the lungs can be identified and the activity within each lung noted. The lung activity on subsequent images can then be expressed as a percentage of the baseline image.

As decay progresses it may be necessary to lengthen the acquisition period to acquire a good quality image. Some studies will require clearance to be measured over 24 hours. For these studies, data will need to be acquired after several half-lives have passed. Data can then be collected by placing the subject between two gamma cameras, or in a whole body counter, and activity in the thorax counted over a longer acquisition period with no image obtained.

The quantitative data obtained will be in the form of a percentage of activity remaining in the lungs at different time points. Clearance over time can then be plotted and analyzed by describing the area under the curve or other indices, such as the time taken to clear 50% of activity or the percentage of activity remaining at a certain time point such as 24 hours (6,9,10).

TOTAL VERSUS REGIONAL DEPOSITION AND CLEARANCE

For many objectives the measurement of total lung clearance is all that is required. However, when assessing the effects of inhaled active agents or disease on mucociliary clearance, it may be considered an advantage to have information on regional clearance as well as total. Currently there are two ways of achieving this. The first is by ascribing regions of interest to a gamma camera image to represent "central" and "peripheral" airways. The second is by the measurement of 24-hour retention/clearance of the radioaerosol and extrapolating that data to clearance from the ciliated airways as opposed to the alveolar compartment.

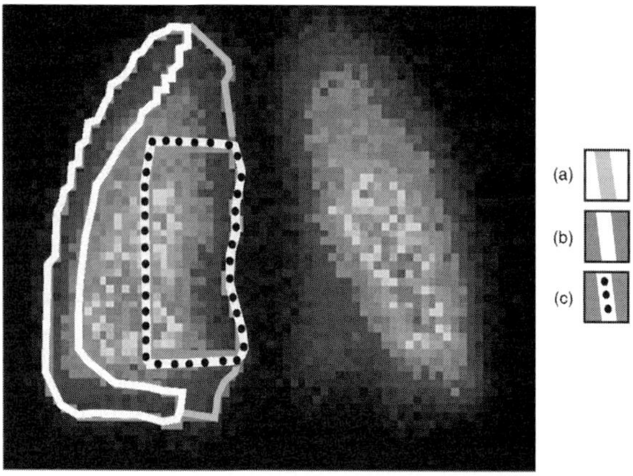

Figure 2 The penetration index. The regions of interest are superimposed on the gamma camera image. The PI is the ratio of counts in the peripheral region to counts in the central region. (A) The total lung zone. (B) The peripheral region. (C) The central region. (From Ref. 4.)

Regional distribution of an inhaled, radio-aerosol is expressed by evaluating "regions of interest" on the gamma camera image chosen to represent the central (C), and peripheral (P), regions of the lung and expressing deposition as a ratio of counts between the two regions. This is commonly referred to as the penetration index (PI) (Fig. 2). The definition of the regions varies between centers and can be seen to be only a loose approximation to anatomy. The central region, for example, is a composite of structures. While the PI has its limitations, it does give some information on regional clearance.

The principle behind 24-hour retention/clearance analysis is that all particles on the mucociliary escalator will be cleared within 24 hours. Any activity left in the lungs after this period must therefore be in the alveolar compartment. This principle has been questioned, with some centers suggesting that particles are retained in the conducting airways after 24 hours (6,11).

THREE-DIMENSIONAL GAMMA CAMERA IMAGING: SINGLE PHOTON EMISSION COMPUTED TOMOGRAPHY

The main disadvantage of gamma camera imaging presented so far is that the lung(s) are viewed two-dimensionally and so may be limited in the amount of spatial information it can provide. The application of novel three-dimensional imaging techniques, such as the use of single photon emission computed tomography (SPECT), holds the promise of greater sensitivity in detecting patterns of deposition in the lung. SPECT gives three-dimensional information on the distribution of a gamma-ray–emitting nuclide. In a similar manner to the two-dimensional studies described above, a radiolabeled aerosol is inhaled and the subject is imaged, but with SPECT this is by a gamma camera(s) that rotates around the patient, acquiring 360° of data. The reconstructed images can then be viewed tomographically in coronal, sagittal, and transverse views (Fig. 3).

SPECT data can be aligned to computed tomography (CT) or magnetic resonance image (MRI) data to provide improved anatomical localization of radionuclide distribution and information on lung volume. In addition, the thoracic CT/MRI data can be converted to represent a map of attenuation of gamma emission by the intrathoracic structures (4,12). Thus the corrected SPECT images represent the true distribution of radionuclide. A SPECT study can now be acquired over a period of 5 minutes (13). This methodology can be used to investigate regional, intrapulmonary clearance with greater sensitivity than the existing two-dimensional gamma camera techniques. The reconstructed three-

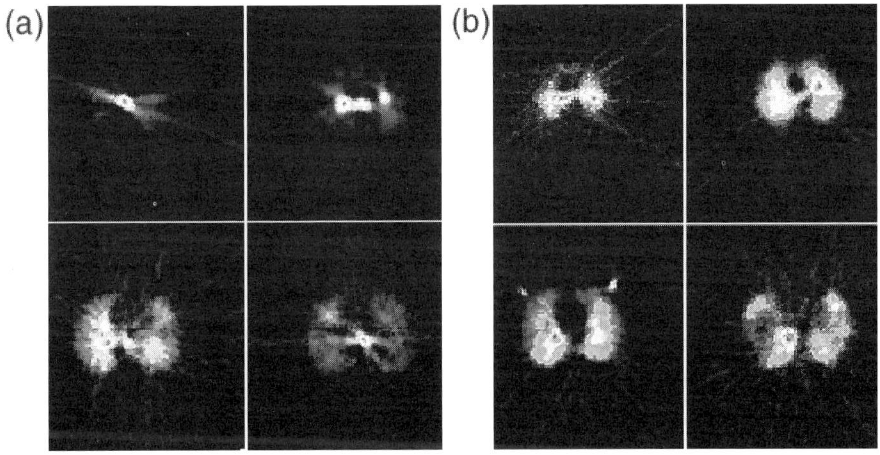

Figure 3 Transverse, tomographic views of SPECT data. The images are of a moderately affected asthmatic after inhaling nebulized Tc-HSA from two different nebulizers. (a) The four images on the left are following the use of a nebulizer, which produces relatively large particles thus tending to deposit more centrally. (b) The four images on the right are from the same subject under identical conditions, except the nebulizer used produces much smaller particles, thus tending to deposit more peripherally. Only four transverse views of a possible 42 have been chosen. Of the four views presented in (a) and (b), the top left view has been taken from high in the thorax, just above the bifurcation of the trachea. The images are then taken from progressively lower points in the thorax. The four images on the left can be directly compared to those on the right.

dimensional data sets can be analyzed by a variety of means, including the use of a mathematical model to describe activity per airway generation (14).

SUMMARY

Mucociliary clearance can be measured in vivo using an inhaled, radioaerosol technique. SPECT may offer some advantages over two-dimensional imaging when regional clearance is the issue. SPECT may help to quantify deposition and clearance in the conducting zone and in the alveoli and so may be useful to investigate the concepts of 24-hour clearance and specific, regional clearance effects of an active agent/intervention.

REFERENCES

1. Dolovich M, Rushbrook J, Churchill E, Mazza M, Powles AC. Effect of continuous lateral rotational therapy on lung mucus transport in mechanically ventilated patients. J Crit Care 1998; 13(3):119–125.
2. Regnis JA, Robinson M, Bailey DL, Cook P, Hooper P, Chan HK, Gonda I, Bautovich G, Bye PT. Mucocilairy clearance in patients with cystic fibrosis and in normal subjects. Am J Respir Crit Care Med 1994; 150(1):66–71.
3. Noone PG, Bennett WD, Regnis JA, Zeman KL, Carson JL, King M, Boucher RC, Knowles MR. Effect of aerosolised uridine-5-triphospate on airway clearance with cough on patients with primary ciliary dyskinesia. Am J Respir Crit Care Med 1999; 160(1):144–149.
4. Fleming JS, Halson P, Conway J, Moore E, Nassim M, Hashish AH, Bailey AG, Holgate ST, Martonen TB. Three-dimensional description of pulmonary deposition of inhaled aerosol using data from multi-modality imaging. J Nucl Med 1996; 37: 873–877.
5. Imai T, Sasaki Y, Ohishi H, Uchida H, Ito S, Mikasa K, Sawaki M, Narita N. Clinical aerosol inhalation cine-scintigraphy to evaluate mucociliary transport system in diffuse panbronchioloitis. J Nucl Med 1995; 36(8):1355–1362.
6. Regnis JA, Zeman KL, Noone PG, Knowles MR, Bennett WD. Prolonged airway retention of insoluble particles in cystic fibrosis versus primary ciliary dyskinesia. Exp Lung Res 2000; 26(3):149–162.
7. Messina MS, O'Riordan TG, Smaldone GC. Changes in mucociliary clearance during acute exacerbations of asthma. Am Rev Repsir Dis 1991; 143(5 Pt 1):993–997.
8. Newman S. Production of radioaerosols. In: Clarke SW, Pavia D, eds. In: Aerosols and the Lung. London: Butterworth and Co. Ltd, 1984:71–91.
9. Saari SM, Vidgren MT, Koskinen MO, Turjanmaa VA, Waldrep JC, Nieminen MM. Regional lung deposition and clearance of 99mTc-labeled beclamethasone-DLPC liposomes in mild and severe asthma. Chest 1998; 113(6):1573–1579.
10. Robinson M, Hemming AL, Regnis JA, Wong AG, Bailey DL, Bautovich GJ, King M, Bye PT. Effect of increasing doses of hypertonic saline on mucociliary clearance in patients with cystic fibrosis. Thorax 1997; 52(10):900–903.

11. Gehr P, Schürch S, Berthiaume Y, Im Hof V, Geiser M. Particle retention in airways by surfactant. J Aerosol Med 1990; 3(1):27–43.
12. Fleming JS. A technique for using CT images in attenuation correction and quantification in SPECT. Nuclear Med Commuun 1989; 10; 83–97.
13. Conway JH, Walker P, Fleming JS, Bondesson E, Borgström L, Holgate ST. Three-dimensional description of the deposition of inhaled terbutaline sulphate administered via the Turbuhaler®. Respir Drug Deliv 2000; 7:607–609.
14. Fleming JS, Hashish AH, Conway JH, Nassim MA, Holgate ST, Halson P, Moore E, Bailey AG, Martonen TB. Assessment of deposition of inhaled aerosol in the respiratory tract of man using three-dimensional multimodality imaging and mathematical modelling. J Aerosol Med 1996; 9:317–327.

30

Regulation of Mucociliary Clearance by Purinergic Receptors

William D. Bennett, Peadar G. Noone, Michael Knowles, and Richard C. Boucher
University of North Carolina at Chapel Hill
Chapel Hill, North Carolina

INTRODUCTION

The P2Y family of receptors comprises seven transmembrane G-protein coupled receptors, for which the naturally occurring ligands appear to be purine and pyrimidine nucleotides (1). The $P2Y_2$ receptor is found on the apical surface of a variety of airway epithelia, including the ciliated epithelial cells and goblet cells of the trachea, bronchi, and bronchioles. Recent in vitro data indicate that extracellular adenosine triphosphate (ATP) and uridine triphosphate (UTP) nucleotides interact with $P2Y_2$ receptors to initiate Cl^- secretion on the apical membrane of human airway epithelia (2,3), principally via calcium-regulated Cl^- channels (4). Benali et al. (3) showed that this enhanced Cl^- secretion resulted in increased hydration of the airway surface with the application of equimolar concentrations of ATP or UTP. ATP has also been shown to stimulate ciliary beat frequency (CBF) via release of intracellular calcium stores both in vitro (5,6) and in vivo (7). Furthermore, Geary et al. (6) showed that the increase in CBF with ATP was twice that observed by equimolar treatment with isoproterenol. Finally, extracellular triphosphate nucleotides (ATP and UTP) have been shown to stimulate goblet cell degranulation and mucin release onto apical surfaces of both normal and cystic fibrosis epithelial explants (4).

Impaired mucociliary clearance may contribute to the pathophysiology of several airway diseases such as cystic fibrosis and chronic bronchitis. These patients have difficulty clearing secretions from their airways, resulting in chronic obstructive changes and repeated pulmonary infections. Similar problems occur in patients with primary ciliary dyskinesia (PCD), a genetic disease characterized by an ultrastructural defect in dynein arms of cilia, which manifests itself as abnormal or absent ciliary movement and thus, impaired mucociliary clearance. Patients with PCD rely on effective cough clearance to clear their airways of secretions (8). The integrated effects of ATP or UTP on the airway epithelium described above may facilitate mucociliary or cough clearance in patients with airways disease. To test the effects of UTP on mucociliary and cough clearance in humans, a series of studies was performed in normal and compromised subjects. Uridine 5′-triphosphate (Prolucin, Inspire Pharmaceuticals, Inc.) was chosen rather than ATP for the human studies because of bronchoconstriction observed with inhalation of adenosine and adenosine 5′-monophosphate (AMP) in patients with asthma (9,10) and chronic bronchitis (11). In addition, intravenous infusions of ATP cause a dose-dependent dilation of pulmonary and systemic vasculature in humans (12), which could potentially lead to systemic hypotension. Studies in anesthetized dogs suggested no adverse effects of aerosolized UTP on airway mechanics, gas exchange, or hemodynamics, whereas infusion of ATP caused hypotension not seen with comparable doses of UTP (13).

Mucociliary clearance rates can be measured in humans by assuming that a nonpermeating, inhaled marker depositing on the airway surface moves out of the lung at the same rate as the airway secretions in which it is immersed. The most common technique is to use inhaled, radiolabeled particles, aqueous or dry, that upon deposition in the lung can be followed by gamma camera or scintillation detectors to determine their rate of egress from the lung. Technetium 99m (^{99m}Tc), generally the radiolabel of choice, may be bound to such impermeable markers as iron oxide, sulfur colloid, red blood cells, albumin, teflon, and polystyrene latex spheres. The aerosol may be generated by a number of means (e.g. nebulizers, spinning top, condensation) for inhalation by the subject. To assess the effects of UTP on mucociliary clearance, we used radiolabeled iron oxide particles (4–5 μm MMAD) generated by a spinning top (14,15). After inhalation of these markers, retention of activity in the lung (as a percent of initial deposition) is monitored as a function of time over a period of 1–6 hours to determine clearance rates (Fig. 1). Whole lung retention is strongly dependent on the sites of particle deposition within the lung and may be confounded by spontaneous coughing episodes during monitoring (15). In the first case, it is important to control particle size characteristics and breathing pattern of the aerosol to repro-

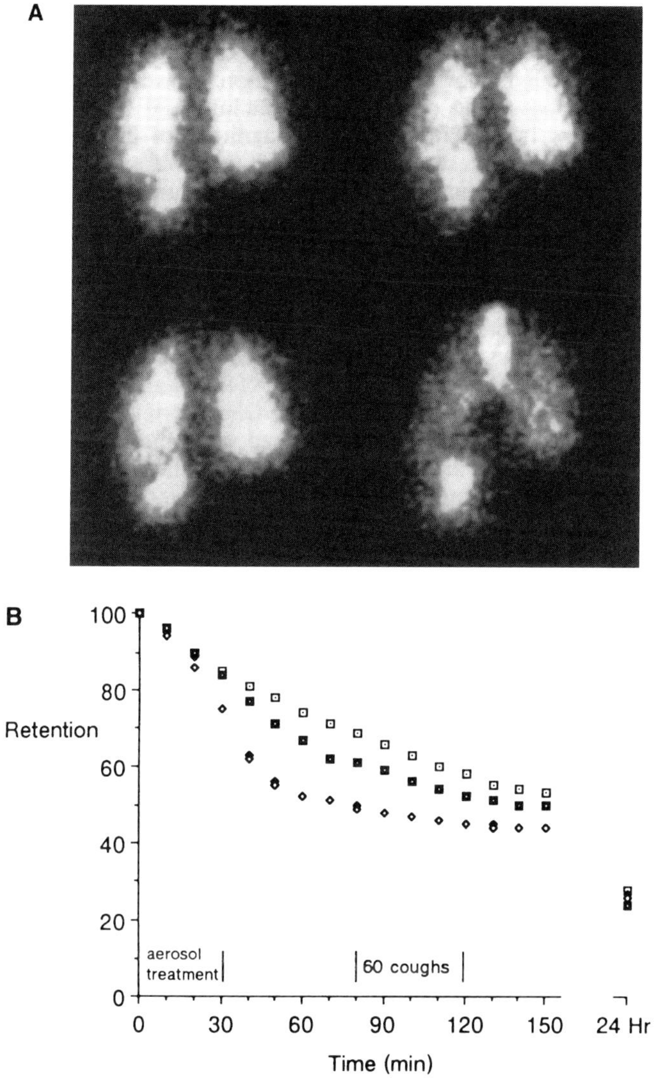

FIGURE 1 (A) Posterior gamma camera images at baseline (left) and 15 minute after inhalation (right) of vehicle (top) and UTP/amiloride (bottom) for a healthy, nonsmoking subject. On the UTP/amiloride study day, a dramatic and rapid movement of particles from the lungs into the trachea has occurred by 15 minutes postinhalation. (B) Whole lung clearance in healthy, nonsmoking subjects illustrated as percent retention vs. time for vehicle (0.12% saline) (open squares), amiloride (closed squares), UTP (open diamonds), and UTP/amiloride (closed diamonds). (From Ref. 14.)

duce deposition patterns as similarly as possible within and between patients. Second, while patients may be advised to avoid coughing during the measurements, spontaneous coughing will inevitably occur. To quantify episodes of spontaneous coughing, the number of coughs should be recorded during the clearance measurements. Then, in the study design the number of spontaneous coughs can be considered as a covariate in the analysis to insure that frequency of these coughs does not overly influence any differences due to therapy effects. Finally, while uncontrolled coughing can confound measures of mucociliary clearance, measuring cough clearance by incorporating controlled, voluntary coughs during the measures of particle clearance may provide a more sensitive indicator of rheological changes in airway secretions (16).

EFFECT OF UTP ON MUCOCILIARY CLEARANCE IN THE NORMAL LUNG

Before testing the effects of aerosolized UTP in patients, we first tested its safety and effects on mucociliary clearance in healthy, nonsmoking adults (14). To achieve optimal clearance in cystic fibrosis via UTP's action on Cl^- secretion it may also be necessary to block the excess Na^+ absorption in these patients (17). Therefore, we tested the effects of UTP alone and in combination with amiloride, an epithelial Na^+ channel blocker, in the healthy subjects as well. The mucociliary clearance protocol design was a double-blind crossover study in 12 healthy male subjects with four aerosol treatment arms: aerosolized saline (0.12%) (vehicle), UTP (10^{-2} M), amiloride (1.3×10^{-3} M), or UTP/amiloride combined. Following inhalation of the ^{99m}Tc-labeled iron oxide particles and an initial deposition scan (two 2 minutes scans) by gamma camera, subjects inhaled the aerosolized treatment (3.5 mL in a Devilbiss 646 jet nebulizer) over a period of 15–20 minutes. Lung scans (2 minutes each) were sequentially acquired over a period of 2 hours, including the treatment period (Fig 1B). During the period of 80–120 minutes, four sets of 15 standardized coughs (16) were interspersed to assess the effect of cough clearance. Subjects returned 24 hours after inhalation of the radiolabeled particles for a half-hour scan to index the slow cleared (alveolar) retention of particles.

The results showed that UTP alone and in combination with amiloride stimulated mucociliary clearance in the healthy subjects (14). Figure 1A shows gamma camera images from a subject before and 15 minutes after aerosolized vehicle or UTP/amiloride illustrating the latter's dramatic effect on particle clearance. Clearance rates through 50 minutes (linear slope through the break point in the biphasic curve: Fig. 1B) for the whole lung were enhanced from 0.47%/min with vehicle to 1.14 and 1.01%/min with UTP/amiloride and UTP, respectively, in normal subjects. On the one hand, the similar enhancement with UTP/amiloride and UTP alone might suggest that the UTP effects on ciliary beating and goblet cell degranulation were more important than any hydration effects for enhancing mucociliary clearance in the normal lung. Interestingly, however, we did see a mild but significant enhancement of clearance with amiloride (to 0.63%/min) in these healthy subjects. Our data mirror that of Benali et al. (Fig 2) (3), who showed that fluid absorption (apical to basal) in cultured nasal epithelial cells was inhibited by amiloride but had no effect in the presence of ATP or UTP. These authors suggested that ATP or UTP may stimulate the same amount of fluid secretion in the presence or absence of amiloride by either (1) ATP/UTP stimulation of calcium-dependent potassium channels that may be enhanced in the absence of amiloride (18,19), or (2) ATP/UTP inhibition of sodium absorption (20). The possibility that increasing periciliary fluid volume on the normal airway surface can enhance mucociliary clearance is supported by recent studies in patients with pseudohypoaldosteronism (PHA) (21), a rare inherited disease characterized by poor Na^+ and volume absortion from the airway surface. These

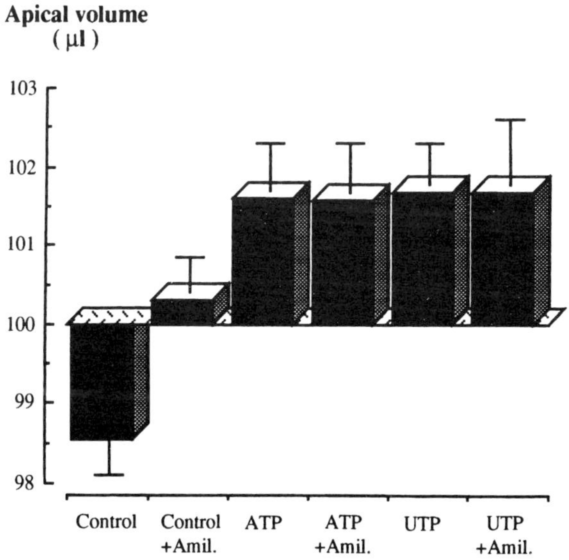

Figure 2 Effect of ATP and UTP on fluid transport in human surface respiratory epithelial cells in culture. At basal state (control) cells absorbed fluid. Pretreatment with amiloride significantly blocked the fluid absorption by cells. The addition of ATP or UTP induced a significant increase in apical volume. In the presence of amiloride, the effects of ATP and UTP were not significantly different when compared with those obtained for ATP and UTP alone. (From Ref. 3.)

patients' airways are ''overhydrated,'' appearing very wet by bronchoscopy. The baseline mucociliary clearance in three adult patients was found to be ''supernormal,'' four times the average normal rate during the first 20 minutes of measurement. The mechanism by which an increased periciliary fluid layer results in enhanced mucociliary clearance is unclear. The efficiency of mucociliary clearance is thought to be dependent on contact of ciliary tips with the inner surface of a layer of mucus. However, the enhanced mucociliary clearance observed in PHA suggests that an increase in the volume of airway surface fluid does not ''float'' mucus off the tips of cilia, slowing mucociliary clearance. On the contrary, it seems likely that the increased fluid facilitates the swelling of mucus, leading to the formation of a highly hydrated mucous blanket with optimal physical properties for an efficient mucociliary transport (3). Thus, it is possible that the enhanced clearance observed with UTP and UTP/amiloride in normal subjects (Fig. 1B) is due to hydration of the normal airway surface in addition to increased mucus and ciliary beating effects.

No enhancement of cough clearance was seen during the 80- to 120-minute period of observation with either UTP or UTP/amiloride. This was likely due to the loss of radiolabeled tracer from the airways by 80 minutes postdeposition (Fig. 1B) following treatment with UTP or UTP/amiloride. It is unlikely, however, that in the normal lung UTP would significantly affect cough clearance, as cough plays a lesser role than mucociliary clearance in the transport of secretions from the normal lung (22). All treatments produced some degree of spontaneous coughing in this study, but the number of coughs was not different between the treatment groups and was not predictive of clearance during the 0 to 50-minute interval where clearance was greatly enhanced by UTP and UTP/amiloride (Fig. 1B).

Immediately following the single treatment of UTP alone (10^{-2} M), there was a small but statistically significant decrement in FEV_1 and maximal midexpiratory flow (MMEF) (14). A small decrease in $Pa{O_2}$ was also noted in subjects immediately following treatment with UTP/amiloride. These changes were clinically insignificant but likely reflected the effect of liquid movement into the airways. Once most of the increased liquid cleared the bronchi (within 15 minutes following treatment; Fig. 1) these small decrements in lung function returned to normal.

Aerosolized UTP was shown to cause a marked increase in mucociliary transport in our normal subjects. It is also possible that endogenous ATP acts in vivo to modulate or accentuate all aspects of mucociliary function, stimulation of CBF, mucus production, and Cl^- and water secretion onto the airway surfaces. Basal levels of ATP on airway surfaces have been found to be in the submicromolar range (slightly less than the ED_{50} for activation of $P2Y_2$ receptors) in both normal subjects and cystic fibrosis (CF) patients (23). The autocrine (or paracrine) secretion of ATP by airway cells may represent one component of mucociliary transport regulation.

EFFECT OF UTP ON MUCOCILIARY CLEARANCE IN CYSTIC FIBROSIS

Cystic fibrosis is a genetic disease characterized by abnormal electrolyte transport by airway epithelia (17,24,25). Reduced Cl^- secretion (25) and excessive Na^+ absorption (17) result in CF patients having thick, viscous airway secretions that are poorly cleared, obstruct airways, and predispose to chronic bacterial infection. In a series of experiments comparing normal and CF airway epithelial cultures in vitro, Matsui et al. (26) showed that CF cultures hyperabsorbed fluid volume from their surface, resulting in a reduced height of airway surface liquid compared to normal. Furthermore, consistent with reduced rates of mucociliary clearance in CF (27), the mucus velocity was reduced to near zero in the CF cultures. Thus, treatment with extracellular nucleotides, alone or in combination with Na^+ channel blockers, aimed at modulating electrolyte and water transport across CF airway surfaces (28) may prove beneficial to CF patients.

Using the same protocol to that described previously for normal subjects, we studied the acute effect of aerosolized UTP alone and in combination with amiloride on mucociliary clearance in adult CF patients with mild-moderate disease (mean FEV_1 (% pred) = 64 ± 2 (se)) (15). In this latter study we focused on clearance from the peripheral regions of the lung as illustrated in Figure 3A. By doing so we were able to minimize effects of mismatched deposition patterns between the CF patients and the healthy subjects, since we eliminated the central region, and spontaneous coughing, since cough is most effective in large, central airways (29). We found that peripheral lung clearance in adult CF patients was about 1/2 of the normal rate (Fig. 3B). This is consistent with the initial pulmonary pathophysiology in CF occurring in the small, peripheral conducting airways (30). Aerosolized UTP plus amiloride produced an acute increase in clearance (0–40 min) from the peripheral lung region in CF (Fig. 3B) to a rate that was similar to the normal baseline clearance rate (i.e., with vehicle). The effect of UTP/amiloride seemed to wane after 40 minutes, i.e., rate not significantly increased through 80 minutes (Fig. 3B). This may reflect either a short duration of action of the administered dose of UTP/amiloride or a general loss of radiolabeled tracer from the airways most affected by the treatment. It may be that the effect of UTP/amiloride on mucociliary clearance is most dramatic in intermediate-size airways—those that clear their deposited particles most rapidly and earliest (i.e. 0–40 min) from the analyzed peripheral region. Unlike the study in healthy subjects (14), however, we found in adult CF patients that only when aerosolized UTP was combined with amiloride was there any improvement in mucociliary clearance. Neither UTP nor amiloride alone appeared to affect peripheral clearance rates (data not shown). This suggests that the mechanism for enhanced clearance by UTP/amiloride in CF may require the synergistic effects of these two compounds on airway surfaces. The combined action of UTP plus amiloride prob-

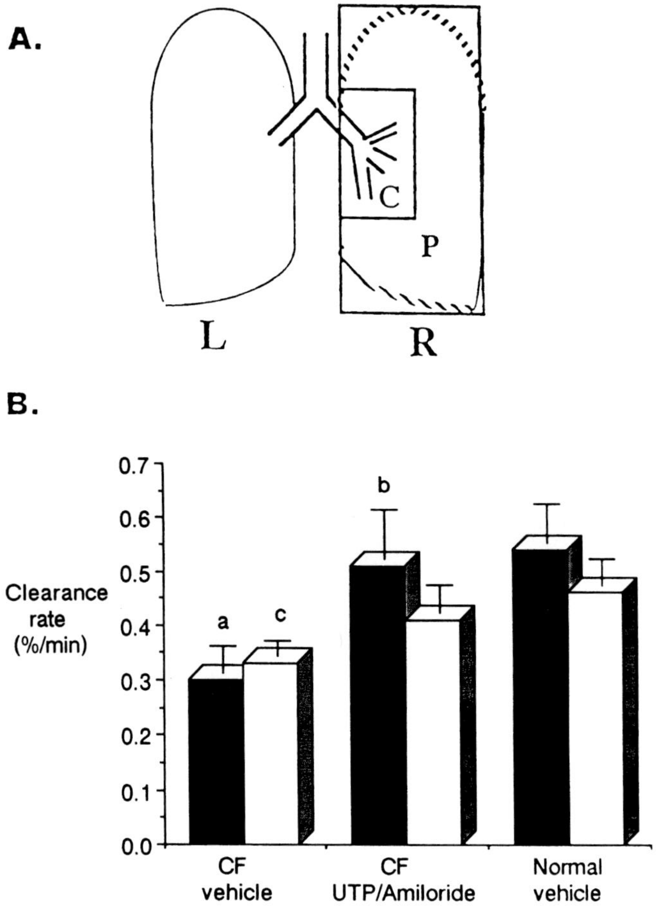

FIGURE 3 (A) Region of interest (ROI) analysis for retention of particles in peripheral (P) region of the lung. The central region includes very large airways (main stem and lobar bronchi). Outline regions are defined from a Xenon 133 equilibrium scan. (B) Mean peripheral clearance rates (%/min) through 40 minutes (closed columns) and 80 minutes (open columns) in adult patients with CF ($n = 14$) for vehicle (0.12% saline) and UTP/amiloride compared with that in normal subjects ($n = 12$). a: Significantly less than normal vehicle ($p = 0.01$); b: significantly greater than CF vehicle ($p = 0.04$); c: significantly less than normal vehicle ($p = 0.03$). (From ref. 15.)

ably reflects an alteration in the rheology of airway surface liquid, through actions on airway epithelial Cl^- and Na^+ transport, which allows for improved clearance by UTP-stimulated ciliary action. It may be that higher doses of UTP or a longer-acting agonist (see below) can be effective alone (i.e., without a Na^+ channel blocker) in CF or in patients with milder airways disease.

The efficacy of extracellular nucleotides in the CF lung may be compromised by their rapid enzymatic degradation in CF sputum. Inspire Pharmaceuticals, Inc., which holds the patent rights for UTP, has recently developed an analog to UTP, INS365, a stabilized uridine nucleotide that produces similar actions as UTP via activation of $P2Y_2$ receptors. A recent study completed in sheep showed significant enhancement of tracheal mucus velocity and whole lung mucociliary clearance (31) with INS365 that was comparable to similar doses of UTP. An investigation of the metabolism of INS365 in chronically infected human airway secretions showed that INS365 was more stable than UTP, with 58% vs. 99% of INS365 vs. UTP, respectively, metabolized after 60 minutes of incubation (M. J. Stutts, personal communication, 1999). Thus, the action of INS365 in the airways of patients with CF may be more prolonged and efficacious than UTP was found to be (15). Current protocols are ongoing to assess the safety and efficacy of INS365 in healthy nonsmokers and smokers, as well as in adult and pediatric CF patients.

Similar to results found in healthy subjects, there was a small but statistically significant decrement in FEV_1, maximal midexpiratory flow (MMEF), and Pa_{O_2} in the CF patients immediately following inhalation of UTP/amiloride (15). Again, these changes were clinically insignificant but likely reflected the effect of liquid movement into the airways.

EFFECT OF UTP ON COUGH CLEARANCE IN PRIMARY CILIARY DYSKINESIA

While the studies described above focus on actions of these agents on mucociliary clearance, CF patients also need effective cough clearance (32) to supplement their abnormal mucociliary clearance. But recent composition studies of CF mucus suggest that cough clearance may be especially compromised in CF compared to other hypersecretory lung disorders (33). Primary ciliary dyskinesia (PCD) is another pulmonary disorder manifested by abnormal mucociliary clearance (34), in this case due to immotile cilia that do not beat in a coordinated fashion and thus are ineffective at propelling mucus out of the lung. As a result it is believed that these patients rely almost entirely on effective cough clearance for removing secretions from their lungs. Interestingly, these patients generally fare better clinically (i.e., lower infection rates and slower decline in lung function) than CF patients (35). Since triphosphate nucleotides are unlikely to stimulate the defective cilia in PCD, these patients provided a genetic model for testing whether

UTP might affect nonciliary aspects (i.e., ion transport, mucins) of clearance during cough. In these patients we tested the effects of UTP on the rate of clearance during cough using predetermined periods of ''controlled'' (directed) coughing by the subjects while monitoring the clearance of the radiolabeled iron oxide particles (36). We studied 12 patients with PCD with mild airways obstruction (mean FEV_1 = 65% predicted). As with previous studies described above, the design was a double-blind crossover conducted on 2 separate days: a UTP treatment day (5 mg/mL, 3.5 mL in Devilbiss 646 nebulizer) and a vehicle treatment day (0.12% saline). Some patients spontaneously coughed during and immediately following either of both treatments. These coughs were counted and the total number of coughs (spontaneous plus directed, controlled) were limited to 90 during the first 60 minutes of gamma camera counting. Controlled coughs were performed by having each patient cough under the direction of the investigators into a spirometer. Peak flow rates for each controlled cough were recorded. In 8 of the 12 subjects we obtained sputum during the cough maneuvers as soon after the aerosolized treatment as possible to measure sputum weights, ion content, and rheology.

We found that aerosolized UTP enhanced cough clearance by about 50% compared to vehicle in the PCD patients (Fig. 4) (36). This effect was only seen

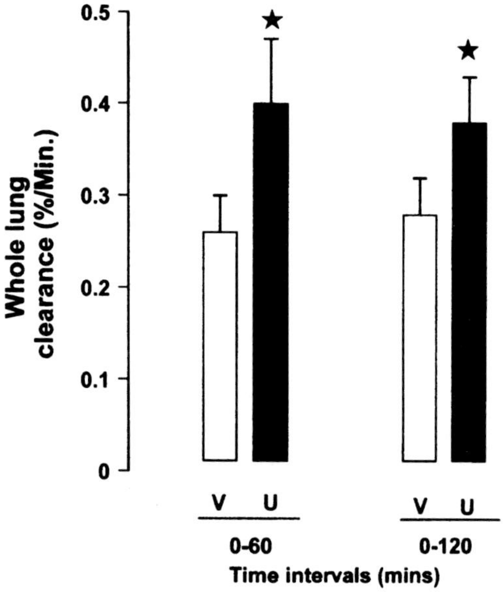

FIGURE 4 Whole lung clearance of radiolabeled particles (% of isotope cleared per minute ± SE) for the periods t = 0–60 min and t = 0–120 min after inhalation of UTP (U) and vehicle (V). *Different from vehicle $p < 0.05$. (From ref. 36.)

in analysis of whole lung clearance, but not from the peripheral region of the lung (Fig. 3A). This was likely due to the preponderance of radiolabeled particle deposition in the central region in these patients and the fact that cough clearance is most effective in the more central, proximal bronchi (29). In those patients for which sputum was collected, there were no differences in ion composition or rheology of sputum between the vehicle and UTP treatment days. Tonicity of the sputum may not be expected to change if water permeability is sufficient for the osmolarity to equilibrate with that of the isotonic submucosal fluid. The rheological measurements were within normal values. The fact that they were unaffected by UTP treatment suggests that the balance between increased water and mucins on the airway surface was maintained. Since rheology was unchanged cough clearance was likely enhanced by an overall increase in the depth of airway surface fluid (29,37) in these patients. The potential for UTP to enhance cough clearance in CF like that observed in PCD has not been investigated, but its potential for doing so may be as important as the effect it had on mucociliary clearance described above (Fig. 3B).

SUMMARY

Triphosphate nucleotides mediate effects on mucociliary clearance via $P2Y_2$ receptors on the apical surfaces of airway epithelial cells. Endogenous ATP, in fact, may serve an autocrine function to help regulate mucociliary clearance. Demonstration of mucociliary and cough clearance changes following aerosolized UTP suggest that this drug or its analogs may correct abnormalities in the mucociliary apparatus in patients with airways disease, leading to long-term improvement in their lung function. It also may be that the effects on cough clearance from these drugs are a more important and sensitive indicator of potential clinical benefit. Studies with more stable analogs of UTP (e.g., INS365) developed by Inspire Pharmaceuticals, Inc. may show greater enhancement of mucociliary/cough clearance in CF. Preliminary studies in mild chronic bronchitis subjects show tolerability and efficacy for enhanced clearance in mild chronic bronchitis with both UTP and INS365. The ability of P2Y2 agonists to enhance mucociliary or cough clearance, as measured by the techniques described above, suggests that such agents could be useful in the treatment of patients with obstructive pulmonary diseases. In contrast to the single dose studies summarized here, the therapeutic utility of these compounds will depend on their ability to improve lung function after repeated treatments over period of time.

ACKNOWLEDGMENTS

We are indebted to Inspire Pharmaceuticals, Inc., for their support of the studies in PCD patients and their review of this manuscript—Kirby Zeman, Kenneth

Olivier, Jeff Regnis, Johnny Carson, and Malcolm King—and to Kathy Hohneker for assisting in some of the studies.

REFERENCES

1. HA Brown, ER Lazarowcki, RC Boucher, TK Harden. Evidence that UTP and ATP regulated phospholipase C through a common extracellular 5′-nucleotide receptor in human airway epithelial cells. Mol Pharmacol 40:648–655, 1991.
2. SJ Mason, AM Paradise, RC Boucher. Regulation of transepithelial ion transport and intracellular calcium by extracellular ATP in human and cystic fibrosis airway epithelium. Br J Pharmacol 103:1649–1656, 1991.
3. R Benali, D Pierrot, M Zahm, S de Bentzmann, Puchelle E. Effect of extracellular ATP and UTP on fluid transport by human nasal epithelial cells in culture. Am J Respir Cell Mol Biol 10:363–368, 1994.
4. LI Clarke, RC Boucher. Chloride secretory response to extracellular ATP in normal and cystic fibrosis nasal epithelia. Am J Physiol 263:C348–C356, 1992.
5. AB Lansley, MJ Sanderson, ER Dirksen. Control of the beat cycle of respiratory tract cilia by Ca^{2+} and cAMP. Am J Physiol 263:L232–242, 1992.
6. CA Geary, CW Davis, AM Paradiso, RC Boucher. Role of CNP in human airways: cGMP-mediated stimulation of ciliary beat frequency. Am J Physiol 268:L1021–L1028, 1995.
7. LB Wong, DB Yeates. Luminal purinergic regulatory mechanisms of tracheal ciliary beat frequency. Am J Respir Cell Mol Biol 7:447–454, 1992.
8. P Camner, B Mossberg, BA Afzelius. Measurements of tracheobronchial clearance in patients with immotile cilia sydrome and its value in differential diagnosis. Eur J Respir Dis 64:57–63, 1983.
9. MJ Cushley, AE Tattersfield, ST Holgate. Inhaled adenosine and guanosine on airway resistance in normal and asthmatic subjects. Br J Clin Pharmacol 15:161–165, 1983.
10. JS Mann, ST Holgate, AG Renwick, MJ Cushley. Airway effects of purine nucleotides and release with bronchial provocation in asthma. J Appl Physiol 61:1667–1676, 1986.
11. Y Oosterhoff, JW DeJong, AM Jansen, GH Koeter, DS Postma. Airway responsiveness to adenosine 5′-monophosphate in chronic obstructive pulmonary disease is determined by smoking. Am Rev Respir Dis 147:553–558, 1993.
12. SJM Gaba, C Prefaut. Comparision of pulmonary and systemic effects of adenosine triphosphate in chronic obstructive pulmonary disease—ATP: a pulmonary vasoregulator? Eur Respir J 3:450–455, 1990.
13. SJ Mason, KN Olivier, D Bellinger, DJ Meuten, PD Pare, MR Knowles, RC Boucher. Studies of absorption and acute chronic effects of aerosolized and parenteral uridine 5′-triphosphate (UTP) in animals (abstr). Am Rev Respir Dis 147:A27, 1993.
14. KN Olivier, WD Bennett, KW Hohneker, KL Zeman, LJ Edwards, RC Boucher, MR Knowles. Acute safety and effects on mucociliary clearance of aerosolized uridine 5′ triphosphate +/− amiloride in normal human adults. Am J Respir Crit Care Med 154:217–223, 1996.

15. WD Bennett, KN Olivier, KL Zeman, KW Hohneker, CA Geary, RC Boucher, MR Knowles. Effect of uridine 5′-triphosphate plus amiloride on mucociliary clearance in adult cystic fibrosis. Am J Resp Crit Care Med 153(6):1796–1801, 1996.
16. WD Bennett, WF Chapman, JM Mascarella. The acute effect of ipratropium bromide bronchodilator therapy on cough clearance in COPD. Chest 103:488–495, 1993.
17. RC Boucher, MJ Stutts, MR Knowles, L Cantley, JT Gatzy. Na+ transport in cystic fibrosis respiratory epithelia. Abnormal basal rate and response to adenylate cyclase activation. J Clin Invest 78:1245–1252, 1986.
18. M Yamaya, T Ohrui, WE Finkbeiner, JH Widdicombe. Calcium dependent chloride secretion across cultures of human tracheal surface epithelium and glands. Am J Physiol 265 (Lung Cell Mol Physiol 9):L170–L177, 1993.
19. L Clarke, T Chinet, RC Boucher. Luminal ATP regulates KCl secretion in human normal CF airway epithelial cultures. Pediatr Pulmonol Suppl 8:167, 1992.
20. DC Devor, JM Pilewski. UTP inhibits Na^+ absorption in wild-type and ΔF508 CFTR-expressing human bronchial epithelia. Am J Physiol 276 (Cell Physiol 45): C827–C837, 1999.
21. F Kerem, T Bistritzer, A Hanukoglt, T Hofmann, Z Zhou, WD Bennett, E MacLaughlin, P Barker, M Nash, L Quittell, R Boucher, MR Knowles. Pulmonary epithelial sodium-channel dysfunction and excess airway liquid in pseudohypoaldosteronism. N Engl J Med 341:156–162, 1999.
22. WD Bennett, WM Foster, WF Chapman. Cough enhanced mucus clearance in the normal lung. J Appl Physiol 69(5):1670–1675, 1990.
23. SH Donaldson, MJ Stutts, RC Boucher, MR Knowles. In vivo regulation of ATP levels in human nasal epithelia. Pediatr Pulmonol (Suppl 13):A286, 1996.
24. PM Quinton. Cystic fibrosis: a disease in electrolyte transport. FASEB J 4:2709–2717, 1990.
25. RC Boucher, EHC Cheng, AM Paradiso, MJ Stutts, MR Knowles, HS Earp. Chloride secretory response of cystic fibrosis human airway epithelia: Preservation of calcium but not protein kinase C- and A-dependent mechanisms. J Clin Invest 84:1424–1431, 1989.
26. H Matsui, BR Grubb, R Tarran, SH Randell, JT Gatzy, CW Davis, RC Boucher. Evidence for periciliary liquid layer depletion, not abnormal ion composition, in the pathogenesis of cystic fibrosis airways disease. Cell 95:1005–1015, 1998.
27. JA Regnis, M Robinson, DL Bailey, P Cook, P Hooper, HK Chan, I Gonda, G Bautovich, PTP Bye. Mucociliary clearance in patients with cystic fibrosis and in normal subjects. Am J Respir Crit Care Med 150:66–71, 1994.
28. MR Knowles, LL Clarke, RC Boucher. Activation by extracellular nucleotides of chloride secretion in the airway epithelia of patients with cystic fibrosis. N Engl J Med 325:533–577, 1991.
29. PW Scherer. Mucus transport by cough. Chest 80:830–833, 1980.
30. PB Davis. Pathophysiology of the lung disease in cystic fibrosis. In: Davis PB, ed. Cystic Fibrosis. New York: Marcel Dekker, Inc. 1993:193–218.
31. JR Sabater, YM Mao, C Shaffer, MK James, TG O'Riordan, WM Abraham. Aerosolization of P2Y(2)-receptor agonists enhances mucociliary clearance in sheep. J Appl Physiol 87(6):2191–2196, 1999.
32. CM Rossman, R Waldes, D Sampson, MT Newhouse. Effect of chest physiotherapy

on the removal of mucus in patients with cystic fibrosis. Am Rev Respir Dis 126: 131–135, 1982.
33. S Girod, C Galabert, A Lecuire, JM Zahm, E Puchelle. Phospholipid composition and surface-active properties of tracheobronchial secretions from patients with cystic fibrosis and chronic obstructive pulmonary diseases. Pediatr Pulmon 13:22–27, 1992.
34. B Mossberg, BA Afzelius, R Eliasson, P Camner. On the pathogenesis of obstructive lung disease: a study on the immotile-cilia syndrome. Scand J Respir Dis 59:55–65, 1978.
35. H Kollberg, B Mossberg, BA Afzelius, K Philipson, P Camner. Cystic fibrosis compared with immotile-cilia syndrome. Scand J Respir Dis 59:297–306, 1978.
36. PG Noone, WD Bennett, JA Regnis, KL Zeman, JL Carson, M King, RC Boucher, MR Knowles. Effect of aerosolized uridine-5′-triphoshate (UTP) on airway clearance with cough in patients with primary ciliary dyskinesia. Am J Respir Crit Care Med 160:144–149, 1999.
37. M King, B Dasgupta, R Tomkiewicz, NE Brown. Clearance of mucus by simulated cough. J Appl Physiol 58:1776–1782, 1985.

31

The Effect of Dry Powder Mannitol on the Clearance of Mucus

Evangelia Daviskas, Michael Robinson, Sandra D. Anderson, and Peter T. P. Bye
Royal Prince Alfred Hospital
Sydney, Australia

INTRODUCTION

In hypersecretory diseases such as bronchiectasis and cystic fibrosis there is abnormal and increased secretion of mucus which the mucociliary system fails to clear. This results in mucus accumulation, cough, bacterial colonization, and infections (1). Treatment of mucociliary dysfunction aims to reduce mucus secretion and to increase its clearance. This treatment includes glucocorticoids, β_2-adrenergic agonists, antibiotics, and mucoactive agents (2).

Osmotic agents have been shown to increase clearance of mucus and are regarded as promising mucoactive agents (3). Knowledge of the effect of the osmotic agents on the mucociliary system originates from the use of hypertonic saline, an ionic substance, which has been shown to increase clearance of mucus in healthy subjects (4) and in patients with bronchitis (5), asthma (4), and cystic fibrosis (6). However, hypertonic saline has the disadvantage that it is rapidly absorbed, and thus its effect may be very short acting. Although salt itself is inexpensive and easily available, delivery of hypertonic saline requires an expensive nebulizer. An additional recent concern is that high salt concentration at the airway surface could inactivate the airway defensins and increase infections (7).

Mannitol, a naturally occurring sugar alcohol ($C_6H_{14}O_6$; MW 182), is a nonionic osmotic agent with low permeability (8) that has been recently shown to increase mucociliary clearance (9). It is not absorbed by the gastrointestinal tract, and it is not metabolized to any appreciable extent when given parenterally. Mannitol, as a dry powder, is a stable substance and can resist moisture resorption even at high relative humidities. Because of these characteristics it is an ideal substance to encapsulate and be delivered conveniently to the patient using portable nonexpensive inhalation devices. Mannitol also has a low permeability, and therefore it is likely to be retained in the airway lumen longer than the hypertonic saline.

Initial investigations of the effect of mannitol on the clearance of mucus were performed in healthy and mild asthmatic subjects (9). These subjects had normal resting lung function and mucociliary clearance, and they had also demonstrated an increase in clearance in response to hypertonic saline (4). These initial studies (9) established that mannitol (1) can stimulate the normal mucociliary system to increase clearance of mucus above the normal baseline level; (2) can as a dry powder have an effect in the small airways; (3) is well tolerated; and (4) induces cough which is not excessive and is acceptable to the subjects.

We have now further investigated the effect of mannitol in patients with abnormal mucociliary clearance—namely, patients with bronchiectasis and cystic fibrosis.

METHODS

Dry Powder Mannitol

The mannitol powder (Mannitol BP, Rhône Poulenc Chemicals Pty., Ltd., Brookvale, NSW, Australia) was specially prepared for inhalation (9–11). Mannitol was delivered from capsules containing 5, 10, 20, and 40 ($\pm$ 0.2) mg using a dry powder inhaler. The dry powder inhaler devices used in these studies were the Dinkihaler (Rhône Poulenc Rorer, Collegeville, PA) and the Inhalator™ (Boehringer Ingelheim, Pty., Ltd., Ingelheim, Germany). Patients inhaled the mannitol with a vital capacity maneuver and a fast inspiratory flow followed by a breath hold of 5 seconds. At a flow rate equal or greater than 60 L/min, the majority of particles are in the respirable range ($<7\ \mu m$).

Measurement of Mucociliary Clearance

In brief, our technique for measuring clearance of mucus involves inhaling a radioaerosol of sulfur colloid tagged with technetium of relatively large droplets (MMAD 6 μm, GSD 1.7). The radioaerosol is delivered with a controlled breathing pattern to ensure deposition in the conducting airways. The initial radioactivity deposited in the lung and the amount subsequently retained was monitored using a gamma camera linked to a computer. Details of our technique have been published elsewhere (4,9–11). Only the right lung was analyzed, which was divided in three regions: central, intermediate, and peripheral.

EFFECT OF MANNITOL ON THE CLEARANCE OF MUCUS

In Patients with Bronchiectasis

Patients with bronchiectasis have impaired mucociliary clearance (12–14). The impairment of mucociliary clearance most likely relates to the excessive load of mucus and to its abnormal rheological properties, both of which can affect the interaction between the cilia and the mucus. These patients require pharmacological and physical treatment daily to help clear secretions.

The effect of mannitol on the clearance of mucus was first investigated in a group of 11 patients with bronchiectasis (10). The patients were 40–62 years of age and had reasonably preserved lung function (mean ± SEM: FEV_1 77 ± 5% predicted, FEF_{25-75} 56 ± 8% predicted). A single dose of mannitol (mean ± SD:320 ± 81 mg) increased clearance acutely, over 75 minutes from the start of the intervention, when compared to control and baseline (mean ± SEM with mannitol was 34.0 ± 5.0% vs. 17.4 ± 3.8% with control and 11.7 ± 4.4% with baseline in the whole right lung) (Fig. 1). The control involved inhaling through the device loaded with an empty capsule and the same number of coughs as on the mannitol day. The mean number of coughs induced by mannitol over the 15-minute inhalation period in this group was 49 ± 11. Clearance was also increased in the central and intermediate region (Fig. 2). The cough control had no significant effect on clearance compared to baseline except in the central region, where cough is known to be most effective.

In this initial study (10), subjects had subjectively commented on the benefits of a single dose of mannitol extending well beyond the acute study period. As mannitol could be retained in the airways for some time because of its low permeability (8), we further investigated, in addition to the acute effect of mannitol, the 24-hour retention of mucus and the clearance rate close to 24 hours following inhalation of a single dose of mannitol (15). This study was performed in eight patients with bronchiectasis 29–70 years of age with well-preserved lung function (FEV_1 80 ± 5% predicted, FEF_{25-75} 50 ± 7% predicted). Mannitol (338 ± 68 mg) was found to increase clearance acutely, as in our previous study (10), and to reduce the 24-hour retention of mucus in the whole right lung and all defined regions when compared to retention without mannitol ($p < 0.02$) (Fig. 3). Mannitol helped patients to clear mucus within 2 hours, which might otherwise have taken up to 24 hours. However, the clearance rate itself 24 hour after

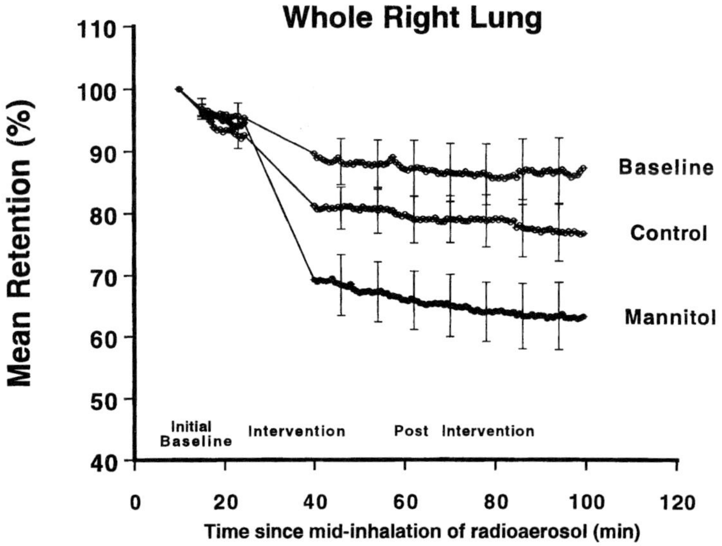

Figure 1 Mean % retention curves on the baseline, control, and mannitol study days in patients with bronchiectasis ($N = 11$) in the whole right lung. Inhalation of dry powder mannitol increased clearance compared to control and baseline ($p < 0.0001$) (From Ref. 10.)

the inhalation of mannitol was the same as on the baseline pretreatment day (clearance at 60 min: 8.7 ± 1.9% vs. 9.7 ± 3.7%; $p > 0.8$).

These results suggest that the effect of a single dose of mannitol, although mostly acute, extends beyond the acute study time of 75 minutes, resulting in a reduction in the 24-hour mucus retention. The findings of the studies in patients with bronchiectasis suggest that if a dose is administered once or twice daily long term, a greater and more prolonged effect in clearance may be achieved than with a single dose.

In Patients with Cystic Fibrosis

Patients with cystic fibrosis have impaired ion transport resulting in abnormal airway secretions, impaired clearance of mucus, and chronic bacterial infections. This leads to global bronchiectasis, respiratory failure, and premature death (16).

The effect of mannitol was investigated in 12 clinically stable patients with cystic fibrosis (11) 16–46 years of age with moderate lung function (FEV_1 60.2 ± 16.5% predicted). In this study the effect of inhaling 300 mg of dry powder mannitol was compared to the effect of inhaling 6% hypertonic saline on both ciliary clearance and cough clearance. Both mannitol and hypertonic saline increased ciliary

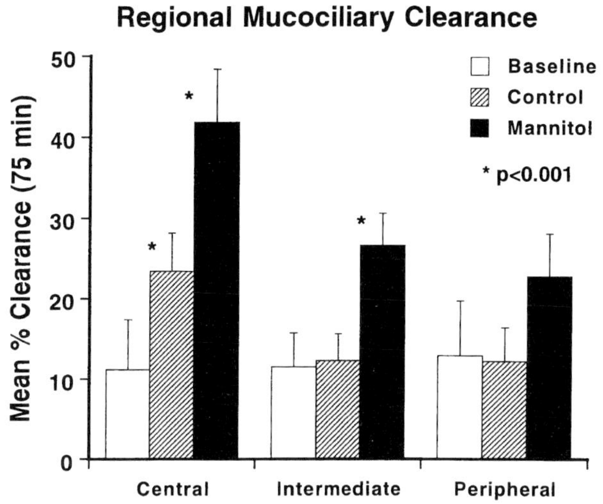

FIGURE 2 Mean % clearance in the defined regions in patients with bronchiectasis ($N = 11$) on the baseline, control, and mannitol study days. Mannitol increased clearance in the central and intermediate region compared to baseline and control. The matched voluntary cough employed as part of the control day had a positive effect on the clearance in the central region only.

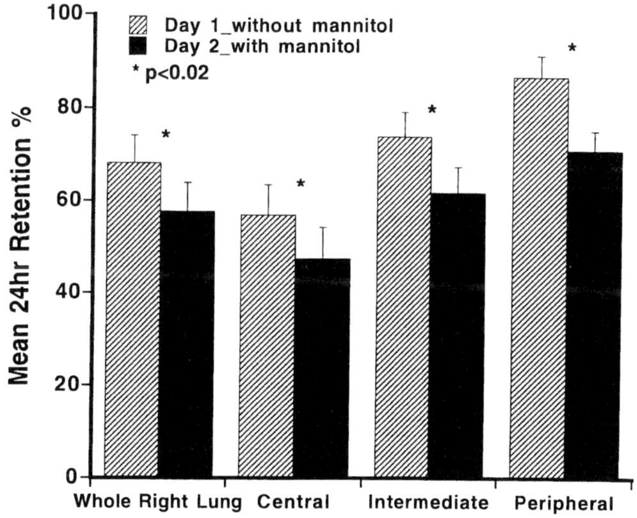

FIGURE 3 Inhalation of dry powder mannitol reduced the 24-hour retention of mucus when compared to the retention without mannitol ($p < 0.02$) in the whole right lung and all defined regions in patients with bronchiectasis ($N = 8$).

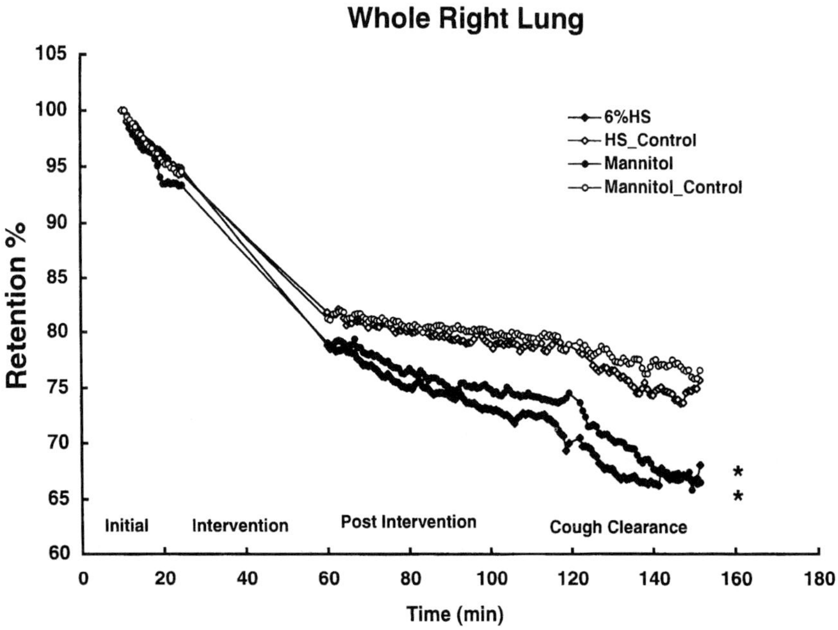

FIGURE 4 Mean clearance curves for the 4 study days involving mannitol, hypertonic saline (6%) (HS), and their respective controls in the whole right lung in patients with cystic fibrosis ($N = 12$). Graphs marked with an asterisk (*) were significantly different from their control day ($p = 0.004$). (Modified from Ref. 11.)

and cough clearance, when compared to their respective controls. There was no difference in efficacy between mannitol and hypertonic saline (Fig. 4).

DISCUSSION

Dry powder mannitol was found to increase clearance of mucus acutely in health and hypersecretory diseases, such as bronchiectasis and cystic fibrosis. In patients with bronchiectasis, a single dose of mannitol decreased the 24-hour retention of mucus compared to baseline, but it did not change the clearance rate 24 hours after inhalation.

Mannitol increased clearance of mucus in patients with cystic fibrosis to a similar extent to hypertonic saline. The magnitude of the effect of mannitol in patients with cystic fibrosis was smaller than that observed in patients with bronchiectasis. This difference in the two disease groups may relate to the differences in the severity of the disease characteristics and suggests that a larger single dose of mannitol may be required in the patients with cystic fibrosis.

The mechanism whereby osmotic agents increase clearance of mucus remains unclear. A combination of factors may be responsible, including changes in the rheology of mucus favoring ciliary and cough clearance (3,17). Osmotic agents may change the viscoelastic properties of the mucus by reducing the number of entanglements that mucin polymers form. While ionic agents may achieve this by shielding the fixed charges along the mucin macromolecules,the nonionic agents may achieve this by disrupting the hydrogen bonds between mucins (3). Osmotic agents also have the potential to increase the water in the airway lumen, so that increased hydration may also contribute to an increase in the transportability of mucus (18,19). Additionally, osmotic agents may increase clearance of mucus by stimulating mucus secretion itself (20). Although an increase in mucus secretion would not seem to be a desirable effect in patients with a hypersecretory disease, the acute stimulation of secretion of fresh mucus may help increase the clearance of viscous and stagnated secretions. Mannitol may also increase clearance of mucus by increasing the frequency of cough.

CONCLUSIONS

Inhalation of a dry powder of mannitol can increase clearance of mucus in patients who have abnormal mucus secretions and impaired airway clearance. A single dose of mannitol increases clearance of mucus acutely. However, its effect can extend beyond the acute stage, resulting in a reduction in the 24-hour retention of mucus as shown in patients with bronchiectasis. The results from these studies suggest that a twice-daily dose may have a greater and more prolonged effect on the clearance of mucus. The initial studies in these patients have provided us with important information and guidance regarding the daily dose and frequency of administration of mannitol for the long-term clinical studies in patients with bronchiectasis and cystic fibrosis.

ACKNOWLEDGMENTS

The authors would like to thank Mr. S. Eberl and the technical staff of the Department of Nuclear Medicine for their help in the studies. The authors would also like to thank Dr. H.-K. Chan from the Department of Pharmacy for preparing the powder of mannitol and the physicians of Royal Prince Alfred Hospital for referring their patients. The studies were supported by grants from the National and Medical Research Council of Australia.

The application for the use of mannitol described in the paper is covered by U.S. Patent No. 5817028, an Australian Patent No. 682756, and an International Patent PCT/AU 95/00086 held by Central Sydney Area Health Service.

REFERENCES

1. A Wanner, M Salathé, TG O'Riordan. Mucociliary clearance in the airways. Am J Respir Crit Care Med 154:1868–1902, 1996.
2. M Salathé, TG O'Riordan, A Wanner. Treatment of mucociliary dysfunction. Chest 110:1048–1057, 1996.
3. M King, BK Rubin. Mucus controlling agents: Past and present. Respir Care Clin North Am 5(4):574–594, 1999.
4. E Daviskas, SD Anderson, I Gonda, S Eberl, S Meikle, JP Seale, G Bautovich. Inhalation of hypertonic saline aerosols enhances mucociliary clearance in asthmatic and healthy subjects. Eur Respir J 9:725–732, 1996.
5. D Pavia, ML Thomson, SW Clarke. Enhanced clearance of secretions from the human lung after the administration of hypertonic saline aerosol. Am Rev Respir Dis 117:199–203, 1978.
6. M Robinson, JA Regnis, DL Bailey, M King, G Bautovich, PTP Bye. Effect of hypertonic saline, amiloride & cough on mucociliary clearance in patients with cystic fibrosis. Am J Respir Crit Care Med 153:1503–1509, 1996.
7. JJ Smith, SM Travis, EP Greenberg, MJ Welsh. Cystic fibrosis airway epithelia fail to kill bacteria because of abnormal airway surface fluid. Cell 85:229–236, 1996.
8. NJ Willumsen, CW Davis, RC Boucher. Selective response of human airway epithelia to luminal but not serosal solution hypertonicity. Possible role for proximal epithelia as an osmolarity transducer. J Clin Invest 94:779–787, 1994.
9. E Daviskas, SD Anderson, JD Brannan, H-K Chan, S Eberl, G Bautovich. Inhalation of dry-powder mannitol increases mucociliary clearance. Eur Respir J 10:2449–2454, 1997.
10. E Daviskas, SD Anderson, S Eberl, H-K Chan, G Bautovich. Inhalation of dry powder mannitol improves clearance of mucus in patients with bronchiectasis. Am J Respir Crit Care Med 159(6):1843–1848, 1999.
11. M Robinson, E Daviskas, S Eberl, J Baker, H-K Chan, SD Anderson, PTP Bye. The effect of inhaled mannitol on bronchial mucus clearance in cystic fibrosis patients: A pilot study. Eur Respir J 14(3):678–685, 1999.
12. RV Lourenço, R Loddenkemper, and RW Carton. Patterns of distribution and clearance in patients with bronchiectasis. Am Rev Respir Dis 106:857–866, 1972.
13. DC Currie, D Pavia, JE Agnew, MT Lopez-Vidriero, PD Diamond, PJ Cole, SW Clarke. Impaired tracheobronchial clearance in bronchiectasis. Thorax 42:126–130, 1987.
14. T Isawa, T Teshima, T Hirano, Y Anazawa, M Miki, K Konno, M Motomiya. Mucociliary clearance and transport in bronchiectasis: global and regional assessment. J Nucl Med 31:543–548, 1990.
15. E Daviskas, SD Anderson, S Eberl, H-K Chan, IH Young. The 24 hr effect of mannitol on the clearance of mucus in patients with bronchiectasis. Chest 119:414–421, 2001.
16. H Matsui, BR Grubb, R Tarran, SH Randell, JT Gatzy, WC Davis, RC Boucher. Evidence for periciliary liquid layer depletion, not abnormal ion composition, in the pathogenesis of cystic fibrosis airways disease. Cell 95:1005–1015, 1998.

17. P Wills, R Hall, W-M Chan, PJ Cole. Sodium chloride increases the ciliary transportability of cystic fibrosis and bronchiectasis sputum on the mucus-depleted bovine trachea. J Clin Invest 99:9–13, 1997.
18. SL Winters, DB Yeates. Roles of hydration, sodium and chloride in regulation of canine mucociliary transport system. J Appl Physiol 83(4):1360–1369, 1997.
19. DB Yeates, BT Chen. Increases in airway surface liquid induced by administration of near-isoosmotic and hyperosmotic mannitol aerosols. J Aerosol Med 12(2):112, 1999.
20. JN Baraniuk, M Ali, A Yuta, S-Y Fang, K Naranch. Hypertonic saline nasal provocation stimulates nociceptive nerves, substance P release and glandular mucus exocytosis in normal humans. Am J Respir Crit Care Med 160(2):655–662, 1999.

32

Neutrophil Elastase and Antigen-Induced Mucociliary Dysfunction

Thomas O'Riordan
State University of New York at Stony Brook
Stony Brook, New York

William M. Abraham
University of Miami at Mount Sinai Medical Center
Miami Beach, Florida

INTRODUCTION

Neutrophil elastase (NE) has long been recognized as a potent degradant of extracellular matrix proteins (1,2), and, as such, its effects have been studied in several animal models of emphysema. However, in recent years it has been recognized that NE has additional properties. Of particular importance is the in vitro data showing that NE is a potent mucus secretagogue (3–6). These findings, in conjunction with renewed interest in the neutrophil's role in the pathogenesis of asthma, suggested that the neutrophil and NE may contribute to the abnormalities in mucus clearance associated with asthma. Therefore, we decided to investigate the possible role of NE in antigen-induced mucociliary dysfunction in an ovine model of allergic asthma.

MUCOCILIARY DYSFUNCTION IN ASTHMA

Before discussing the potential role of NE in antigen-induced mucociliary dysfunction, we will briefly review the evidence that mucociliary dysfunction is important in asthma (7). The principal evidence supporting the importance of mucociliary dysfunction comes from autopsy data published over the past 40 years, which revealed mucus plugging of small airways in patients dying of acute asthma. In addition, patients admitted to hospital with status asthmaticus were found to have severely impaired mucociliary clearance (as measured by clearance of radiolabeled aerosols) within the first 48 hours after hospital admission (9) (Fig. 1). When studied 3–6 weeks following discharge there was a marked improvement in mucociliary clearance. Similar techniques were used to study patients with chronic asthma (10). Patients with severe asthma as defined by the presence of expiratory flow limitation during tidal breathing were compared to patients with milder asthma as defined by the absence of tidal flow limitation. Not surprisingly, the patients with severe asthma had marked impairment of mu-

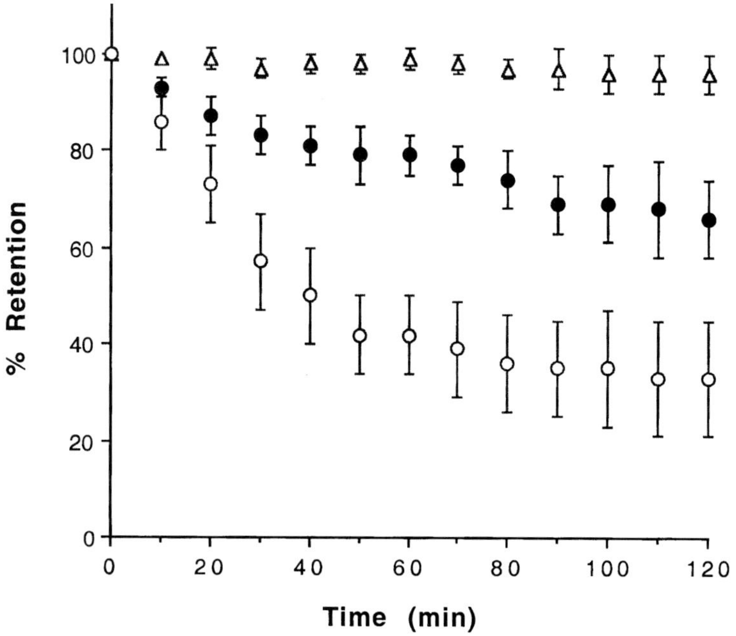

FIGURE 1 Retention of radioactivity vs. time; patients in hospital (open symbols), after discharge (closed symbols). Error bars indicate standard errors. Numerical values are *p*-values. (From Ref. 9.)

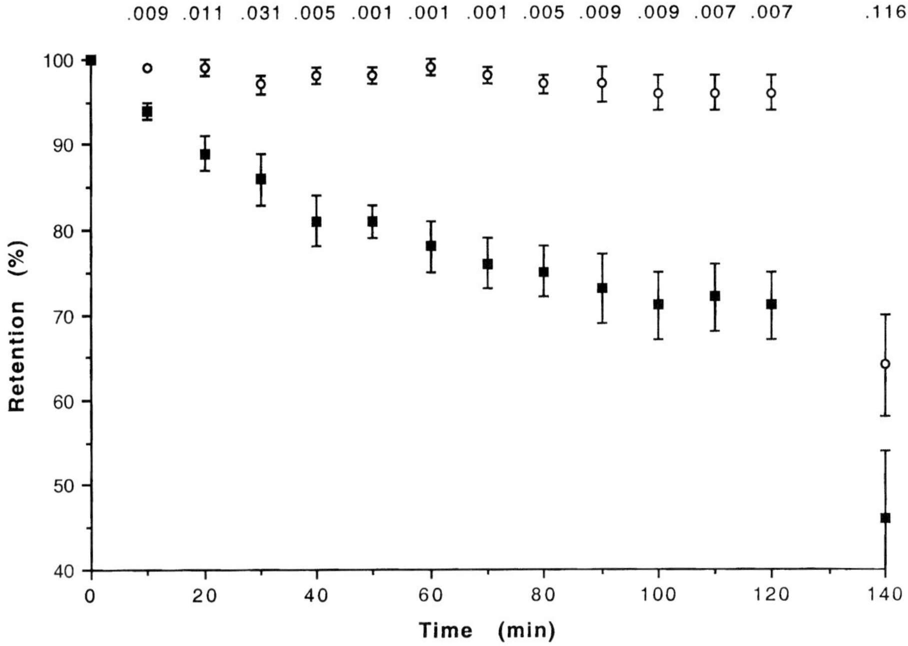

FIGURE 2 Whole lung radioaerosol mucociliary clearance data in patients with chronic stable asthma(10) with superimposed results from patients with acute exacerbations of asthma (9). The stable patients with tidal flow limitation are represented by closed circles, the stable patients without tidal flow limitation by open circles, and the acutely hospitalized patients by triangles. (From Ref. 10.)

cociliary clearance. However, even though mucociliary clearance was impaired in the patients with severe asthma, when these data were superimposed on the data from the patients studied while in status asthmaticus (9), the patients with severe but stable asthma had a much less severe degree of impairment of mucociliary clearance (Fig. 2). These findings suggest that the changes in clearance rates parallel a patient's clinical deterioration and recovery.

THE ROLE OF THE NEUTROPHIL IN ASTHMA

Because asthma is believed to be caused by airway inflammation, it is worth considering possible mechanisms by which inflammatory cells may contribute to mucociliary dysfunction. In deciding whether or not to study NE as a cause of mucociliary dysfunction asthma, we should first briefly review the evidence supporting the potential role of the neutrophil in asthma, a role that has in the past tended to be a subject of controversy (11). Endobronchial biopsies and bronchoalveolar lavage fluid (BALF) obtained in patients with mild stable asthma reveal no evidence of neutrophilia; instead the dominant cell is clearly the eosinophil (12). However, neutrophilia and elastase activity have been noted in airways secretions of patients with acute exacerbations of asthma, especially those that had a preceding viral illness (13). Increased neutrophils have also been detected in BALF obtained in patients with nocturnal worsening of asthma, in patients developing a late response to inhaled allergen, and in some patients with occupational asthma (11). In addition, increased neutrophils have been noted in autopsies of patients who died acutely within a number of hours of developing an exacerbation of asthma (14). Finally, recent studies of patients with severe asthma not adequately controlled with inhaled steroids revealed the increased numbers of neutrophils in BALF. Collectively, these findings indicate that the neutrophil may have a role in severe persistent asthma (15).

OVINE MODEL OF ANTIGEN-INDUCED MUCOCILIARY DYSFUNCTION

To further investigate whether neutrophil elastase can play a role in antigen-induced mucociliary dysfunction, experiments were conducted in an ovine model (16–18). Sheep, which are naturally allergic to *Ascaris suum*, respond to inhalation of this allergen with the development of an early bronchoconstrictor response. In over 50% of these animals, there is also a late bronchoconstrictor response that is associated with airway inflammation (19). The vast majority of studies on this model have focused on the mechanisms by which airway inflammation contributes to allergen-induced bronchoconstriction. However, these animals also respond to exposure to allergen by developing a marked decrease in mucociliary clearance (20). In these animals mucociliary clearance is measured by measuring the tracheal mucous velocity (TMV). The animals are intubated and teflon discs are then insufflated into the airway, the movement of which can be measured by fluoroscopy. Decreases in TMV are usually detectable within 30 minutes after allergen exposure and are then measured at hourly intervals for the duration of the experiment. Impairment of TMV in some animals may last for some days. Similar fluoroscopic techniques have been used to measure TMV in human subjects (21,22) exposed to ragweed allergen. Interestingly, the findings in humans are similar to those seen in sheep.

We utilized this model to determine if the neutrophil and/or NE contribute to antigen-induced mucociliary dysfunction. To do this, three inhibitors of NE were studied: α-1, PI, SLPI (secretory leucoprotease inhibitor), and ICI 200,355, a synthetic inhibitor of NE. In addition, because glucocorticoids are the mainstay of current anti-inflammatory therapy of asthma, we studied the effects of the potent topical corticosteroid budesonide in this model of antigen-induced impairment of mucociliary clearance (16–18).

RESULTS

Prevention of Antigen-Induced Bronchoconstriction

Figure 3 demonstrates that treatment of animals with α-1PI or a synthetic inhibitor of NE (ICI 200,355) prevented the antigen-induced decrease in TMV. In contrast, α-1 PI that had been denatured was ineffective in protecting against the effects of antigen, indicating that the protection afforded by α-1 PI is not a nonspecific response to the inhalation of a protein. Budesonide administered one hour prior to allergen exposure was also effective at preventing decreases in TMV (17). When the α-1 PI, SLP1, and ICI 200,355 and the budesonide are administered in the absence of antigen, there is no effect on TMV (16–18).

Effect of NE Inhibitors and Budesonide When Administered One Hour After Antigen Exposure

It is generally believed that the early effects of antigen on the airway are mediated through the degranulation of mast cells, with the later effects being mediated

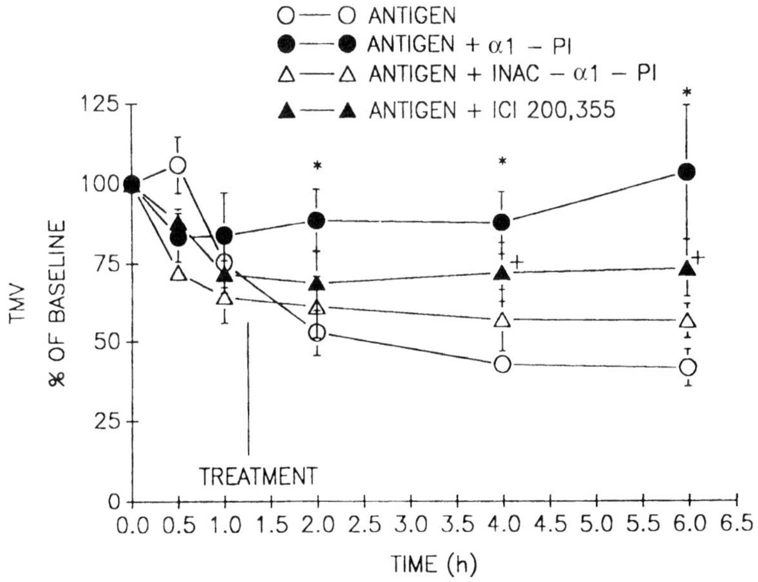

FIGURE 3 Challenge with antigen produced a significant decrease in TMV at 2, 4, and 6 hours, which was prevented by pretreatment with α-1 PI and ICI 200,355 but not by inactivated α-1 PI. Values are mean± SE. *$p < 0.05$ versus antigen alone. (From Ref. 16.)

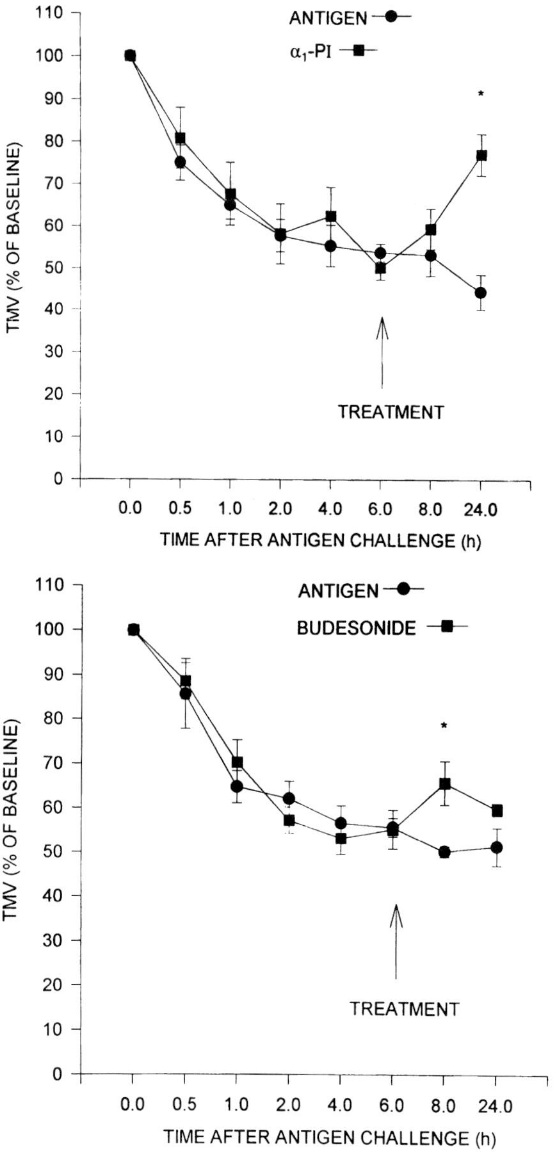

Figure 4 (a) Effect of α-1 PI ($n = 5$) administered 6 hours after antigen challenge on the antigen-induced fall in TMV. α-1 PI attenuated the antigen-induced fall in TMV at 24 hours. $^{*}p < 0.05$ versus antigen alone. (b) Effect of budesonide ($n=6$) administered 6 hours after antigen challenge on the antigen-induced fall in TMV. Budesonide significantly improved clearance at 8 hours but had no effect at 24 hours after antigen. $^{*}p < 0.05$ versus antigen alone. (Figure 4a from Ref. 16, Figure 4b from Ref. 17.)

through second messengers cells such as the neutrophil and the eosinophil. These following experiments were designed to see if NE inhibitors would be effective when given after antigen exposure and after the presumed mast cell degranulation. We found that α-1 PI, SLPI, ICI 200,355, and budesonide were all effective in reversing antigen-induced mucociliary dysfunction when given one hour after antigen challenge. These results suggest that the mechanism by which these agents improve TMV is not via the stabilization of mass cells (16–18).

Effects of NE Inhibitors and Budesonide When Administered 6 Hours Following Antiallergen Exposure

After 6 hours the antigen-induced reduction in TMV is maximal. At this time, BALF, neutrophils, NE, and other inflammatory markers are significantly increased. Given these observations, it was of interest to determine if α-1 PI and budesonide, administered 6 hours after allergen exposure, could affect TIMV. As seen in Figure 4, there was a small improvement in TMV with both agents at the 8-hour point. However, with α-1 PI there was a further marked improvement in TMV at the 24-hour time point, but no significant improvement was noted with budesonide. These findings demonstrate that α-1 PI can reduce even well-established antigen-induced mucociliary dysfunction. The long duration of action is consistent with the known pharmacokinetics of inhaled α-1 PI, whereas the lack of effect with budesonide may be due to its shorter duration of action.

Direct Effects of Elastase on TMV

It was important to confirm that ovine elastase could itself reduce TMV and that this effect could be blocked by α-1 PI. We confirmed that the elastase activity contained in supernatants from PMA and opsonized zymosan stimulated ovine neutrophils was inhibited by both ICI 200,355 and α-1 PI in vitro (16). In another series of experiments, we showed that when sheep inhaled elastase, the decrease in TMV was similar to that achieved with antigen, and this decrease was prevented by pretreatment with α-1PI. This experiment provides further evidence for the hypothesis that NE contributes to the decrease in TMV observed after antigen challenge (Fig. 5).

DISCUSSION

Summary of Observations

Collectively, the results of these studies indicate that elastase may mediate mucociliary dysfunction that follows antigen challenge and that agents that have anti-

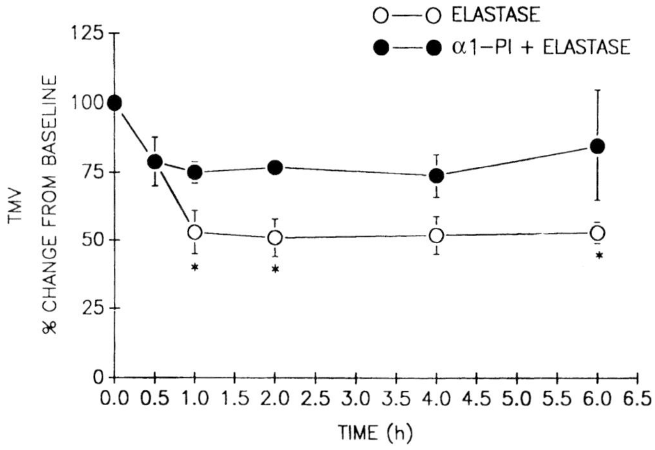

FIGURE 5 Effect on TMV, of administration of degranulation supernatants from isolated ovine neutrophils containing elastase activity, alone (open circles), and after pretreatment with α1-PI (closed circles). Values are mean = SE for three sheep. *$p < 0.05$ versus elastase alone. (From Ref. 16.)

elastase activity may be beneficial in blocking this mucociliary impairment. This conclusion is supported by the following observations:

1. The antigen-induced reduction in TMV was prevented by pretreatment with three elastase inhibitors (α-1 PI, ICI 200,355, and SLPI).
2. Administration of the elastase inhibitors one hour after antigen also reversed the reduction in TMV.
3. A single dose of α-1 PI administered 6 hours after antigen challenge significantly attenuated the severity of mucociliary dysfunction present 24 hours after antigen challenge.
4. Antigen challenge was associated with neutrophilia and detectable free elastase activity in BALF.
5. Inhalation of degranulation supernatants containing elastase activity from isolated ovine neutrophils caused a reduction in TMV similar to that seen after antigen challenge, and this impairment was attenuated by α-1 PI.
6. α-1 PI and ICI 200,355, inhibitors of human neutrophil elastase, inhibited ovine elastase activity in vitro.
7. A single dose of a topical glucocorticosteroid administered 30 minutes before or 1 hour after antigen challenge can prevent or reverse the antigen-induced reduction in TMV.

8. Treatment with a single dose of budesonide 6 hours after antigen challenge resulted in a transient increase in TMV at 8 hours after challenge, but TMV was not significantly improved at 24 hours after challenge, indicating that the protease inhibitor had a more sustained effect that the corticosteroid.

Mechanism of Action: NE as a Secretagogue

The experiments described above provide evidence that NE may be important in causing impairment of mucociliary clearance after antigen exposure. However, the mechanism of action by which NE causes clearance to be decreased is not known. Based on studies that demonstrate that NE is a potent secretagogue (3–6), the most likely explanation for a change in TMV is an alteration in the composition of airway secretions, which results in impaired mucociliary clearance.

Our findings of increased numbers of neutrophils and increased neutrophil elastase activity in the airways after antigen challenge and its association with abnormal mucociliary function are consistent with previous findings reported by Tabachnik et al. (5) in ragweed-sensitive dogs. In that study, perfusion of an isolated tracheal segment with ragweed for 8 hours resulted in increased submucosal gland secretion as indicated by increased lysozyme levels that correlated with increased elastase in the perfusate, and the effects were blocked by ICI 200,355. Although we showed that antigen challenge caused an increase in measurable elastase activity in the BALF, this increase in activity was disproportionate to the numbers of neutrophils recruited to the airway. One potential explanation for the apparent dichotomy between the two measured parameters (i.e., neutrophilia and elastase activity) may be that the elastase released by stimulated neutrophils may be bound and inhibited by mucus (23).

Other Possible Mechanisms of Action

It is also possible that NE may slow clearance through interaction with other inflammatory cells. Recent data suggest, for example, that NE can cause the degranulation of eosinophils (24). Eosinophilic products such as major basic protein have been shown to impair ciliary beat frequency and injure the epithelium. NE may also be chemoattractant and may cause migration of the airway of cells that may alter mucociliary function (7). Finally, there is also evidence from in vitro preparations that NE can decrease the rate of ciliary beat frequency and evidence that it may be directly toxic to ciliated epithelium (25). However, this possibility is somewhat unlikely, because if destruction of ciliated epithelium were the primary mechanism of action by which elastase affected mucus clearance administration, elastase inhibitors would not be expected to reverse this in a matter of hours.

Distinguishing Between the Effects of Antigen on Airway Mechanics and Mucociliary Clearance

The results of this study are indicative of the dichotomy between the mechanisms leading to mucociliary dysfunction and allergic bronchoconstriction following antigen challenge. Pretreatment of sheep with 10 mg α-1 PI was effective in blocking the antigen-induced fall in TMV, whereas this same dose of α-1PI had no effect on the antigen-induced early and late responses (26). Thus, our findings suggest that elastase has a role in antigen-induced mucociliary dysfunction, but not on the bronchoconstrictor response seen following antigen challenge.

The observation that treatment after the antigen challenge can reverse the antigen-induced fall in TMV is consistent with the hypothesis that the effects on TMV are mediated by secondary cells or mediators recruited to the airway as a result of antigen-induced mast cell degranulation. That mast cell degranulation is an initiating event in this process was demonstrated previously, when we showed that pretreatment with the anti allergic agent cromolyn sodium prevented the *Ascaris suum* antigen-induced fall in TMV in dogs and sheep (27,28). Cromolyn also inhibited *Ascaris suum* antigen-induced effects on ciliary beat frequency in vitro. Given this evidence, it was not surprising to us that treatment with α-1 PI, administered after the antigen challenge, was also effective, because it coincides with the timing of the neutrophil influx into the airways following allergen challenge in this model. Our previous studies have shown that neutrophil numbers begin to increase in the airways between 1 and 2 hours (19) after antigen challenge and that they continue to be present through 96 hours, which is similar to the time course of the antigen-induced impairment in TMV. The effectiveness of the posttreatment strategy also reinforces the belief that the fall in TMV is likely linked to an infiltrating inflammatory cell because the stimulus for cell recruitment, i.e., mast cell degranulation, was not affected.

Controls

An important control for the present experiments was the failure of inactivated α-1 PI to prevent or reverse the antigen-induced impairment of mucociliary clearance. This indicates that the protective effect of α-1 PI was not a nonspecific effect that resulted from the inhalation of a protein. Furthermore, the lack of an effect of aerosolized α-1 PI and ICI 200,355, alone, on TMV removes the concern that the protective effects that we observed were due to a direct stimulatory action of these agents.

Although our previous studies indicate that antigen-induced mast cell degranulation is primarily responsible for the slowing of TMV following *Ascaris* inhalation, in light of our recent studies it is important to rule out possible contamination by endotoxin (29). We found that TMV was not affected by endotoxin challenge. The extract of *Ascaris suum* used in this study contained endotoxin at a level of 50 EU/mL (17). While endotoxin can induce neutrophilia and elastase activity in human subjects, we believe that it is unlikely that this would have been the primary cause of the mucociliary impairment in this model. This opinion is based on the fact that challenge with a concentration of 30,000 EU/mL failed to induce bronchoconstriction in the model.

Effects of Budesonide on Mucociliary Clearance

The effectiveness of budesonide in preventing or reversing antigen-induced mucociliary dysfunction is most likely due to its anti-inflammatory properties. This is supported by the observation that the drug does not appear to have a direct stimulatory effect on TMV. A previous study in the sheep model with another topical glucocorticosteroid, beclomethasone, also reported that there was no direct effect on TMV (30). Previous studies have suggested that the anti-inflammatory effects of glucocorticosteroids may enhance mucociliary clearance in asthma. Treatment of asthmatic patients with oral prednisolone (15 mg for 2 weeks, followed by 30 mg for a further 2 weeks) resulted in some mucociliary improvement (31). In another study of patients with status asthmaticus treated with high-dose systemic glucocorticosteroids, significant improvement in mucociliary clearance occurred after hospital discharge compared to their acute exacerbation, leading the authors to speculate that the glucocorticosteroids may have induced a reduction in airway inflammation that may have resulted in improved mucociliary clearance (9). The mechanism by which budesonide prevented antigen-induced mucociliary dysfunction in the present study is not known, but it is likely to be multifactorial because glucocorticosteroids suppress the inflammatory effects of antigen challenge through many mechanisms.

CONCLUSION

The data presented here suggest that NE may contribute to mucociliary dysfunction in asthma but that the effects on mucociliary dysfunction may be ameliorated by treatment with elastase inhibitors. These agents may be worthy of further investigation as potential therapies for mucociliary dysfunction in asthma.

REFERENCES

1. NG McElvaney, RC Crystal. Proteases in lung injury. In: Crystal RG, West, JB, Weibel ER, Barnes PJ, eds. The Lung: Scientific Foundations. Philadelphia: Lippincott-Raven, 1997, pp 2205–2218.
2. RA Stockley. Neutrophils and protease/antiprotease imbalance. Am J Respir Crit Care Med 160:549–552, 1999.
3. JA Nadel, K Takeyama. Mechanisms of hypersecretion in acute asthma, proposed cause of death, and novel therapy. Pediatr Pulmonol (Suppl) 18:54–55, 1999.

4. JV Fahy, A Schuster, I Ueki, HA Boushey, JA Nadel. Mucus hypersecretion in bronchiectasis. The role of neutrophil proteases. Am Rev Respir Dis 146:1430–1433, 1992.
5. F Tabachnik, A Schuster, WM Gold, JA Nadel. Role of neutrophil elastase in allergen induced lysoszyme secretion in the dog trachea. J Appl Physiol 73:695–700, 1992.
6. K Takeyama, C Agusti, I Ueki, J Lausier, LO Cardell, JA Nadel. Neutrophil-dependent goblet cell degranulation: role of membrane-bound elastase and adhesion molecules. Am J Physiol 275:L294–302, 1998.
7. A Wanner, M Salathe, TG O'Riordan. State of the art. Mucociliary clearance in the airways. Am J Respir Crit Care Med 154:1868–1902, 1996.
8. JB MacDonald, ET MacDonald, DA Seaton Williams. Asthma deaths in Cardiff: 1963–74: 53 deaths in hospital. Br Med J 2:721–723, 1976.
9. M Messina, TG O'Riordan, GC Smaldone. Changes in mucociliary clearance during acute exacerbations of asthma. Am Rev Respir Dis 143:993–997, 1993.
10. TG O'Riordan, J Zwang, G Smaldone. Mucociliary clearance in adult asthma. Am Rev Respir Dis 146:598–603, 1992.
11. PM Henson, LC Borish. Neutrophil mediators in asthma. In: WW Busse, ST Holgate, eds. Asthma and Rhinitis. Boston: Blackwell, 1995, pp 367–380.
12. R Djukanovic, WR Roche, J Wilson, CR Beasley, OP Twentyman, PH Howarth, ST Holgate. State of the art. Mucosal inflammation in asthma. Am Rev Respir Dis 142:434–457, 1990.
13. JV Fahy, KW Kim, J Liu, HA Boushey. Prominent neutrophilic inflammation in sputum from subjects with asthma exacerbation. J Allergy Clin Immunol. 95:843–852, 1995.
14. S Sur, TB Crotty, GM Kephart, BA Hyma, TV Colby, CE Reed, LW Hunt, GJ Gleich. Sudden onset fatal asthma. A distinct entity with few eosinophils and relatively more neutrophils in the airway submucosa. Am Rev Respir Dis 148:713–719, 1993.
15. S Wenzel, SJ Szefler, DYM Leung, SI Sloan, MD Rex, RJ Martin. Bronchoscopic evaluation of severe asthma. Persistent inflammation associated with high dose glucocorticoids. Am J Respir Crit Care Med 156:737–743, 1997.
16. TG O'Riordan, R Otero, Y Mao, I Lauredo, WM Abraham. Elastase contributes to antigen-induced mucociliary dysfunction in the airway. Am J Respir Crit Care Med 155:1522–1528, 1997.
17. TG O'Riordan, Y Mao, R Otero, J Lopez, JR Sabater, WM Abraham. Budesonide affects allergic mucociliary dysfunction. J Appl Physiol 85:1086–1091, 1998.
18. CD Wright, AM Havil, SC Middleton, MA Kashem, PA Lee, DJ Dripps, TG O'Riordan, MP Bervilicqua, WM Abraham. Secretory leukoprotease inhibitor prevents allergen-induced pulmonary responses in animal models of asthma. J Pharmacol Exp Ther 289:1007–1014, 1999.
19. WM Abraham, MW Sielczak, A Wanner, AP Perruchoud, L Blinder, JS Stevenson, A Ahmed, LD Yerger. Cellular markers of inflammation in the airways of allergic sheep with and without allergen induced late responses. Am Rev Respir Dis 136: 1565–1571, 1986.
20. L Allegra, WM Abraham, GA Chapman, A Wanner. Duration of mucociliary dys-

function following allergen challenge in allergic sheep. J Appl Physiol 55:726–730, 1985.
21. RJ Mezey, MA Cohn, RJ Fernandez, AJ Januszkiewicz, A Wanner. Mucociliary transport in patients with antigen-induced bronchospasm. Am Rev Respir Dis 118: 677–684, 1978.
22. T Ahmed, DW Greenblatt, S Birch, B Marchette, A Wanner. Abnormal mucociliary transport in allergic patients with antigen-induced bronchospasm: role of slow reacting substance of anaphylaxis. Am Rev Respir Dis 124:110–114, 1981.
23. C Nadziejko, I Finkelstein. Inhibition of neutrophil elastase by mucus glycoprotein. Am J Respir cell Mol Biol 11:103–107, 1994.
24. H Liu, SC Lazarus, GH Caughey, IV Fahy. Neutrophil elastase and elastase-rich cystic fibrosis sputum degranulate human eosinophils in vitro. Am J Physiol 276(1 Pt 1):L28–34, 1999.
25. F Frigas, DA Legering, GO Solley, GM Farrow, GJ Gleich. Elevated levels of eosinophil granule major basic protein in the sputum of patients with bronchial asthma. Mayo Clin Proc 56:345, 1981.
26. R Forteza, Y Botvinnikova, A Ahmed, A Cortes, RH Gundel, A Wanner, WM Abraham. The interaction of α-1 proteinase inhibitor and tissue kallikrein in controlling allergic ovine airway hyperresponsiveness. Am J Respir Crit Care Med 154:36–42, 1986.
27. A Wanner, S Zarecki, J Hirsch, S Epstein. Tracheal mucous transport in experimental asthma. J Appl Physiol 39:950–957, 1975.
28. WM Abraham, A Wanner, JS Stevenson, GA Chapman. The effect of an orally active leukotriene D4 and E4 antagonist, LY 171883, on antigen-induced airway responses in allergic sheep. Prostaglandins 31:457–461, 1986.
29. LW Hun, GJ Gleich, T Ohnishi, DA Weiler, ES Mansfield, H Kita, S Sur. Endotoxin contamination causes neutrophilia following allergen challenge. Am J Respir Crit Care Med 149:1471–1475, 1994.
30. MA Sackner, M Reinhart, B Arkin. Effects of beclomethasone diproprionate on tracheal mucous velocity. Am Rev Respir Dis 115:1069–1070, 1977.
31. JF Agnew, JRM Bateman, D Pavia, SW Clarke. Peripheral airways mucous clearance in stable asthma is improved by oral glucocorticosteroid therapy. Bull Eur Physiopathol Respir 20:295–301, 1984.

33

Mechanisms of Clearance of Soluble Substances from the Intrathoracic Airways

W. Michael Foster* and Elizabeth M. Wagner
The Johns Hopkins University
Baltimore, Maryland

INTRODUCTION

Information on regional dosimetry of substances deposited within the respiratory tract is essential for calculating the dose/tissue burden at epithelial surfaces and identifying the pathways for diffusion and clearance of these substances. This information is key to the design strategies for delivery of therapeutic agents and for understanding toxicology and injury from bioaerosols. Although numerous studies have focused on the fractional deposition of inhaled aerosols within the conducting airways as a function of particle diameter (1), only a few investigations have examined in vivo the fate of soluble substances inhaled or placed onto the epithelial surface of the airway (2,3). The rate of diffusion of a substance into submucosal tissue depends upon its lipid and water solubilities, size, and shape. Lipophilic molecules pass mainly via transcellular routes, and hydrophilic molecules pass via paracellular routes (4). Since the bronchial circulation anastomoses with pulmonary veins (5) and drains into the left heart, the absorption of

* *Current affiliation*: Duke University Medical Center, Durham, North Carolina.

soluble substances into the blood perfusing the subcarinal airways is difficult to assess in vivo separately from the pulmonary vascular component (6).

When direct in vivo assessment of adsorption into the airway vasculature has been evaluated, the uptake of soluble tracers by the tracheal venous outflow was inversely related to the level of airway perfusion (7). Therefore, when tracheal perfusion was altered by vasoactive agonists or mechanical means and blood flow increased, the adsorption of soluble tracer by the vasculature decreased, and conversely when tracheal inflow decreased, the concentration of tracer in venous blood increased. This suggested that when tracheal blood flow is diminished, the uptake of soluble tracer into the venous outflow would increase due to a prolonged dwell time.

The purpose of the present study was to investigate the time course and pathway(s) for clearance of soluble tracers from the surface of intrathoracic subcarinal airways and evaluate whether this process was influenced by airway perfusion. We expected that uptake of a soluble tracer into bronchial blood would parallel vascular clearance phenomenon reported for the extrathoracic tracheal airway (7,8). We utilized an instrumentated sheep model for our in vivo investigation and employed a soluble, low molecular weight, hydrophilic radiolabeled tracer, ^{99m}technetium labeled diethylene triamine pentaacetic acid (^{99m}Tc-DTPA), to gauge vascular uptake. Two-dimensional radio-imaging of the bronchial airway assisted our systemic blood measurements of ^{99m}Tc-DTPA after site-specific delivery of the tracer directly onto the bronchial airway surfaces using a flexible bronchosocope. Two major pathways of clearance, by a vascular route and by epithelial mucociliary function, were observed.

METHODS

Animal Preparation

The animal protocol was approved by the Johns Hopkins Animal Care and Use Committee. Anesthesia was induced in mixed breed sheep (25–35 kg) with intramuscular ketamine (30 mg/kg) and subsequently maintained with intravenous pentobarbital sodium (20 mg/kg/h). The animal was placed onto a surgical table in a prone position and tracheal intubation was performed. Animals were paralyzed with pancuronium bromide (2 mg iv), and the lungs were mechanically ventilated (10–12 mL/kg) at a rate (12–15 breaths/min) sufficient to maintain normal blood gases. Five cm H_2O positive end-expiratory pressure was applied. The left thorax was opened at the 5th intercostal space, and heparin (20,000 U iv) was administered. The esophageal and thoracic branches of the bronchoesophageal artery were ligated as previously described (9). The bronchial branch of the bronchoesophageal artery was isolated, cannulated, and perfused (0.6 mL/

min/kg) with autologous blood that was withdrawn from the descending aorta and pumped through a variable speed roller pump.

Radioaerosol Administration

Local delivery of the small (492 daltons), soluble, hydrophilic tracer ^{99m}Tc-DTPA onto the bronchial epithelium was employed to gauge absorption of soluble substances from the airway surface. DTPA was freshly prepared as ^{99m}Tc-DTPA (Medi-Physics, Arlington Heights, IL). ^{99m}Tc-DTPA was randomly sampled predelivery and assayed for unbound ^{99m}Tc with silica gel media and thin-layer chromatography to verify the labeling procedures (10). Local airway delivery was performed to ensure deposition of the DTPA exclusively onto bronchial airway surfaces. A fiberoptic bronchoscope (OlympusBF type P10, New Hyde Park, NY) was advanced into the trachea and passed beyond the carina and mapped into a fourth-generation bronchus. A polyethylene catheter with a microspray nozzle (11) at its tip that delivers spray radially onto the epithelial surface was advanced through a channel of the bronchoscope and visualized beyond the end of the bronchoscope within the center of the bronchial lumen. Ventilation was withheld and 6–10 μL of ^{99m}Tc-DTPA was deposited through the nozzle and sprayed directly onto the bronchial wall using 0.7 mL of air to clear the catheter tip (an additional 0.7 mL of air cleared any residual activity from the catheter). Controlled ventilation was resumed and serial images of ^{99m}Tc-DTPA retention were recorded every 120 seconds over a 30-minute period using a gamma scintillation camera (MaxiCamera, General Electric Medical Systems, Pittsburgh, PA) set with a 15% window around the peak energy of 140 keV and shielded with a parallel hole collimator. Images were acquired from the ventral aspect of the thorax and the fraction of sprayed droplets initially delivered within the lung was normalized to 100%. All images were background subtracted and decay corrected to time zero (time immediately after droplet delivery) for radioisotopic decay of the ^{99m}Tc. Systemic venous blood was sampled (0.5 mL) every 6 minutes during the 30-minute imaging period in order to assess take-up of the ^{99m}Tc-DTPA into the blood. Once ^{99m}Tc-DTPA is absorbed into the circulation, it is cleared from the blood by the kidneys (12). Blood samples were analyzed by a gamma scintillation counter (GammaTrac, Tm Analytic, Tampa, FL). Total absorption of the ^{99m}Tc-DTPA into blood was estimated by the calculation of multiplying the activity in blood (counts/mL) by a nominal blood volume equal to 8.5% of body weight (13).

Experimental Protocol

The design of the protocol was to use each animal preparation as its own control and thus to perform measures of ^{99m}Tc-DTPA clearance with bronchial perfusion intact (control condition), followed by redelivery of ^{99m}Tc-DTPA and measure clearance with bronchial perfusion interrupted (experimental condition). We did not randomize the order of this sequence, as previously we have demonstrated the stability of the preparation over the time period of the measurements (14). Our concern was that if the experimental condition was evaluated first, then the time course of recovery of the preparation from the no flow condition might extend the measurement period beyond the period during which the preparation was physiologically stable, adding uncertainty to any control clearance measures acquired at a late time point. To eliminate issues of nonhomogeneous tissue attenuation of radioactivity due to regional differences in airway geometry, all sprayed deliveries of ^{99m}Tc-DTPA were performed as paired comparisons (control vs. experimental) in the same airways. Residual activity from the initial delivery (in our experience usually $<2\%$) that was not cleared by the delivery time of the second administration of ^{99m}Tc-DTPA was considered as nonclearable and added to the room and animal background corrections that were subtracted from acquired images. With the present sheep model and other models, we have had good success in using the bronchoscopic method to map into a specific bronchial airway and deliver radiolabeled tracers directly onto the airway surface and at a later time point in the protocol remap into the airway and redeliver isotopic tracer for a second measurement period of tracer clearance (15).

Additional sheep were evaluated with protocols to investigate blood uptake of unbound label, $^{99m}TcO_4$, and clearance of ^{99m}Tc-DTPA after vagotomy. Isotopic tracer was delivered in these additional protocols in a manner identical to the methods used above in control and interrupted bronchial blood flow protocols. For the vagotomy protocol the vagus nerves were isolated bilaterally and sectioned as previously described (9).

Data Management and Statistics

Radioisotope delivery and clearance data were quantitated with techniques modified from Foster and Freed (15). The initial bronchial image acquired immediately after delivery of the ^{99m}Tc-DTPA was stored on a video screen, and this enabled a region of interest to be selected by cursor manipulation and drawn to cover the airway site of ^{99m}Tc-DTPA delivery. For the clearance of ^{99m}Tc-DTPA, activity time plots were constructed for the region of interest, and the retention of radioactivity within the region during the 30-minute washout was corrected for background and radioactive decay and expressed as a percentage of the ^{99m}Tc-DTPA delivered to the region at time zero (immediately after the nozzle catheter and bronchoscope were withdrawn from the bronchial airway). The natural logarithm of the proportion of radioactivity remaining within an airway region was plotted as a function of time. The semilogarithmic regression line for the interval from peak radioactivity after delivery to the end of the observation time point was determined by a least-squares fit. The slope of the regression line was determined according to the equation, $A = Aoe^{-kt}$, where Ao is the y intercept and A is the count rate at any time t. The slope of the line is the rate constant (k) for the clearance of ^{99m}Tc-DTPA from the bronchus, and it can be converted to a clearance half-time (t_{50}, clearance index) by $t_{50} = 0.693/k$ (15). Corrections to the clearance analysis of ^{99m}Tc-DTPA were not made for nonairway epithelial radioactivity, because it has been demonstrated that such corrections over the 30-minute imaging period do not significantly affect the measured clearance rate (6,16,17).

Comparisons of right to left bronchial clearance, and normal bronchial blood flow to interrupted bronchial blood flow treatment on the ^{99m}Tc-DTPA clearance half-times were accomplished by a paired t-test analysis. Effects of interrupted bronchial blood flow on the bronchial retention levels of ^{99m}Tc-DTPA were analyzed with analysis of variance for repeated measures and a Newman-Keuls post hoc test for significance of the differences. A p-value of <0.05 was considered significant.

RESULTS

A total of 12 sheep were investigated and the baseline bronchial artery pressure was 87 ± 5 mmHg. The mean pressures were obtained during perfusion at the control flow (17 ± 1 mL/min), which had been set based on sheep body weight (27.5 ± 1.7 kg). Mean systemic arterial pressure for the group of sheep evaluated was 95 ± 2 mmHg. Peak inspiratory pressure was 17 ± 2 cm H_2O.

Table 1 shows ^{99m}Tc-DTPA clearance half-times for right and left lobar bronchi that were measured in 10 sheep during the 30-minute imaging period of control bronchial perfusion. The mean half-times for the right and left lobar bronchi were 14.8 (SEM ± 2.7) minutes and 8.6 (SEM ± 1.7) minutes, respectively. The difference in mean clearance half-times between right and left bronchial regions was significant. In eight of the sheep during control bronchial perfusion the corresponding absorption of ^{99m}Tc-DTPA into the systemic blood was assessed, and the time course of this event is demonstrated in Figure 1. The activity present in systemic venous blood peaked on average at 18.5 minutes after delivery of the ^{99m}Tc-DTPA and decreased thereafter to plateau levels that ranged between 60 and 80% of the maximum level. To further characterize the clearance pathways, an estimate of the total ^{99m}Tc-DTPA present in blood was extrapolated from the amount in a 0.5 mL sample and blood volume of the animal (estimated as 8.5% of the mass in kg) (13). Thus, the blood sample with the maximum amount of radioactivity during the sampling period (0–36 min postdelivery) was used to calculate the fraction of ^{99m}Tc-DTPA delivered to the airway that cleared to blood; on average this occurred at the 18.5-minute time point postdelivery. Knowing the initial amount of ^{99m}Tc-DTPA delivered to each bronchial segment,

TABLE 1 ^{99m}Tc-DTPA Clearance Half-times (t_{50}) for Lower Right and Left Bronchi

Animal number	Right bronchus (min)	Left bronchus (min)
02	24.2	21.7
03	15.2	12.0
04	10.4	7.1
06	6.2	5.0
08	15.9	8.4
09	8.0	4.9
10	10.8	4.2
11	34.3	12.1
14	12.7	3.8
18	10.6	7.2
Mean	14.8[a]	8.6[a]
±SE	2.7	1.7

[a] Clearance haftimes (t_{50}): difference between means is significant, $p < 0.01$.

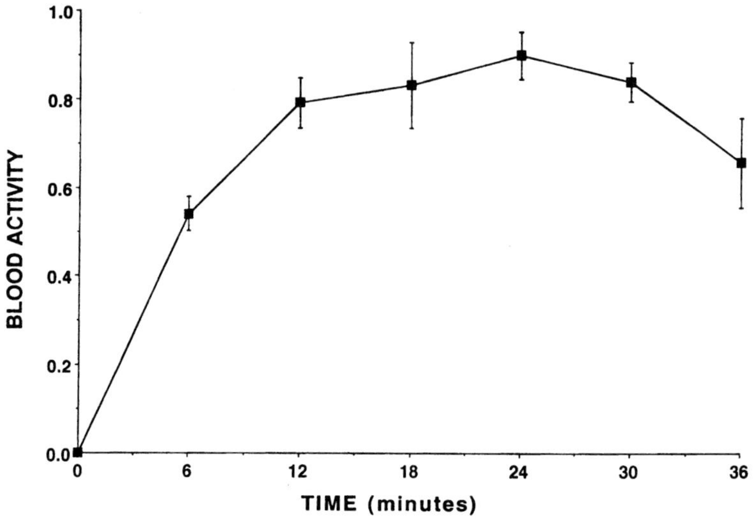

FIGURE 1 Course of ^{99m}Tc activity in systemic venous blood samples. Activity was normalized to maximum level and expressed as a fraction of the maximum activity level over the course of 36 minute. Mean ± SEM activity are indicated for each respective time point after delivery (time = 0) of ^{99m}Tc-DTPA onto the bronchial surface ($n = 8$).

the amount retained, and the fraction cleared to the blood enabled an estimate of the amount cleared by a second pathway for clearance, i.e., mucociliary function, from the start (delivery of isotope onto the bronchial surface) to the point at which concentrations of the tracer were maximal in blood. Using this mass balance approach, the fraction of delivered activity absorbed into the vasculature, cleared by mucociliary function, or retained at the delivery site was on average 0.179, 0.432, and 0.389, respectively. For comparison the mean (± SEM) of each component as a fraction of the amount sprayed onto the airway surface are presented in Figure 2. Thus, at the time point when ^{99m}Tc-DTPA levels in blood were maximal, approximately 18% of the ^{99m}Tc-DTPA delivered to the airway had cleared into the circulation and 43% followed a mucociliary pathway, and this difference was significant.

In additional control sheep we investigated with a separate protocol the absorption into blood of the unbound label, ^{99m}Tc-pertechnetate ($^{99m}TcO_4^-$), delivered directly to the bronchial surface in the same manner as the ^{99m}Tc-DTPA.

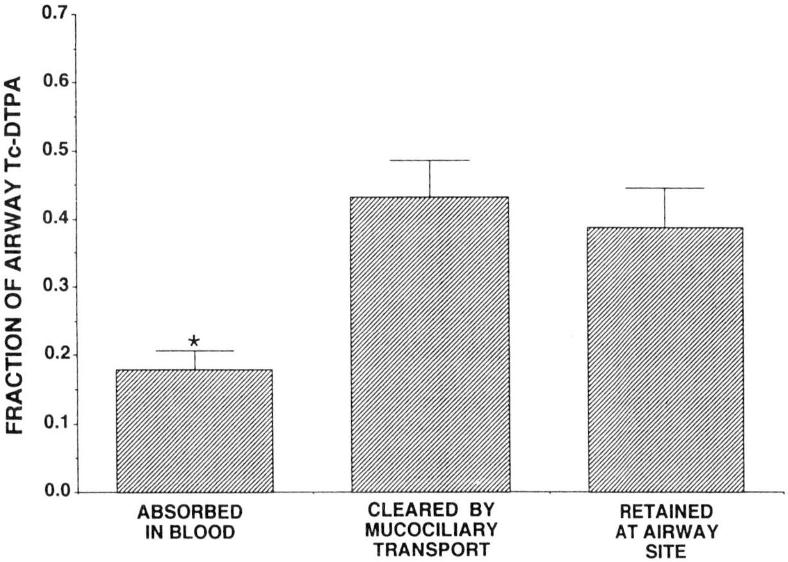

FIGURE 2 Distribution of ^{99m}Tc-DTPA activity in blood, retained at the airway delivery site, and attributed to mucociliary clearance. Means ± SEM are expressed as fraction of the total ^{99m}Tc-DTPA activity delivered to the bronchial surface and represent distribution of activity at the time point after delivery when blood ^{99m}Tc activity was maximal (see Fig. 1). *Significant difference between fractions absorbed into blood and cleared by mucociliary function, $p < 0.05$ ($n = 8$).

Maximal blood uptake for $^{99m}TcO_4$ in three sheep studied with the protocol were similar to the blood level activities found after ^{99m}Tc-DTPA had been delivered to bronchial surfaces within contralateral lobes. Expressed as a fraction of the $^{99m}TcO_4$ or ^{99m}Tc-DTPA delivered, the fractional means were 0.15 and 0.18, respectively. As a further characterization of the factors with potential to influence the clearance processes in one additional animal, the influence of acute bilateral vagotomy on the kinetics of ^{99m}Tc-DTPA adsorption and mucociliary clearance was evaluated. Vagotomy did not appear to influence the clearance kinetics of ^{99m}Tc-DTPA, and clearance half-times for lower right and left bronchi after vagotomy were 7.5 and 6.5 minutes, respectively, and comparable to the range in half-times found under control conditions (Table 1).

In seven animals the interruption of bronchial artery perfusion to the lower lung segmental bronchi led to decreased clearance of ^{99m}Tc-DTPA from the bronchial segments. Paired curves are presented in Figure 3 for retention of ^{99m}Tc-DTPA within lower right bronchi under conditions of control bronchial blood

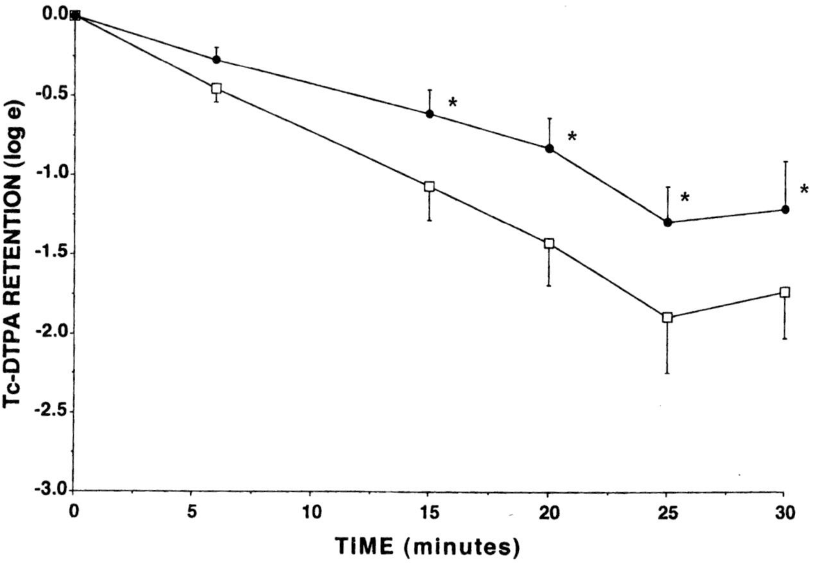

FIGURE 3 Influence of interrupted bronchial perfusion on bronchial airway retention of ^{99m}Tc-DTPA. Mean =SEM retentions of ^{99m}Tc-DTPA are represented for paired measurements made under conditions of control perfusion (□) vs. interrupted perfusion (●). Retention points are expressed as logarithm of fraction of ^{99m}Tc-DTPA activity delivered into the lower right bronchus at time 0. *Significantly different than corresponding retention level for control perfusion, $p < 0.05$ (n = 7).

Table 2 Distribution of Clearance Pathways and Airway Retention of ^{99m}Tc-DTPA

	Animal No.	Absorbed in blood	Cleared by mucociliary transport	Retained at airway site
Normal perfusion				
	08	0.206	0.334	0.460
	09	0.141	0.569	0.293
	16	0.151	0.399	0.450
Mean		0.166	0.434	0.401
Interrupted perfusion				
	08	0.306	0.304	0.390
	09	0.058	0.429	0.513
	16	0.126	0.215	0.659
Mean		0.163	0.316	0.521

flow versus no flow (comparisons were only made for the right bronchial segment). The mean clearance half-times during perfusion, i.e., 13.9 (SEM ± 2.4) minutes, increased significantly to 41.1 (SEM ± 21.8) minutes, representing an approximate twofold increase in the clearance half-time for soluble ^{99m}Tc-DTPA tracer when bronchial perfusion was interrupted. Under the conditions when bronchial blood flow was stopped, we acquired ^{99m}Tc-DTPA blood levels in only three of the seven animals evaluated. Pathway clearance and bronchial retention levels of these three animals under no flow conditions at the time point when ^{99m}Tc-DTPA blood levels were maximal (on average 20 min postdelivery) are presented in Table 2. The fractional amounts of ^{99m}Tc-DTPA delivered to the airway that cleared into vascular and mucociliary pathways or retained had mean values of 0.163, 0.316, and 0.521, respectively. Thus, compared to control flow conditions interruption of bronchial blood flow appeared to predominantly influence the mucociliary clearance pathway, with 43.2% clearing by this pathway when bronchial blood flow was intact (Fig. 2) versus 31.6% when blood flow was interrupted.

DISCUSSION

We have developed in the laboratory sheep model methodology to evaluate the clearance pathways for soluble tracers placed directly onto the bronchial epithelial surface. Although our measurements were accomplished in an anesthetized animal that was paralyzed, intubated, and mechanically ventilated, we believe the results provide information relevant to understanding the pathways and time course for absorption and airway clearance of soluble substances that deposit onto intrathoracic airway surfaces. We chose to use ^{99m}Tc-DTPA as a soluble tracer because of its molecular radius (~0.57 nm) and its γ-imaging capabilities. We found under conditions of control bronchial perfusion that clearance was different between lower right and left bronchial segments and that the difference was significant (Table 1). In additional animals we measured the influence of perfusion to the lung airway segments on the clearance of DTPA. Interruption of bronchial blood flow caused a significant slow down in the clearance (twofold increase in the clearance half-time as compared to controlled perfusion conditions). In separate investigations, the addition of vagotomy did not appear to alter the clearance kinetics for ^{99m}Tc-DTPA, and when $^{99m}TcO_4^-$ was compared to ^{99m}Tc-DTPA, the fractions taken up by blood were similar for both tracers, as were the time courses for absorption into blood.

The interplay between fluid transfer across the airway epithelium (permeability) and the efficiency of particle clearance by the mucociliary apparatus along with respiratory secretions and cellular debris atop the ciliated epithelium have often been postulated as essential components of epithelial barrier function (18). An association between permeability and the kinetics of mucociliary clearance of the airway epithelial surface has not been firmly established in vivo; however, in this connection it is known that smokers have increased parenchymal permeability (19) and abnormal mucociliary function within small peripheral airways (20). Results of two human studies suggest that when an aerosol of ^{99m}Tc-DTPA is deposited onto the airways, it not only permeates the airway wall and enters the blood, but is removed by mucociliary transport as well (21,22). In both these studies systemic blood levels of radiolabel were not assessed, but rather clearance of radiolabel was compared on separate test days using tracers with widely differing molecular weights and calculating clearance half-times for each. In animal studies, however, clearance of ^{99m}Tc-DTPA delivered directly to the bronchial mucosa did not appear to follow a mucociliary pathway. Wolfe and colleagues placed 50 μL of soluble tracer directly onto airway surfaces at two intrathoracic airway sites (fifth- and tenth-generation airways) in a laboratory dog model and showed that there were no significant differences in clearance rates of the instilled tracer from these two discrete regions of the tracheobronchial airway (2). Interestingly, these investigators did not observe any mucociliary clearance of the tracer and removal appeared to be entirely via a vascular pathway. Since the animal model was anesthetized with halothane, an anesthetic known to reduce mucociliary transport (23), the authors suggested that this might explain the lack of a mucociliary component in their clearance measurement. Lay et al. (3), also using a dog model anesthetized with halothane, evaluated the clearance of a small volume (6 μL) of soluble tracer that was sprayed directly onto the surface of a third-generation airway. In this study, only a small fraction of the tracer appeared to follow a mucociliary clearance pathway, with approximately 90% being absorbed into the circulation and about 5% retained at the delivery site to the airway.

Our results clearly demonstrate co-pathways for clearance of a small molecular weight tracer placed directly onto the surface of an intrathoracic airway, with the mucociliary component responsible for about 43% and absorption by blood for approximately 18% of clearance under conditions of control bronchial perfusion. As controls for radiolabel leaching and potential for local irritant effects of the procedures on clearance pathways, we measured the absorption into blood of unbound $^{99m}TcO_4$ and in addition evaluated the influence of bilateral vagotomy on the clearance of ^{99m}Tc-DTPA. Blood uptake and clearance protocols using the unbound label and after vagotomy, respectively, gave results similar to the clearance of ^{99m}Tc-DTPA observed from the right and left bronchi during control bronchial perfusion.

The finding that kinetics of clearance (combination of mucociliary mechanisms and absorption) can differ significantly on a regional basis between lower right and left bronchi was not surprising (Table 1). For example, in the dog, and in sheep as well, we have found regional differences in mucociliary function to be present (14,24). The factors responsible for these differences are not well understood and may be related to the topography of the cellular epithelium (density of ciliated and secretory cells) and completeness of the liquid lining. The size and amount of DTPA droplets delivered to the lung can influence the adsorption rate of the DTPA (25). This effect was not influential in our measurements because the size, volume, and delivery time of DTPA droplets were controlled with equal amounts of DTPA delivered locally onto the bronchial surface at baseline and during retesting of clearance under experimental conditions. Others have shown using a dog model that when ^{99m}Tc-DTPA liquid was instilled onto airway surfaces (trachea and fifth- and tenth-generation airways), the clearance rate of the DTPA tracer into the vasculature was uniform between sites. However, using a bronchoscope to deliver an aerosol of ^{99m}Tc-DTPA in the dog, we found regional differences in absorption can exist within sublobar bronchial airways (15). Thus, with respect to permeability of the airway epithelium to soluble tracers, local differences may exist, which parallels the situation widely found for the alveolar epithelium in both the human and dog lung (15,26). Regional differences in vascular absorption are believed to reflect variation in epithelial permeability rather than nonuniform endothelial permeability or a local disparity in the capillary network serving the tissue.

With respect to the influence of perfusion on the clearance of soluble tracer from the airway surface, interruption of bronchial blood flow significantly delayed clearance. It was not surprising that the interruption of blood flow decreased the component cleared by mucociliary function (Fig. 3). In a previous report in the sheep model, we demonstrated the dependency of the mucociliary function system on bronchial blood supply (14). Although the precise mechanism(s) leading to delays in mucociliary clearance are not certain, some level of bronchial perfusion seems likely to be required to maintain homeostasis (heating/humidification) of the airway environment and normal ciliary and periciliary fluid interaction. For three animals in which we had measures of ^{99m}Tc-DTPA blood levels, when bronchial blood flow was interrupted the absorption of ^{99m}Tc-DTPA into the vascular clearance pathway remained comparable to the vascular clearance levels when bronchial perfusion was under normal flow conditions (Figs. 2 and 3). This result is not in agreement with the report by Hanafi and colleagues (7), in which decreased perfusion resulted in an increase in the venous uptake and clearance of a soluble tracer. This difference may be explained by their model in which absorption was limited to the tracheal airway and in addition their approach (liquid-filled lumen) precluded any assessment of mucociliary function as a component pathway of airway clearance of a soluble tracer. We do not have a definitive explanation why blood levels of ^{99m}Tc-DTPA remained approximately the same even when bronchial blood flow was suspended. It would appear that even without a bronchial blood supply, ^{99m}Tc-DTPA delivered to the bronchi gets into the systemic venous blood pool. Activity that cleared from the bronchi by mucociliary function was transported through airways such as the trachea where, in our model, we did not have control of the blood supply. Thus it is likely that ^{99m}Tc-DTPA was absorbed into the blood supply serving the trachea. Blood activity levels of ^{99m}Tc-DTPA represent an active process of both absorption (from the bronchial site, and from airway surfaces along the mucociliary clearance pathway as ^{99m}Tc-DTPA is transported to the trachea) and clearance by the kidney (12). As demonstrated in Figure 1, uptake of ^{99m}Tc-DTPA into blood attained a maximum level at about 18 minutes after delivery onto the bronchial airway surface, and after the maximum is attained it is presumed that clearance by the kidney begins to predominate over absorption processes. Since some losses in blood activity occur during the initial attainment of the maximum, these losses represent a slight underestimation of the amount absorbed into blood (Figure 2). This component is small and when added to the amount of ^{99m}Tc-DTPA cleared by mucociliary function represent the difference between the total amount of ^{99m}Tc-DTPA initially delivered to the bronchial surface minus the activity that is (1) found in blood and (2) retained at the airway delivery site.

In summary, we have demonstrated that a soluble, low molecular weight substance cleared the intrathoracic airway surface by two primary pathways: directly absorbed into blood and by mucociliary transport into the tracheal airway. Regional right to left bronchial differences in clearance appear to exist, but clearance from the lung of small liquid volumes of soluble tracer directly placed onto the bronchial surface does not appear to be influenced by bilateral vagotomy. When the bronchial blood supply serving the intrathoracic airways is interrupted, bronchial clearance is delayed, and this response in large measure reflects an impairment in mucociliary function.

ACKNOWLEDGMENTS

Dr. Michael Foster is supported from awards of the Center for Indoor Air Research (#97-11, Linthicum, MD) and the National Institute of Environmental Health Sciences (#ES-03819, Washington, DC). The research was supported by an award from the National Heart Lung and Blood, NIH, Bethesda, MD (#RO1HL58577).

REFERENCES

1. WM Foster. Deposition and clearance of inhaled particles. In: S Holgate, J Samet, H Koren, eds. Air Pollution and Health. New York: Academic Press, 1999, pp 295–324.
2. R Wolff, JV Kitzman, BA Muggenburg, JL Mauderly. Clearance of ^{99m}Tc-DTPA from four sites in the respiratory tract of dogs. J Aer Med 1:371–377, 1988.
3. JC Lay, CR Berry, CS Kim, WD Bennett. Retention of insoluble particles after local intrabronchial deposition in dogs. J Appl Physiol 79:1921–1929, 1995.
4. MP Barrowcliffe, JG Jones, JE Agnew, RA Francis, SW Clarke. The relative permeability of human conducting and terminal airways to ^{99m}Tc-DTPA. Eur J Respir Dis 71(Suppl 153):68–77, 1987.
5. EM Wagner, WA Mitzner, RH Brown. Site of functional bronchopulmonary anastomoses in sheep. Anat Rec 254:360–366, 1999.
6. NW Rizk, JM Luce, J Hoeffel, DC Price, JF Murray. Site of deposition and factors affecting clearance of aerosolized solute from canine lungs. J Appl Physiol 56:723–729, 1984.
7. Z Hanafi, DR Corfield, SE Weber, JG Widdicombe. Tracheal blood flow and luminal clearance of ^{99m}Tc-DTPA in sheep. J Appl Physiol 73:1273–1281, 1992.
8. Z Hanafi, SE Weber, JG Widdicombe. Permeability of ferret trachea in vitro to ^{99m}Tc-DTPA and [^{14}C]antipyrine. J Appl Physiol 77:1263–1273, 1994.
9. EM Wagner, DB Jacoby. Methacholine causes reflex bronchoconstriction. J Appl Physiol 86:294–297, 1999.
10. MW Billinghurst. Chromatographic quality control of ^{99m}Tc-labeled compounds, J Nucl Med 28:903–906, 1973.
11. MD Hoover, JR Harkema, BA Muggenburg, JW Spoo, P Gerde, HJ Staller, JA Hotchkiss. A microspray nozzle for local administration of liquids or suspensions to lung airways via bronchoscopy. J Aer Med 6:67–72, 1993.
12. JW Stather, H Smith, MR Bailey, A Birchall, RA Bulman, FEH Crawley. The retention of ^{14}C-DTPA in human volunteers after inhalation or intravenous injection. Health Phys 44:45–52, 1983.
13. S Schermer. The Blood Morphology of Laboratory Animals. Philadelphia: FA Davis, 1967.

14. EM Wagner, WM Foster. Importance of airway blood flow on particle clearance from the lung. J Appl Physiol 69:1878–1883, 1996.
15. WM Foster, AN Freed. Regional clearance of solute from peripheral airway epithelia: recovery after sublobar exposure to ozone. J Appl Physiol 86:641–646, 1999.
16. G Coates, H O'Brodovich. Extrapulmonary radioactivity in lung permeability measurements. J Nucl Med 28:903–906, 1987.
17. G Oberdorster, MT Utell, DA Weber, M Ivanovich, RW Hyde, PE Morrow. Lung clearance of inhaled ^{99m}Tc-DTPA in the dog. J Appl Physiol 57:589–595, 1984.
18. I Nathanson, JA Nadel. Movement of electrolytes and fluid across airways. Lung 162:125–137, 1984.
19. BD Minty, C Jordan, JG Jones. Rapid improvement in abnormal pulmonary epithelial permeability after stopping cigarettes. Br Med J 282:1183–1186, 1981.
20. WM Foster, EG Langenback, EH Bergodsky. Disassociation in the mucociliary function of central and peripheral airways of asymptomatic smokers. Am Rev Respir Dis 132:633–639, 1985.
21. JS Ilowite, WD Bennett, MS Sheetz, ML Groth, DM Nierman. Permeability of the bronchial mucosa to ^{99m}Tc-DTPA in asthma. Am Rev Respir Dis 139:1139–1143, 1989.
22. WD Bennett, JS Ilowite. Dual pathway clearance of ^{99m}Tc-DTPA from the bronchial mucosa. Am Rev Respir Dis 139:1132–1138, 1989.
23. AR Forbes. Halothane depresses mucociliary flow in the trachea. Anesthesiology 45:59–63.
24. WM Foster, AN Freed. Methodology for delivery and kinetics of clearance of insoluble particles from sublobar lung segments. Inhal Tox 12(Suppl. 1):99–105, 2000.
25. S Groth, J Mortensen, P Lange, S Vest, N Rossing, D Swift. Effect of change in particle number on pulmonary clearance of aerosolized ^{99m}Tc-DTPA. J Appl Physiol 66:2750–2755, 1989.
26. WM Foster, PT Stetkiewicz. Regional clearance of solute from the respiratory epithelia: 18–20 h postexposure to ozone. J Appl Physiol 81:1143–1149, 1996.

34

The Role of Cough in Lung Mucus Clearance

Amir Hasani and John E. Agnew
Royal Free and University College Medical School
London, United Kingdom

Mucus secretion, cellular debris, and any insoluble deposited particles can be cleared from the conducting airways of the human lungs by two main clearance mechanisms: mucociliary and cough clearance. These clearance mechanisms help to keep the lungs clear and free from infection and pollutants.

Cough is an important defense mechanism of the lungs. It is a reserve mechanism for mucus clearance (1). Cough rarely occurs in healthy subjects except in emergency situations, following the inhalation of a foreign body or bronchial irritants. It can come dramatically into action in airway diseases where mucociliary transport is often compromised (2) and the increased amount of secretions cannot be adequately removed by mucociliary clearance.

RADIOAEROSOL TECHNIQUE FOR MEASURING MUCUS TRANSPORT

The radioaerosol technique is now widely used for measuring lung mucus transport in humans (3). This technique involves the inhalation of an insoluble aerosol firmly labeled with a gamma-emitting radioisotope. The radiation from the subject's chest can be monitored by external scintillation counters or a gamma camera, and therefore the initial radioaerosol deposition in the lungs can be determined. Sequential counting of the lung radioactivity shows a progressive fall due to two factors: biological clearance of the radioaerosol and physical decay of the radionuclide. Correction for the physical decay can readily be made knowing the physical half-life of the radionuclide and a lung clearance curve can be obtained for the radioactivity remaining in the lungs (as a percentage of the initial count) against time.

It is conventionally accepted that the lung retention of radioaerosols 24 hours postinhalation indicates the proportion of the inhaled aerosols deposited in the alveolar region of the lungs (4). Even though this assumption may not be totally valid, the estimate of alveolar deposition so obtained when subtracted from the lung clearance curve gives a tracheobronchial clearance curve. This curve represents the amount of radioaerosol that was cleared from the conducting airways.

Aerosol particles can be produced by a jet or ultrasonic nebulizer (5) or by a spinning disc generator (6). The last apparatus is capable of producing monodisperse aerosols of any size in the range of 1–10 μm. A variety of aerosols are being used for measuring lung mucus clearance including polystyrene, iron oxide, Teflon, Lucite, and resin (7). These insoluble aerosols are labeled with several types of gamma-emitting radionuclides (5). The most widely used radioisotope for labeling aerosol particles is technetium-99m (^{99m}Tc).

The radioaerosols should be administered to human subjects under controlled conditions to minimize variability in their initial topographical distribution and hence in the measurement of mucus clearance. The system for the generation of radioaerosols and their delivery needs to be reliable and simple so as to get the required deposition within the lungs. Deposition of inhaled particles in the lungs in inevitably affected by the physical properties and the mode of inhalation of the aerosol as well as the patency of the airways (8).

WHOLE LUNG CLEARANCE

Many studies evaluating the efficacy of cough on the transport of inhaled deposited radioaerosol particles from the human lungs have been reported. Toigo and associates (9) suggested that cough increased the clearance of labeled carbon particles (40–70 μm in diameter) in eight healthy subjects and eight patients with chronic lung disease. They observed that after each cough the amount of radiation, measured over the carina, dropped sharply. However, Yeates and associates (10) studying 42 healthy nonsmoking adults reported that coughing did not greatly affect the movement of a local concentration (bolus) of microspheres in the trachea, whereas coughing was the major clearance mechanism from the trachea in patients with cystic fibrosis (11). These results were confirmed by Puchelle and associates (12) in patients with chronic bronchitis. The mean percentage of bronchial radioactivity cleared after one hour from inhalation by mucociliary clearance in 10 healthy subjects was 30%, which was about twice that eliminated by 27 chronic bronchitics (14%). At the end of the one-hour period, the healthy subjects and patients were asked to cough and the percentage of retained bronchial radioactivity was measured again. The chronic bronchitics eliminated a further 20% compared to the healthy subjects, who cleared only 2.5%. This study demonstrated that in the chronic bronchitics a high percentage of deposited particles was cleared by coughing, while in the healthy subjects this was not so. The effect of cough was also studied in eight patients with respiratory tract disease and six healthy subjects by Camner and associates (13). The healthy subjects did not produce any sputum and did not clear any test particles on instructed vigorous coughing, whereas the patients who produced sputum (6 of 8) were able to clear test particles from their lungs by coughing.

It thus appears that the presence of an increased amount of mucus is an essential prerequisite for cough to be effective as a clearance mechanism. We have reported that tracheobronchial clearance of deposited radioaerosol was faster in chronic bronchitic patients who produced a high volume of sputum and coughed frequently (14). We also suggested that cough may be a very important clearance mechanism in asthma (15). During the first 2 hours of the study, the asthmatic patients who coughed frequently cleared more than twice as much radioaerosol as the asthmatic patients who coughed less despite a slightly higher initial radioaerosol penetration.

Voluntary coughing in 12 patients with immotile-cilia syndrome was also found to be an important clearance mechanism (16); on average, 30% of the deposited particles were removed after coughing. Directed coughing was evaluated in 10 patients with obstruction and copious sputum by Sutton and associates (17). Each patient underwent a 30-minute treatment period in which the patients were asked to cough and another 30 minutes as a control period after inhaling radioaerosol particles. The percentage of radioactivity remaining in the lungs after the treatment period was less than the control period, during which the reduction in radioactivity was due to mucociliary clearance and spontaneous coughing.

The effect of a controlled coughing maneuver (10 coughs every 10 minutes for 1 hour) compared to a control period in 12 nonsmoking healthy subjects was investigated by Bennett and associates (18). The controlled cough consisted of forceful exhalation against a closed solenoid valve, which automatically opened after a threshold airway pressure was reached. The amount of radiolabeled particles retained after the coughing maneuver were significantly less than that retained during the control period. The same results were observed with rapid inhalations (90 inhalations per hour) rather than exhalations (coughs) when compared to a control period in eight healthy subjects. Therefore, they postulated that the observed enhancement of mucus clearance by controlled coughing might be due to a stimulation of the mucociliary mechanism. However, using the same protocol, 10 young asymptomatic smokers were unable to enhance their rate of mucus clearance by coughing or rapid inhalations suggesting a change in the mucociliary transport from normal (19).

REGIONAL LUNG CLEARANCE

Estimation of regional mucus clearance has been carried out by relatively few investigators. There has been general agreement that cough significantly enhances the clearance of mucus from the lungs as a whole and from the central region and that it almost fully compensates for a defective mucociliary clearance in patients with hypersecretion. However, attempts to study lung mucus clearance from the peripheral region have yielded contradictory results.

Oldenburg and colleagues (20), studying eight patients with obstructive chronic bronchitis (mean % predicted FEV_1 was 53) and producing sputum volumes ranging from 10 to 120 mL/day, found that coughing produced a very significant effect on whole lung and peripheral lung mucus clearance. Using similar technique, Bateman and associates (21) studied six patients (three chronic bronchitics and three bronchiectatics) who were more severely obstructed (mean % predicted FEV_1 was 37) and were producing larger sputum volumes (50–300 mL/day). Mucus clearance during cough was significantly increased from the whole lung and central and intermediate regions. However, no significant effect on clearance from the peripheral region was noted, and that differs from the study of Oldenburg and colleagues (20). It is possible in the Oldenburg study that the significant mucus clearance reported from the peripheral region reflected clearance from larger conducting airways since the proximal boundary of the peripheral region corresponded approximately to 2–4 mm airways (i.e., in the peripheral region there must have existed large airways). Furthermore, in the study of Oldenburg the initial radioaerosol deposition was more central than in the Bateman study (21).

Six patients with cystic fibrosis were studied by Rossman and associates (22). The mean % predicted FEV_1 was 38 and the mean sputum volume was 67 mL/day. The effect of cough on mucus clearance was similar to that reported by Oldenburg and associates study (20), where after coughing the mucus transported from the peripheral region of the lungs was increased as well as from the central region.

We have shown that cough is effective in clearing lung secretions from central and intermediate regions in a group of 19 patients with airways obstruction with mean % predicted FEV_1 of 52 and sputum wet weight of 37 mL/day (23). This is in agreement with the Bateman and colleagues study (21) but in contrast with those reported by Oldenburg and associates (20) and Rossman and colleagues (22). The apparent discrepancy may have arisen because of differences in the selection of regions of interest between those studies and because more of the radiotracer (20% approximately) was deposited in the outer region in the studies of Bateman and ours than in the other two studies (12.5% approximately). However, in another group of 14 patients with airways obstruction (mean % predicted FEV_1 was 54 and sputum production was 9 mL/day) we were able to demonstrate that cough resulted in movement of secretions proximally from all regions of the lungs (24). These findings were supported by the study of Bennett and associates (25), which also demonstrated that voluntary cough was effective in clearing lung secretions from the central as well as the peripheral regions of the lungs. The reason that cough in our study (23) did not achieve significant level in clearing lung secretions from the peripheral airways could be attributed to the amount of daily sputum produced by the patients.

Recently we evaluated the effect of different airway diseases on lung mucus transport during instructed coughing (26). Cough was able to significantly enhance the movement of mucus from the central and peripheral regions of the lungs in eight patients with chronic bronchitis (mean ± SEM % predicted FEV_1 of 41 ± 3). However, in two groups of eight patients with asthma and eight patients with bronchiectasis (Mean ± SEM % predicted FEV_1 of 65 ± 6 and 57 ± 10, respectively), cough was only able to significantly enhance clearance from the central region of the lungs (Fig. 1). These results were control corrected where

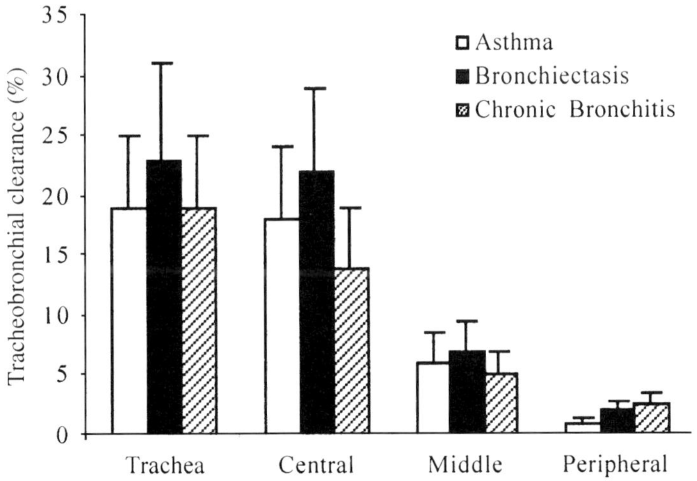

FIGURE 1 Mean ± SEM tracheobronchial clearance of radioaerosol particles from four regions of the lungs during cough in patients with asthma, bronchiectasis, and chronic bronchitis.

patients did not cough, and thus mucus clearance was primarily achieved by mucociliary transport mechanism. The difference in cough action in chronic bronchitic, asthmatic, and bronchiectatic patients could be due to differences in physiochemical properties of mucus in each group.

CONCLUSION

Cough in patients with airways obstruction results in an improvement of whole lung clearance and movement of secretions proximally from all regions of the lungs. As such coughing (productive or unproductive) must be partly compensating for the well-documented compromised mucociliary transport in such patients. It therefore follows that allowance should be made for coughing when interpreting lung mucociliary clearance curves.

REFERENCES

1. A Hasani, D Pavia. Cough as a clearance mechanism. In: PC Braga, L Allegra, eds. Cough. New York: Raven Press, 1989, pp 37–52.
2. E Houtmayers, R Gosselink, G Gayan-Ramirez, M Decramer. Regulation of mucociliary clearance in health and disease. Eur Respir J 13:1177–1188, 1999.
3. A Hasani, D Pavia, S Rotondetto, SW Clarke, MA Spiteri, JE Agnew. Effect of oral antibiotics on lung mucociliary clearance during exacerbation of chronic obstructive pulmonary disease. Respir Med 92:442–447, 1998.
4. WD Bennett, WF Chapman, JC Lay, TR Gerrity. Pulmonary clearance of inhaled particles 24 to 48 hours post deposition: effect of beta-adrenergic stimulation. J Aerosol Med 6:53–62, 1993.
5. SP Newman. Production of radioaerosols. In: SW Clarke, D Pavia, eds. Aerosols and the Lung. Boston: Butterworths, 1984, pp 71–91.
6. KR May. An improved spinning top homogeneous spray apparatus. J Appl Physiol 20:932–938, 1949.
7. D Pavia, JRM Bateman, NF Sheahan, JE Agnew, SP Newman, SW Clarke. Techniques for measuring lung mucociliary clearance. Eur J Respir Dis 61(suppl 110): 157–177, 1980.
8. JE Agnew. Physical properties and mechanisms of deposition of aerosols. In: SW Clarke, D Pavia, eds. Aerosols and the Lung. Boston: Butterworths, 1984, pp 49–70.
9. A Toigo, JJ Imarisio, H Murmall, MN Lepper. Clearance of large carbon particles from the human tracheobronchial tree. Am Rev Respir Dis 87:487–492, 1963.
10. DB Yeates, N Aspin, H Levison, M Jones, AG Byran. Mucociliary tracheal transport rates in man. J Appl Physiol 39:487–495, 1975.
11. DB Yeates, J Sturgess, S Khan, H Levison, N Aspin. Mucociliary transport in trachea of patients with cystic fibrosis. Arch Dis Childhood 51:28–33, 1976.
12. E Puchelle, JM Zahm, F Girard, A Bertrand, JM Polu, F Aug, P Sadoul. Mucociliary

transport in vivo and in vitro: relation to sputum properties in chronic bronchitis. Eur J Respir Dis 61:254–264, 1980.

13. P Camner, B Mossberg, K Philipson, K Strandberg. Elimination of test particles from the human tracheobronchial tract by voluntary coughing. Scand J Respir Dis 60:56–62, 1979.
14. JE Agnew, F Little, D Pavia, SW Clarke. Mucus clearance from the airways in chronic bronchitis- smokers and ex-smokers. Bull Eur Physiopath Resp 18:473–484, 1982.
15. JE Agnew, JRM Bateman, NF Sheahan, AM Lennard-Jones, D Pavia, SW Clarke. Effect of oral corticosteroids on mucus clearance by cough and mucociliary transport in stable asthma. Bull Eur Physiopath Resp 19:37–41, 1983.
16. B Mossberg, BA Afzelius, R Eliasson, P Camner. On the pathogenesis of obstructive lung disease: a study on the immotile-cilia syndrome. Scand J Respir Dis 59:55–65, 1978.
17. PP Sutton, RA Parker, BA Webber, SP Newman, N Garland, MT Lopez-Vidriero, D Pavia, SW Clarke. Assessment of the forced expiration technique, postural drainage and directed coughing in chest physiotherapy. Eur J Respir Dis 64:62–68, 1983.
18. WD Bennett, WM Foster, WF Chapman. Cough-enhanced mucus clearance in the normal lung. J Appl Physiol 69:1670–1675, 1990.
19. WD Bennett, WF Chapman, TR Gerrity. Ineffectiveness of cough for enhancing mucus clearance in asymptomatic smokers. Chest 102:412–416, 1992.
20. FA Oldenburg, MB Dolovich, JM Montgomery, MT Newhouse. Effect of postural drainage, exercise and cough on mucus clearance in chronic bronchitis. Am Rev Respir Dis 120:739–745, 1979.
21. JRM Bateman, SP Newman, KN Daunt, NF Sheahan, D Pavia, SW Clarke. Is cough as effective as chest physiotherapy in the removal of excessive tracheobronchial secretions? Thorax 36:683–687, 1981.
22. CM Rossman, R Waldes, D Sampson, MT Newhouse. Effect of chest physiotherapy on the removal of mucus in patients with cystic fibrosis. Am Rev Respir Dis 126:131–135, 1982.
23. A Hasani, D Pavia, JE Agnew, SW Clarke. Regional lung clearance during cough and forced expiration technique (FET): effect of flow and viscoelasticity. Thorax 49:557–561, 1994.
24. A Hasani, D Pavia, JE Agnew, SW Clarke. Regional mucus transport following unproductive cough and forced expiration technique in patients with airways obstruction. Chest 105:1420–1425, 1994.
25. WD Bennett, KL Zeman. Effect of enhanced supramaximal flows on cough clearance. J Appl Physiol 77:1577–1583, 1994.
26. A Hasani, N Toms, JE Agnew, D Pavia, SW Clarke. Regional mucus clearance following cough and forced expiration technique in patients with asthma, chronic bronchitis and bronchiectasis (abstr). Eur Respir J 8(suppl 19):297s, 1995.

35

Cough Clearance of Mucus Simulants in Endotracheal Tubes: Patterns of Catastrophic Separation at Controlled Linear Velocities

Peter Krumpe and Bruce Denney
University of Nevada and VA Sierra Nevada Health Care System
Reno, Nevada

Amgad Hassan, Ross Albright, and Cahit Evrensel
University of Nevada
Reno, Nevada

INTRODUCTION

Cough requires coupling of air flowing over airway mucus to develop surface waves on the mucus. The shear causing the formation of standing waves is created by linear velocity of airflow. Mucus adheres to underlying airway surfaces so that adhesive forces oppose airflow-induced shear forces. In order for mucus to be coughed out, a separation of the mucus from the airway surface must occur (1). Such a separation occurs suddenly and is rather like an avalanche (2). The case of cough-induced catastrophic separation of mucus from airway can be thought of as a ''mucolanche.''

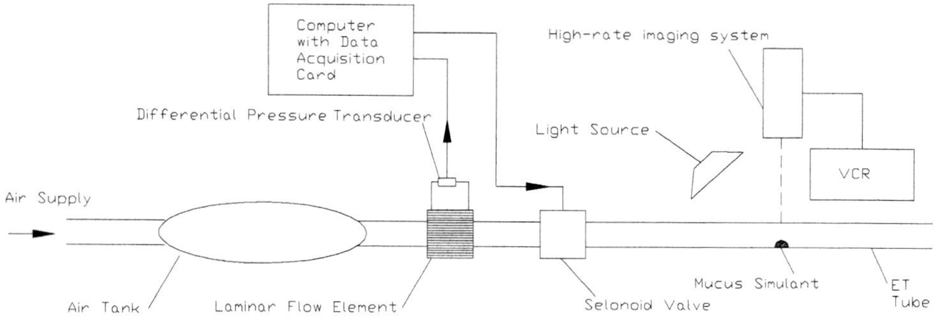

FIGURE 1 Experimental design.

Previous studies have demonstrated that mucus simulants (MS) with lower viscoelastic (VE) ratios (given by η'/G') are able to develop standing waves on their surfaces when studied in Plexiglas channels (3). Mucus simulants undergo catastrophic separation at lower linear velocity (U) than did mucus simulants with higher VE ratios (4).

We postulated that similar conditions would occur when mucus simulants were studied in dry endotracheal (ET) tubes. We used dry ET tubes to minimize the wetting effect of water or surface-active materials that would tend to allow MS globs to spread along rather than separate from the tube wall. The ability to cough clear secretions from endotracheal tubes is of interest because these tubes are used for airway support in mechanically ventilated patients. We reasoned that the flow velocities required to create standing waves on the surface of mucous globs and to clear mucus simulants would relate to the rheologic properties of the simulants (5). Hypothetically, simulants with lower VE ratios should develop deeper standing waves and exhibit catastrophic separation from the underlying ET tube at lower linear velocity than would simulants with higher VE ratios.

METHODS

A simulated cough machine was developed that provided a 0.5-ms burst of square wave flow pattern at flow rates between 60 and 400 L/min (Fig. 1). These flows are characteristic of adult male expiratory flows, ranging from flows seen during exacerbations of airway disease to those of normal health. They were forced into ET tubes having internal diameters of 6–12 mm, thereby creating cough linear velocities that would likely be seen in intubated patients. Mucus simulants were prepared using locust bean gum cross-linked with Borax (3, 6, 9, and 12%) as previously described (B. Rubin, personal communication, 1995). MS rheological studies were performed using an S5 Rheometer (Rheometric Scientific, Piscataway, NJ). Each simulant underwent calculation of VE ratio; elasticity and viscosity were plotted against frequency of oscillation.

Cough testing was performed by placing a 0.5 cc glob of MS in the ET tube. Air-mucus interactions were videotaped during 0.5-second bursts of known airflow (V) from 60 to 400 L/min in real time (Sony, Hi-8), and with a high-speed camera (Kodak, Ektapro), so as to view wave initiation on the MS surface. The high-speed camera (capable of resolving 2000 frames per second) and a standard VCR camera were placed so as to observe surface changes at the initiation of cough flow and the bulk movement of the MS glob after the cough.

RESULTS

Rheological studies are shown in Figures 2–4. The 12, 9, and 6% Borax MS preparations behaved more or less the same but were different from the 3% MS. VE ratios were higher with the 3%, and although VE ratios decreased for all MS borax concentrations, the 3% exceeded the others. The 12% MS preparation

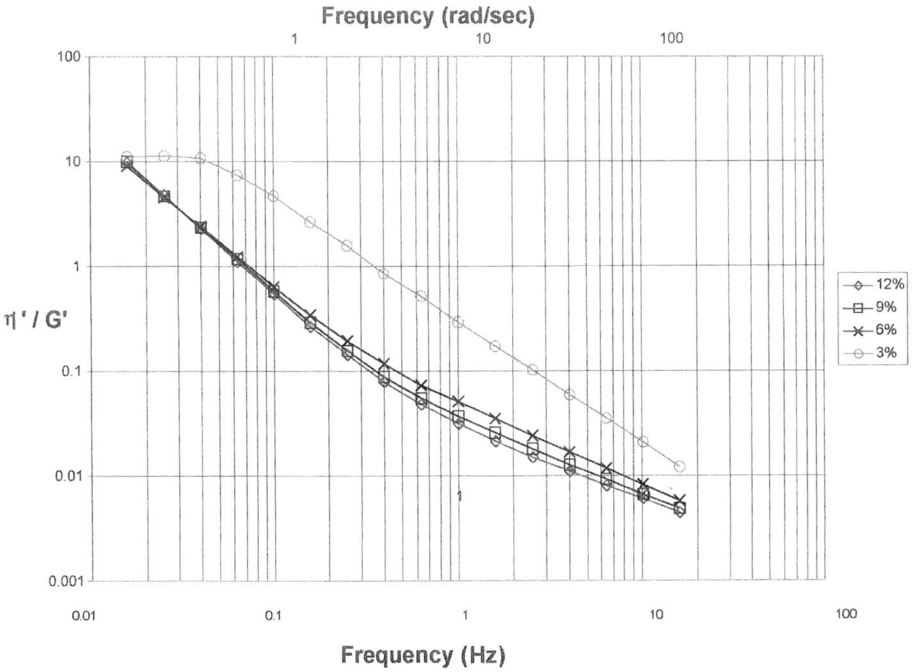

Figure 2 Mucus simulant rheological measurements—viscoelasticity ratio.

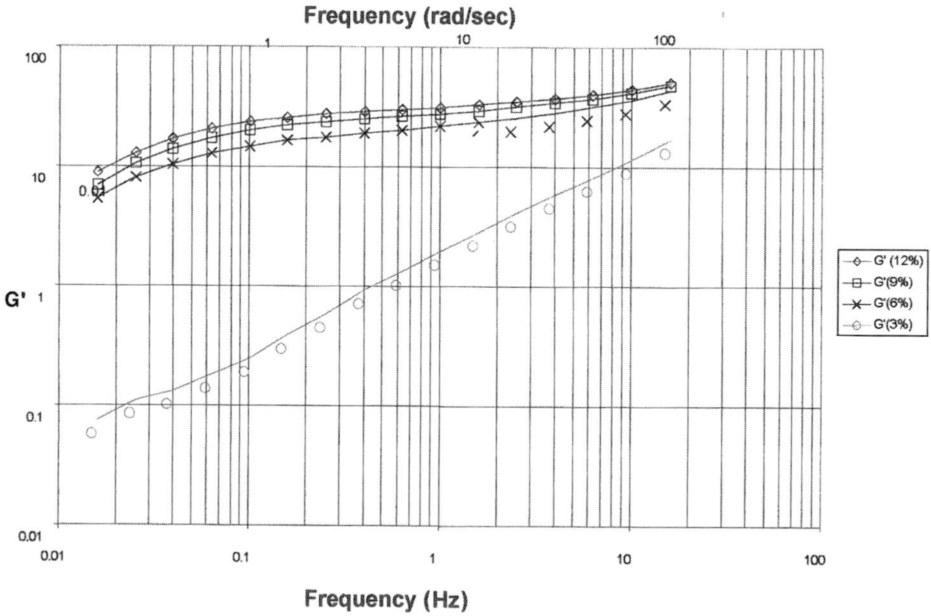

Figure 3 Mucus simulant rheological measurements—elasticity vs. frequency.

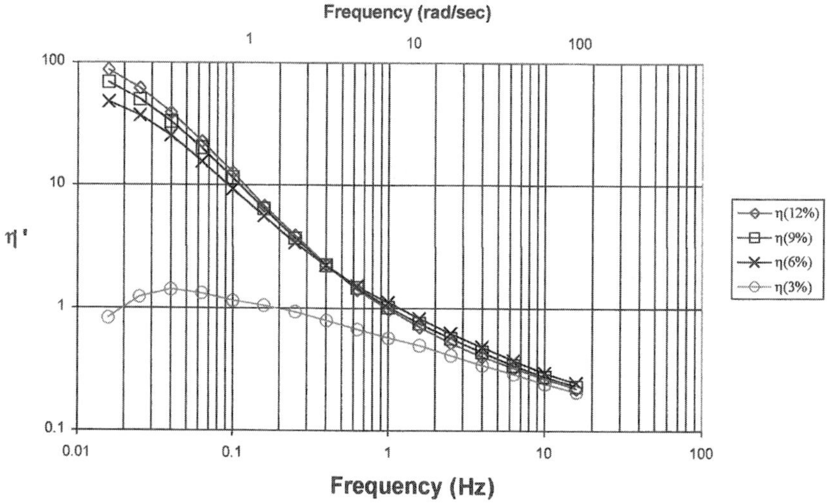

Figure 4 Mucus simulant rheological measurements—viscosity vs. frequency.

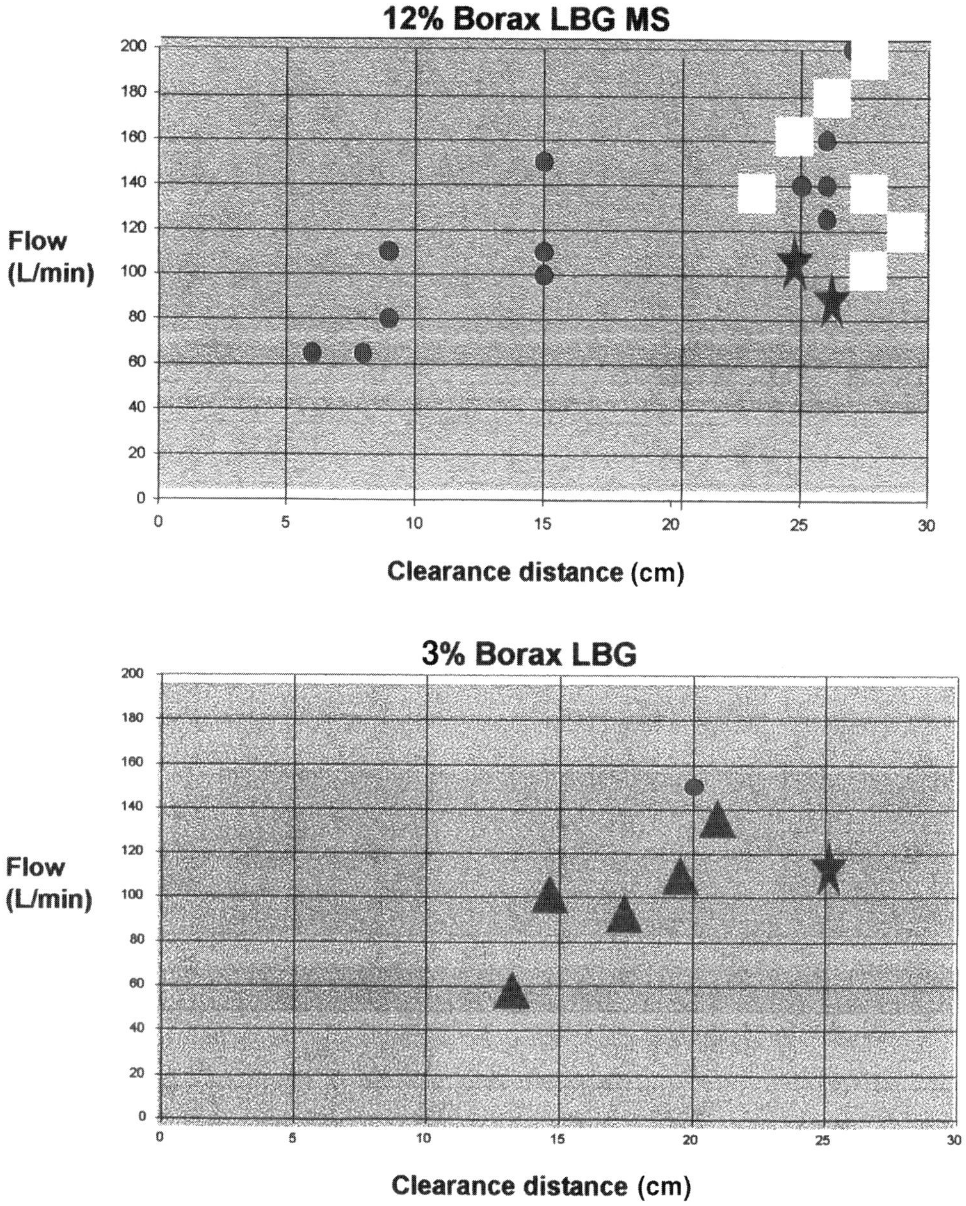

FIGURE 5 Patterns of cough clearance from dry #10 ID endotracheal tube (see text for explanation of symbols).

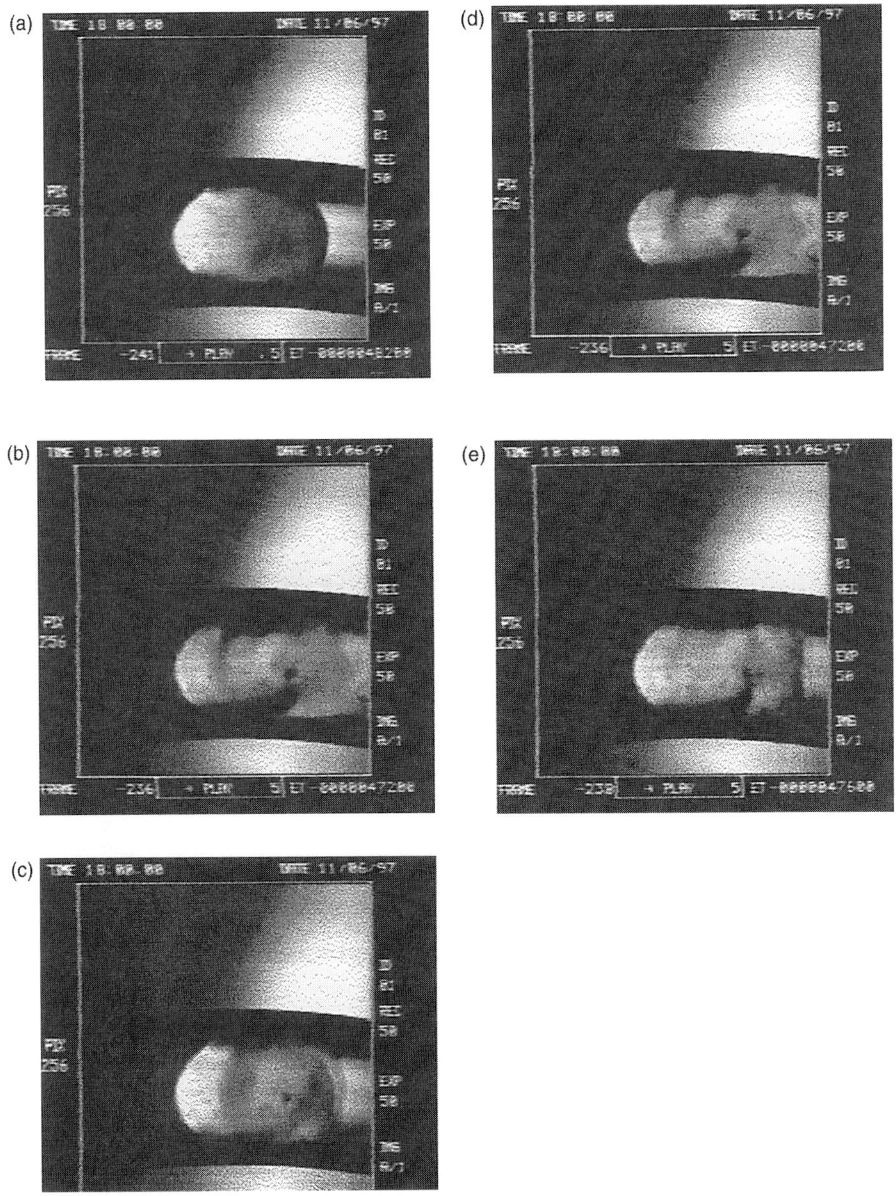

FIGURE 6 High-speed photography of mucus clearance from dry endotracheal tube. [Micrographs are sequential from (a) through (e).]

was more gel-like and more coherent than the 3%; because of the similarities of the 6, 9, and 12% solutions, cough comparisons were made between the 3% and 12% preparations.

Behaviors of wave initiation and patterns of movement of MS during simulated cough were graded as follows: MS spreads out along the ET tube (triangles, Fig. 5); MS scatters along the ET tube (stars, Fig. 5); MS separates as a coherent glob and moves along ET tube (circles, Fig. 5); and MS clears out of ET tube (squares, Fig. 5).

During cough experiments the 12% MS exhibited more coherent separation and catastrophic separation and clearance from the ET tube than did the 3% MS (Fig. 5). The 3% MS, however, smeared along the tube and typically produced a longer pattern of spreading out along the tubes (Fig. 5). Examples of wave initiation on the surface of the MS are shown in the high-speed camera series of photos (Fig. 6). Wave formation starts at the leading edge of the MS glob and spreads over the surface, reflecting retrograde and interacting with oncoming waves. At low flow rates (<60 L/s) in the 10 mm tube using the high VE (3% Borax) MS, surface waves were either not observed or were shallow and not complex.

DISCUSSION

As % Borax increased, the MS viscosities decreased and elasticities decreased. MS globs at higher % Borax were more gel-like and less syrupy. However, the VE ratios fell, as did the change in VE ratio with increasing frequence when studied with the rheometer.

The linear velocities studied in these experiments were chosen so that they approximated the flows that normals and patients with obstructive airway disease might generate during coughing. Catastrophic separation did not occur except at the highest linear velocities (high flows through small tubes). It was clear that at lower U, initiation of wave, movement of MS was difficult. Deeper surface waves with more complex surface undulations occurred with the 12% MS than the 3% MS preparations. The adhesive forces between MS globs and the dry surfaces of the Tigon ET tube plastic exceeded the shearing forces produced by the cough machine at flows below 60 L/min. The finding that lower VE simulants cleared at lower cough U is similar to findings of King (4) and Galstaldi et al. (6).

These data support the clinical observation that older and sicker patients who are unable to generate high cough flow through their ET tubes are able to clear airway mucus only when their secretions are moved centrally as a coherent (low VE ratio) glob (7). Furthermore, the data argue that liquefying secretions by pouring saline into the ET tube (effectively increasing VE ratios of their airway mucus and decreasing adhesive forces) might further impair the ability to cough.

Lower VE ratio secretions do not remain in place but slither toward dependent lung regions, where they are inaccessible to cough clearance and even airway suctioning. Indeed, airway remodeling (goblet cell metaplasia and mucous gland hyperplasia) that produce mucus with lower VE ratios may serve to improve survival in inflammatory airway states because lower VE ratio mucus may enhance cough clearance (7,8).

REFERENCES

1. CA Evrensel, MR Kahn, S Elli, PE Krumpe. Viscous airflow through a rigid tube with a compliant lining: a simple model for the air-mucus interaction in pulmonary airways. J Biomech Eng 115:263–270, 1993.
2. C Evrensel, K Korver, P Krumpe. High-speed visualization of the airflow/viscoelastic layer interface. In: AE Engin, ed. Proceedings of the Third Biannual Joint Conference on Engineering Systems Design and Analysis (ESDA). 1996, pp 183–188.
3. CA Evrensel, A Hassan, PE Krumpe. An experimental investigation of interactions of airflow with mucus simulant in a channel. In: B Simon, ed. 1997, pp 39–40.
4. M King. The role of mucus viscoelasticity in cough clearance. Biorheology 24:589–597, 1987.
5. C Evrensel, A Hassan, P Krumpe. An experimental investigation of interaction of airflow with a viscoelastic layer. Advances in Bioengineering 39:93–94, 1998.
6. A Gastaldi, JR Jardim, PH Saldiva. Factors influencing mucus displacement in a simulated cough machine. Am J Respir Dis Crit Care Med 158:366A, 1998.
7. P Krumpe. Overview of respiratory system clearance mechanisms. In: B Simon, ed. Proceedings of the IMECE 97 Bioengineering Conference (ASME). 1997, pp 37–38.
8. PE Krumpe. The evolutionary biology of airway clearance. Advances in Bioengineering, 43:51–52, 1999.

Index